Current Topics in Microbiology and Immunology

255

Editors

Springer-Verlag Berlin Heidelberg GmbH

Marek's Disease

Edited by K. Hirai

With 27 Figures and 12 Tables

Springer

Professor Dr. KANJI HIRAI†
Tokyo Medical and Dental University
Department of Tumor Virology
Medical Research Institute
Yushima 1-5-45
Bunkyo-ku 113-8510 Tokyo
Japan

Cover Illustration: Jozef Marek (1868–1952)
Picture by courtesy of Dr. Lápis Karoly

ISSN 0070-217X
ISBN 978-3-540-67798-7 ISBN 978-3-642-56863-3 (eBook)
DOI 10.1007/978-3-642-56863-3

Originally published by Springer-Verlag Berlin Heidelberg New York in 2001

Library of Congress Catalog Card Number 15-12910

Cover Design: *design & production GmbH*, Heidelberg
Typesetting: Scientific Publishing Services (P) Ltd, Madras
Production Editor: Angélique Gcouta
Printed on acid-free paper SPIN: 10718362 27/3020GC 5 4 3 2 1 0

Preface

Marek's disease virus (MDV) is a herpesvirus which causes a lymphoproliferative disorder of domestic chickens worldwide. Marek's disease (MD), is named after Jozef Marek (portrait on the cover), who first described it as polyneuritis in 1907. The serious economic problems caused by MD were mostly solved by the development of an effective vaccine against this disease. The MD vaccine, introduced in the late 1960s, is the first practical vaccine against a tumor disease of any type in any species. In addition, there are obvious similarities among MDV, Epstein-Barr virus, and Kaposi's sarcoma-associated herpesvirus in humans. Therefore, MD has received considerable attention as an experimental model in viral oncology. However, it has become apparent recently that the virulence of MDV in the field is continuing to increase, gradually resulting in a vaccine break. There has also been a rapidly expanding understanding of the pathogenesis of this disease along with some of the factors that dictate the outcome of infection in a given chicken, and more recently, the molecular bases of various features of the infection have been studied in some detail. In particular, international symposia on Marek's disease held in Berlin (1978), Ithaca (1983), Osaka (1988), Amsterdam (1992), East Lansing (1996), and Montréal (2000) have contributed greatly to the progress of MD research and the promotion of international friendship and collaboration among the participants. The meetings succeeded in forming the global "MDV network". Without these meetings, I would not have met the leading investigators who contributed their areas of expertise to this book.

This volume provides an overview of many aspects of MDV research and summarizes recent advances in the field. The topics include history and biology of MDV, molecular biology, pathogenesis, vaccinal immunity, immune response, and genetic resistance and development of recombinant polyvalent vaccines. The volume can be divided into two sections. The first three chapters, by Drs. Peter Biggs, Bruce W. Calnek, and Richard L. Witter, provide an overview of the biology and pathology of MDV and

MD vaccines. The period from 1960 to the early 1970s provided most of the fundamental knowledge about MD and the production of effective MD vaccines. The above persons contributed greatly to key events in the early period and have been leaders in MD research to date. During the period between 1968 and 1974, I was investigating the simian tumor virus SV40 at Wistar Institute in Philadelphia and was surprised to hear that cancer could be prevented by live vaccines. Interestingly, MD vaccines prevent tumor formation but not viral infection.

Dr. Peter M. Biggs at Houghton Poultry Research Station established the name of Marek's disease and identified a herpesvirus as the causative agent of MD. The history of MD research in his chapter is an essential part of this book, describing who contributed to MD research during the fruitful period. Dr. Bruce W. Calnek at Cornell University clarified mainly the process from initial infection to tumor formation and spread of MDV. It can safely be said that he established the outline of MD pathology. Dr. Richard L. Witter at Avian Disease and Oncology Laboratory of USDA isolated herpesvirus of turkeys (HVT) and developed an HVT vaccine against MD. He has played a central role in the development of MD vaccines. Many talented researchers who collaborated with these distinguished researchers or trained as postdoctoral fellows in their laboratories have also contributed greatly to MD research. The history of MD in Chap. 1 also introduces the achievements of these people. From the first three chapters, young researchers in the fields of oncology and immunology as well as veterinary medicine will learn that MD research covers numerous interesting topics.

The "legacy" of MD research during the early period has to be succeeded by the next generation, who are well aware of modern molecular biology and immunology. In fact, introduction of various techniques in molecular biology such as recombinant DNA and monoclonal antibody techniques accelerated MD research from the late 1970s onwards. The establishment of lymphoblastoid cell lines derived from MD tumors also resulted in a new phase of MD research. Dr. Shiro Kato at Osaka University was an initial key contributor by establishing an MD cell line, MSB-1, which has been used in many laboratories around the world. I could not have initiated MD research without his help. The new phase of MD research has progressed mainly at the laboratories of contributors in the second part of this volume. Unfortunately, I could not ask all of the talented researchers to contribute to this volume because I could not contact them or their specialty overlapped that of the present authors. Among them, Dr. Leland F. Velicer at Michigan State

University, Dr. Keyvan Nazerian at Avial Disease and Oncology Laboratory, and Dr. Norman L.J. Ross at the Institute for Animal Health Compton Laboratory contributed to progress in molecular biology of MDV. Furthermore, most MD investigators do not forget that the late Dr. Meihan Nonoyama made major advancements in this area. Among his achievements, he provided evidence for a circular plasmid state of latent MDV DNA as well as EBV DNA in transformed cells and succeeded in constructing a restriction map of MDV type 1 DNA. Because of this map, we have been able to locate many viral genes and identify their functions. I would like to dedicate this volume to the memory of Dr. Nonoyama who always encouraged and supported my work.

Thus, the second part of this volume, from Chap. 4 on, is an up-to-date review of fascinating topics in MD research written by authoritative figures in this area. The chapter by Drs. Karel A. Schat and C.J. Markowski reviews the innate and acquired immune responses to MDV infection and vaccinal immunity. Despite its importance, few investigators have performed immunological studies of MD. Among them, Dr. Schat has played a leading part in the field. Dr. Larry D. Bacon and colleagues reviewed the parameters influencing MD resistance, especially MHC and non-MHC genes controlling it. In two successive chapters, Drs. Robert F. Silva, and Lucy F. Lee and colleagues provide a comprehensive overview of the structure and gene functions of MDV genome in comparison with other herpesviruses. Mr. Yoshihiro Izumia and colleagues in Dr. Takeshi Mikami's group summarize their study of the MDV2 genome. This is the first complete genomic sequence of MDV2 DNA. They willingly provided the complete sequence for publication here, although it has not previously been published elsewhere. Dr. Robin W. Morgan and colleagues concentrate on the latency of MDV with special emphasis on latency-associated transcripts. Dr. Hsing-Jien Kung summarizes his study of the Meq protein, which is the only candidate oncoprotein of MDV at present. Finally, the volume ends with a review, written by Mr. Masashi Sakaguchi and myself, of recombinant MDV vaccine capable of protecting commercial chickens with maternal antibodies from both MD and Newcastle disease by one-time inoculation.

It is hoped that the reviews and new results in this volume will not only provide encouragement to researchers in this area but also contribute greatly to progress in the poultry industries and medical science. The research will extend further in the twenty-first century.

I gratefully appreciate the help of many distinguished colleagues who have contributed to this volume. I wish to thank Dr. Hilary Koprowski for encouragement and support. I very much appreciate the thoughtful assistance in the preparation of this volume provided by Ms. Doris Walker at Springer-Verlag in Heidelberg. Finally, I should acknowledge many people who have supported and encouraged my MD research for more than 25 years. I would especially like to thank Dr. Kazuyoshi Ikuta, a professor of Osaka University, who has made a significant contribution to molecular biological studies of MD in the initial stage, for his friendship and collaboration.

November 2000, Tokyo KANJI HIRAI

List of Contents

List of Contributors

(Their addresses can be found at the beginning of their respective chapters.)

The History and Biology of Marek's Disease Virus

P.M. Biggs

Willows, London Road, St. Ives, PE27 5ES, UK

1 Introduction

Marek's Disease (MD) was the term chosen in 1960 for a common lymphoproliferative disease principally of the domestic chicken (*Gallus domesticus*), which is now known to be caused by a cell-associated alpha herpesvirus. The lymphoproliferative process in MD can involve most organs and tissues including peripheral nerves. The most commonly affected organs and tissues are peripheral nerves, iris, gonads, spleen, heart, lung, liver and muscle. Cytolytic changes and atherosclerosis can be a manifestation of MD virus (MDV) infection as can a clinical syndrome described as transient paralysis.

MD can occur at any age from a few weeks old, but is most common between 2 and 6 months of age. The incidence of disease before vaccination was introduced could vary between a few percent of a flock to as high as 60% in exceptional circumstances, but 20–30% was not unusual. Involvement of organs and tissues in addition to peripheral nerves occurs in the more severe outbreaks of disease. In general, involvement of tissues and organs, other than peripheral nerves, increases with reducing age at which disease occurs and increasing incidence of disease.

MD occurs in all poultry-producing countries and, before vaccination was introduced for its control, was responsible for serious economic loss to the poultry industries throughout the world. Infection with MDV is ubiquitous in the domestic chicken and is also present in jungle fowl (*Gallus gallus*) (Weiss and Biggs 1972).

It is the intention to cover in this chapter the highlights of the history of the disease and its definition, its transmissibility and identification of its causative agent and to describe the biological properties of the MDV group.

2 History of Marek's Disease

2.1 The Disease and Its Definition

2.1.1 First Description of Marek's Disease

The story starts with a publication in 1907 by Dr. Joseph Marek, an eminent veterinary clinician and pathologist and Professor and Head of the Department of Veterinary Medicine at the Royal Hungarian Veterinary School in Budapest. This paper entitled "Multiple Nervenentzündung (Polyneuritis) bei Hühnern" (Marek 1907) describes a disease in four adult cockerels which were affected by paralysis of the legs and wings. In the cockerel he examined in detail, he noted thickening of the sacral plexuses and spinal routes which were infiltrated by mononuclear cells. He described the disease as a "neuritis interstitialis" or a "polyneuritis". A study of his description leaves little doubt that this was the first published account of what was to be later called MD.

A similar condition was described by KAUPP in the USA (1921) and VAN DER WALLE and WINKLER-JUNIUS (1924) in The Netherlands. These early descriptions of MD suggested that the pathological changes only occurred in the central and peripheral nervous system. The disease was known to poultrymen as "fowl paralysis" or "range paralysis".

2.1.2 Recognition of the Lymphoproliferative Nature of Marek's Disease

The next landmark was the important and seminal studies by PAPPENHEIMER et al. (1926, 1929a). They made a detailed study of the epidemiology and distribution of the disease and concluded that it was present in varying prevalence in most states of the USA. However, the most important observation was that of 60 cases studied in detail; they found that in addition to lesions in nerves and the central nervous system, 10% of the chickens had lymphoid tumours. Although these principally involved the ovary, tumours were also seen in the liver, kidneys, lungs, adrenals and muscle. For various reasons they considered these tumours to be a manifestation of the disease. For that reason they considered the terms previously used to describe the disease to be unsatisfactory. Because the outstanding pathological features of the disease were lymphoid infiltration of peripheral nerves (although they did recognise that inflammatory lesions did occur) which was frequently accompanied by lymphomatous growths in other tissues and organs, they suggested the term neurolymphomatosis gallinarum for the disease and used "visceral lymphomatosis" and "visceral lymphomata" to describe the visceral tumours.

The significance of these findings was the recognition that the disease was not one of the nervous system alone but was a lymphoproliferative process resulting in lesions in peripheral nerves which were essentially similar in nature to the visceral tumours that occurred in a proportion of cases.

2.1.3 Confusion with Lymphoid Leukosis

At about the same time that Marek was studying his four cockerels, Ellerman and Bang were undertaking their classic studies of a group of neoplastic conditions of the haemopoietic system of the domestic chicken, which they called leukoses. The culmination of this work was a book published in 1922 (ELLERMANN 1922). Three forms of leukosis were described: lymphoid, myeloid and erythroid, the former being the most common under field conditions.

In the 1930s, 1940s and particularly the 1950s, the poultry population expanded and the incidence of MD and lymphoid leukosis increased. Because diagnosis relied on pathological examination and there was difficulty in differentiating between the visceral lymphomata of MD and lymphoid leukosis, confusion arose between these two diseases. This difficulty led to proposals for classifications based on pathology under the heading of the Avian Leukosis Complex with the term lymphomatosis used to describe all lymphoproliferative disease in the domestic chicken (ANON 1941; COTTRAL 1952). Although it was pointed out that this classification was not based on aetiology, the concept that all lymphoid tumours were a manifestation of a single disease became commonplace. This view persisted in the USA until well into the 1960s. Although this classification "had the object of facilitating uniformity in terminology and interpretation of data", in practice its use led to difficulty in interpreting experimental results.

In Europe some scientists continued to hold the view that there were two distinct and unrelated diseases, e.g. FRITZSCHE (1939) and CAMPBELL (1945, 1956). This conflict of views led the World Veterinary Poultry Association to select the subject of the "Classification of the Avian Leucosis Complex and Fowl Paralysis" for discussion at its first conference in 1960. Papers were presented by Campbell and by Biggs (BIGGS 1961; CAMPBELL 1961) and both advocated the separation of the leukoses from "fowl paralysis". Biggs suggested the term of MD for the latter because pathological terms had been responsible for the confusion. The conclusion of the conference was the recommendation that these two conditions be separated and that "fowl paralysis" be termed MD (BIGGS 1962).

This recommendation was vindicated soon after by the successful and convincing transmission of a disease conforming to the definition of MD (SEVOIAN et al. 1962; BIGGS and PAYNE 1963). Also, comparative studies using lymphoid leukosis virus and the agent of MD showed that each of these agents produced characteristic and distinct disease patterns. The diseases differed in latent period, in the distribution and cytopathology of their lymphoid tumours and in the susceptibility of lines of chicken to the two agents (BIGGS and PAYNE 1964).

2.1.4 Changes in Marek's Disease Through the Years

The disease described by MAREK (1907) and by VAN DER WALLE and WINKLER-JUNIUS (1921) in Europe only involved the nervous system as did the disease described by KAUPP (1921), which he first saw in the USA in 1914. By 1922, the disease was present in most parts of the USA and it appeared to have become more severe with outbreaks involving up to 20% of a flock with a significant proportion of affected birds having lymphoid tumours in organs and tissues (PAPPENHEIMER et al. 1926, 1929a).

A high incidence of "visceral lymphomatosis" in addition to neural involvement was reported in the 1950s in the USA in young chickens between 8 and 10 weeks of age (BENTON and COVER 1957); a condition that became widely known as "acute leukosis". A similar condition was reported in Great Britain (BIGGS et al. 1965). On the basis of studies of the disease in the field and transmission experiments, it was concluded that it was a form of MD and that it would more appropriately be called acute MD than "acute leukosis" (BIGGS et al. 1965; PURCHASE and BIGGS 1967). The features of acute MD were the explosive nature of outbreaks, the young age at which the disease could first appear, high mortality of 30% or more and the high incidence of visceral lymphoid tumours in addition to neural lesions. This form of MD was clearly more severe than the disease that was present in the 1920s, 1930s and 1940s, which BIGGS (1966) described as classical MD to differentiate it from the acute form. Acute MD became the predominant form by the 1960s in most countries that had a well-developed poultry industry.

After the introduction and widespread use of vaccines in the early 1970s the disease was at first well controlled, although vaccines do not protect against super infection with wild type virus (CHURCHILL et al. 1969b; WITTER et al. 1976). However, strategies and type of vaccine had to be changed in response to disease appearing in vaccinated chickens during the late 1970s. A similar situation arose in the 1990s. In both cases it was shown that changes in the virulence of the circulating field virus had occurred (WITTER 1996) (see Sect. 3.1). Although the disease was essentially the same as that which had been present since the 1960s, it was appearing commonly in vaccinated flocks. Some outbreaks had unusual clinical and pathological aspects, such as a high incidence of ocular lesions (FICKEN et al. 1991; SPENCER et al. 1992), early mortality without visceral lymphomata resulting from severe cytolytic changes in lymphoid organs, or an encephalitis which sometimes led to transient paralysis (IMAI et al. 1992; KROSS 1996; VENUGOPAL et al. 1996; WITTER 1996).

2.2 Experimental Transmission of Marek's Disease

An examination of the older publications shows that there was widespread belief that MD was infectious. Attempts to transmit the disease started with the early studies of KAUPP (1921) and continued over the years until unequivocal transmission was achieved in the 1960s (SEVOIAN et al. 1962; SEVOIAN and CHAMBERLAIN 1962; BIGGS and PAYNE 1963, 1967).

The early studies were difficult to interpret and all, for one reason or another, were not accepted as providing irrefutable evidence of the infectious nature of the disease. A close examination of these reports show why this was so. In some the criteria used for diagnosis in the transmission experiments were not satisfactory; lymphoid leukosis and even other manifestations of avian leukosis were confused with MD (PATTERSON et al. 1932; JOHNSON 1934; JUNGHERR 1937). In these studies and others, there was always an incidence of the disease in control groups since there were not adequate isolation facilities, for an infection we now know to be highly contagious, to prevent spread from the experimental to the control chickens. In many reports there was also inadequate detail presented or the number of chickens used was too small to provide convincing evidence of transmission (VAN DER WALLE and WINKLER-JUNIUS 1924; WARRACK and DALLING 1932; LERCHE and FRITZSCHE 1933; SEAGAR 1933; FURTH 1935; FRITZSCHE1938; DE OME 1943). It is probable that MD was transmitted by some of these authors because an examination, in the light of current knowledge, of the more notable reports of the period (PAPPENHEIMER et al. 1926, 1929b; BLAKEMORE 1939; DURANT and MCDOUGLE 1939, 1945) suggests that the disease was transmissible. Belief in this view was not possible at the time because of either a relatively low incidence of MD in experimentally treated chickens or high incidence in the controls. For example, PAPPENHEIMER et al. (1926, 1929b), who made the first extensive and well-documented report on transmission of MD, produced 25% incidence of MD in inoculated chickens as compared with 6% in the controls. BLAKEMORE (1939) published similar results with a lower incidence in the controls. Durant and McDougle (1945) were able to induce a much higher incidence of MD (78%) but also had a higher incidence (34%) in their controls.

Convincing evidence for the transmissibility of MD came from the studies of Sevoian et al. (1962) and Biggs and Payne (1963, 1967). Three factors led to this success. The availability of secure isolation facilities and highly susceptible lines of chicken and the use of living cells as the inoculum. Biggs and Payne used a crude homogenate of an ovarian lymphoid tumour to initiate their studies, but whole blood for subsequent passage. Whole blood was chosen because of the high incidence of MD produced by Durant and McDougle (1945) using whole blood. Biggs and Payne (1963, 1967) and Sevoian et al. (1963) were able to show conclusively that the disease could spread by direct and indirect contact and therefore was caused by a contagious infectious agent.

2.3 Properties of Infectious Material Used to Transmit Marek's Disease

Sevoian et al. (1962) described the successful transmission of MD with a 0.3-μm Millipore filtrate of a suspension of JM isolate. Biggs and Payne (1963, 1967) described transmission of the disease with plasma but they were unable to transmit the disease with a 0.45-μm Millipore filtrate. Biggs (1965) made reference to unpublished studies that indicated that successful transmission was dependent on whole cells and that their destruction results in loss of infectivity. Proof of this suggestion came from qualitative and quantitative studies (Biggs and Payne 1967; Biggs et al. 1968). For example, homogenization with a mechanical homogenizer or ultrasonic waves destroyed all or nearly all infectivity of whole blood. Further studies showed that all infectivity of blood and a tumour suspension with 50% infectivity titres of $10^{3.8}$ and $10^{4.0}$ per millilitre, respectively, was destroyed by lyophilization. Also, there was no infectivity in supernatants of tumour cells, or washed tumour cells, disrupted by a high-speed blender or sonication, respectively, and then centrifuged at 2000 *g* for 20 min. These studies suggested that viable cells were necessary for infectivity. This view was supported by studies showing that both dimethyl sulfoxide and glycerin, which are recognised as substances that protect viability of frozen cells, prevented the destruction of infectivity caused by freezing of tissue (Spencer and Calnek 1967).

It was concluded from these studies that the infectious unit was either a living avian cell or a free-living organism with similar physical properties. The latter possibility was discounted by studies which found that none of the variable assortment of microorganisms isolated from cultures of blood and tumours produced MD when inoculated into young chicks (Biggs et al. 1968).

The association of infectivity with living cells suggested that cell transplants or an avidly cell-associated virus was responsible for the infectivity of blood and tumour cells. The transplant possibility seemed unlikely since the disease had been shown to spread not only by direct contact but also indirect contact between infected and susceptible chickens. In addition, studies using the sex chromosome as a cell marker found that tumours in 4 out of 4 and 18 out of 20 chickens that had received either blood or tumour suspension were made up of cells of host origin (Owen et al. 1966; Biggs et al. 1968).

The conclusion from this series of studies on the properties of the infectious material was that the causative agent of MD was an avidly cell-associated virus.

2.4 Isolation of a Cell-Associated Herpesvirus

Because the behaviour of MD in the field suggested that it was an infectious disease and that its cause was likely to be a virus, attempts were made to isolate a virus from the early days of research on the disease, but without success. It was not until cell culture techniques became available, the right choice of cell type chosen for primary isolation and the recognition of the cell-associated nature of the causative agent that its isolation and identification were successful. An avidly cell-associated herpesvirus was isolated from blood and from tumour cells of chickens with MD in chick kidney cells (CKC) (Churchill and Biggs 1967) and in duck embryo fibroblasts (DEF) (Nazerian et al. 1968; Solomon et al. 1968). In each cell type the inocula produced a cytopathic effect which had characteristics of changes produced by herpesviruses. Inclusion bodies were present in the areas of cytopathic effect and herpes-type virus particles were seen in infected cell cultures. The cytopathic effect was in the form of plaques and these could be used to quantify the dose in the inoculum.

Immediately following these reports, there were many publications from several areas of the USA reporting the isolation of a herpesvirus from field cases of MD and from recognised isolates of the causative agent of the disease (Ahmed and Schidlovsky 1968; Bankowski et al. 1969; Calnek and Madin 1969; Eidson et al. 1969; Sharma et al. 1969).

2.5 Evidence that the Herpesvirus Is the Aetiological Agent of Marek's Disease

Whether the herpesvirus was the aetiological agent of MD or not, could not, at first, be formally answered. Koch's postulates are difficult to fulfil for viruses because of the difficulty of guaranteeing the purity of cell cultures. In the case of a cell-associated virus it is impossible. Also, because herpesviruses are common and widespread in animal populations it was suggested that the herpesvirus isolated from cases of MD could be a common virus infection of the domestic chicken that was co-isolated as a passenger with the causative agent. Other questions that were asked at the time were: isn't it more likely that a retrovirus is the cause of a lymphoproliferative condition like MD?; and how can a highly contagious disease be caused by a cell-associated agent? These questions were answered in a number of ways.

2.5.1 Is the Aetiological Agent of Marek's Disease a Retrovirus?

If the causative agent of MD was a retrovirus, it must be a defective or partially defective virus of the avian leukosis/sarcoma group because of its cell-associated nature. The question of whether the agent of MD was a retrovirus was addressed by testing homogenates of tumours for the *gs* antigen. Sixteen out of 24 tumours collected from chickens infected with a classical or acute isolate of MD had no titre for *gs* antigen. Because there had never been a failure to transmit MD with suspensions of tumours, it was concluded that the aetiological agent of MD was not a retrovirus (Biggs et al. 1968).

2.5.2 Herpesvirus Passenger or Aetiological Agent of Marek's Disease?

The question of whether the herpesvirus was a passenger or the aetiological agent of MD was addressed in a number of ways. By examining the association between the herpesvirus and MD, by examining the quantitative relationship between infectivity responsible for CPE in cultured cells and for lesions of MD and by comparing the properties of the agent of MD with that of the herpesvirus.

An association between the herpesvirus and MD was provided by the consistent isolation of the cytopathic agent from experimental birds with MD produced by a wide range of recognised isolates of the causative agent of the disease and from field cases of MD. Cells from cultures showing characteristic cytopathic effect (CPE) produced MD and cells from cultures in which the CPE was not present did not. Quantitative evidence for the association was provided by studies showing that the ability of blood or infected cell cultures to produce lesions of MD was diluted out at the same point as the ability to produce the characteristic CPE in cell culture. Most importantly, the agent present in cell cultures was highly cell-associated and no infectivity was detectable in cell culture supernatants. Cells were found to be necessary to passage in cell culture infectivity for chickens. Supernatant of infected cultured cells centrifuged twice at 2000 *g* for 20 min and filtrates of 0.45-µm Millipore filters were not able to produce MD. Infectivity of cell cultures was destroyed using the same techniques to disrupt cells already referred to as destroying the disease-causing ability of blood and tumour cells taken from cases of MD (Churchill and Biggs 1967; Biggs et al. 1968; Churchill and Biggs 1968; Solomon et al. 1968; Bankowski et al. 1969; Eidson et al. 1969; Sharma et al. 1969; Witter et al. 1969a).

Clearly, the presence of the herpesvirus was closely associated with MD and disease-causing materials and it had the same cell-associated property as the agent of Marek's disease. These studies provided strong evidence that the herpesvirus was not a passenger but the aetiological agent of MD.

2.5.3 Explanation for the Contagious Nature of Marek's Disease

The question of how a disease caused by a cell-associated agent could be highly contagious was soon answered by the demonstration by Calnek and his co-workers that antigens of the herpesvirus were consistently present in the superficial layers of the feather follicle epithelium and that enveloped herpesvirus particles were present at this site (CALNEK and HITCHNER 1969). They also showed that cell-free extracts of skin were infectious both for cultured cells and the chicken (CALNEK et al. 1970a,b). These observations not only provided further evidence that the herpesvirus is the causative agent of MD, but also provided an explanation of how the virus could spread. Poultry dust and litter, which contain desquamated epithelial cells, has been shown to be infectious (BEASLEY et al. 1970; JURAJDA and KLIMEŠ 1970; CARROZZA et al. 1973) and remain infectious for over a year and for at least 16 weeks, respectively (WITTER et al. 1968; CARROZZA et al. 1973; HLOZÁNEK et al. 1973). These observations provided a satisfactory explanation as to why and how the aetiological agent of MD is highly contagious.

2.5.4 Conclusive Evidence for the Herpesvirus Aetiology of Marek's Disease

Conclusive evidence for the herpesvirus aetiology of MD was the demonstration that an attenuated pathogenic herpesvirus and a herpesvirus isolated from turkeys closely related antigenically to the chicken herpesvirus were both capable of protecting chickens from MD (CHURCHILL et al. 1969b; OKAZAKI et al. 1970).

3 Biology of the Marek's Disease Virus Group

This section will describe the main biological features of viruses of the MDV group. Details of the molecular biology and proteins of MDV will be covered in other chapters.

Antigenically related viruses varying in pathogenicity from highly oncogenic to nononcogenic have been isolated from domestic chickens, and together with an antigenically related virus, isolated from turkeys (HVT), they form the MDV group.

3.1 Historical Aspects of Marek's Disease Virus Isolates

A characteristic of the early isolates was that they varied in their pathogenicity and oncogenicity. For example, HPRS-B14 produced mainly neural lesions with only 18% of 451 cases with visceral tumours in a susceptible line of chicken and only 3.3% out of 61 cases in a resistant line of chicken, and none out of 118 cases in another resistant line (BIGGS and PAYNE 1967). Whereas HPRS-16 produced a higher incidence of disease and proportion of visceral tumours in the same strains of chicken: 85% of 42 and 64% of 44 cases of MD in susceptible and resistant lines of chicken, respectively (PURCHASE and BIGGS 1967). The JM and GA strains were similar to the HPRS-16 strain (SEVOIAN et al. 1962; EIDSON and SCHMITTLE 1968), and the Conn-A and HPRS-17 strains to the HPRS-B14 strain (CHOMIAK et al.

1967; Purchase and Biggs 1967). Classical and acute MD were terms introduced to describe these two types of disease (Biggs et al. 1965). The terms were also used to describe strains of the causative agent that produced these two types of MD; however, it was recognised that there may be strains in the field with all grades of virulence between "severe acute and mild chronic forms" (Purchase and Biggs 1967). Interestingly, the disease produced by most isolates made up until the late 1970s generally fitted one or other of these types. In the late 1970s, an acute type of disease appeared in HVT-vaccinated flocks. The characteristic of viruses isolated from such flocks (Md/5, Md/11, ALA-8, RB1B) was that they produced more severe disease and a significant incidence of disease in HVT-vaccinated experimental chickens (Witter et al. 1980; Eidson et al. 1981; Schat et al. 1982). In order to control disease in the field, a change in vaccine strategy was made and bivalent and trivalent vaccines were introduced (see the chapter by R.L. Witter, this volume). During the 1990s, MD was seen at a significant incidence in flocks vaccinated under the new regimes. This suggested that a further change in virulence had occurred. A series of isolates was made and a proportion were found to produce, under standard test conditions, a higher incidence of MD in bivalently vaccinated (HVT + SB-1) chickens than those isolated during the 1980s (Witter 1996, 1997). These isolates are also characterised by their ability to produce a significant early mortality. Isolates with a similar property or other novel properties have been made during the 1990s from the United States, Europe and Japan (Ficken et al. 1991; Imai et al. 1992; Kross 1996; Venugopal et al. 1996).

In addition to the isolation of pathogenic strains of varying virulence, isolates were made that were found to be nononcogenic and in most cases apathogenic. Examples of early isolates are HPRS-24 and -27 (Biggs and Milne 1972; see Phillips and Biggs 1972), HN (Choy and Kenzy 1972) and SB-1 (Schat and Calnek 1978). In a study in which isolates from a number of flocks with varying incidence of MD were categorised, it was found that "apathogenic" MDV was widespread in poultry populations (Biggs and Milne 1972; Jackson et al. 1976). These viruses which may have a very low level of pathogenicity but no oncogenicity are most frequently referred to as nononcogenic MDVs.

A herpesvirus was isolated from cell cultures of kidney of normal turkeys (Kawamura et al. 1969). Witter et al. (1970) isolated from turkeys an apparently identical virus, which they showed to be antigenically related to MDV. It was classified as a herpesvirus of turkeys (HVT) and the strain was designated FC-126. It was cell-associated and non-pathogenic for turkeys and chickens.

3.2 Classification of the Marek's Disease Group

Initially, the MD viruses were classified as either classical or acute MDV (Biggs et al. 1965) on the basis of the type of disease they produced. HVT was added to the group because it was shown to be antigenically related to MDV (Witter et al. 1970), and this was followed by apathogenic MDV which was also shown to be antigenically related to pathogenic MDV (Biggs and Milne 1972). As time progressed, the classification of pathogenic MDVs into classical and acute became inadequate due to the isolation of viruses of greater virulence (see Sect. 3.1).

3.2.1 Serotypes

It has been shown that HVT is antigenically related to MDV and that antibody to HVT could be distinguished from antibody to MDV by the distribution of antigens in HVT-infected cell cultures detected by immunofluorescence (WITTER et al. 1970; PURCHASE et al. 1971). BÜLOW and BIGGS (1975a) found no difference in the distribution of MDV-induced and HVT-induced antigens regardless of whether homologous or heterologous antisera were used. This could have been due the use of different cell types. However, they did find differences between HVT and MDV in quantitative immunofluorescence studies.

In a series of studies using immunofluorescence, immunodiffusion and neutralisation, it was found that there were antigenic differences between pathogenic and apathogenic MDV and between both of these and HVT (BÜLOW and BIGGS 1975a,b; BÜLOW et al. 1975). On the basis of these results, BÜLOW and BIGGS (1975a,b) and BÜLOW et al. (1975) suggested that the virus strains they examined (HPRS-16, JM, and GA acute MDV and their attenuated derivatives, HPRS-B14 and VC classical MDV, HPRS-24 and -27 apathogenic MDV and FC-126 HVT) could be divided into three serotypes:

1. Pathogenic strains of MDV and their attenuated derivatives.
2. Apathogenic strains of MDV.
3. HVT.

The MDV group consists of antigenically closely related viruses with many antigens in common. Quantitative differences in immunofluorescence, immunodiffusion and neutralisation studies have shown that the difference in titres between homologous and heterologous serotype reactions to be only two- to eightfold in most instances.

The division of the MDV group into three serotypes has been supported by subsequent studies (BüLOW 1977; SCHAT and CALNEK 1978; KING et al. 1981; IKUTA et al. 1982; LEE et al. 1983) and by the demonstration that genomes of viruses of the three serotypes differ in their restriction endonuclease digestion patterns (HIRAI et al. 1979; ROSS et al. 1983; SILVA and BARNETT 1991). Monoclonal antibodies have been developed which have confirmed the specificity of the three serotypes and aided in the serotyping of viruses (IKUTA et al. 1982; LEE et al. 1983). The classification of MDV group viruses into three serotypes is now generally accepted and widely used.

3.2.2 Pathotypes

All pathogenic (oncogenic) MDVs belong to serotype-1, but they vary greatly in their pathogenic and oncogenic potential. Although this variation is likely to be a continuum, it has been found useful to classify isolates and strains into categories of

Table 1. Serotypes and pathotypes of the Marek's disease virus group

Virus	Serotype	Pathotype	Old terms
MDV	1	Very virulent plus (vv+) MDV	
		Very virulent (vv) MDV	
		Virulent (v) MDV	Acute MDV
		Mild (m) MDV	Classical
MDV	2	Nononcogenic	
HVT	3	Nononcogenic	

MDV, Marek's disease virus; HVT, herpesvirus of turkeys.

pathogenicity or virulence. The original classification was based on the terms classical and acute which were used to distinguish between two distinct forms of MD (Biggs et al. 1965), and these terms are still useful for that purpose. Because virus isolates could be described as being isolated from and/or producing, under experimental conditions, the classical or acute forms of disease, they were used to categorise virus strains (Biggs and Milne 1972; Jackson et al. 1976). With the isolation of viruses that have characteristics of greater virulence than acute MDV (Witter et al. 1980; Schat et al. 1982), these terms became inadequate for the classification of MDV viruses. An alternative nomenclature was proposed by Witter (1983, 1985). In this system, classical strains become mild MDV (mMDV), acute strains, virulent MDV (vMDV) and strains isolated in the 1980s from vaccinated flocks, very virulent MDV (vvMDV). Even this nomenclature has difficulties because when isolates made in the 1990s were found to be more virulent than vvMDV it was difficult to decide what term should describe them. Very virulent plus (vv+MDV) was chosen (Witter 1996). A system and criteria for assigning strains of MDV to one of these categories have been proposed by Witter (1997).

Table 1 shows the currently most widely used serotype and pathotype classification.

3.3 Recombination and Mutation in Marek's Disease Virus

There has been and there is ample opportunity for recombination amongst the three serotypes of MDV group of viruses. Superinfection with field virus of chickens vaccinated with attenuated serotype-1 and with serotype-3 HVT virus has been reported (Churchill et al. 1969b; Biggs et al. 1970; Purchase and Okazaki 1971), and clearly is a widespread phenomenon. Viruses of serotype-1 and -2 have been isolated from single pens of chickens, from the same chicken in sequential studies (Biggs et al. 1972; Jackson et al. 1976) and from the same tissues in dually infected chickens (Cho 1977) and from a cell line in culture (Nii et al. 1988). Despite these observations, there has been no report of spontaneous recombination between the three serotypes.

Spontaneous mutation apparently does occur. All three serotypes alter in their biological and molecular properties on passage in cell culture (Churchill et al. 1969a; Witter and Offenbecker 1979; Hirai et al. 1981; Maotani et al. 1986; Witter et al. 1990; Silva and Barnett 1991). A temperature-sensitive mutant of

MDV and a phosphonoacetate mutant of HVT have also been described (LEE et al. 1978; WITTER and OFFENBECKER 1979). Changes in the virulence of MD over the years, which are described in Sect. 3.1, also provide support for the view that mutation under suitable conditions is part of the biology of the MDV group.

3.4 Morphology

There have been many ultrastructural studies describing the morphology and morphogenesis of MDV and HVT, and these have been reviewed by PAYNE et al. (1976), KATO and HIRAI (1985) and SCHAT (1985). These studies have shown essentially similar results and, in general, viruses of all three serotypes have characteristics typical of other herpesviruses. Most studies have used infected cell cultures because the presence of particles is infrequent in the chicken in tissues other than the feather follicle epithelium (SCHIDLOWSKY et al. 1969; CALNEK et al. 1970a,b; UBERTINI and CALNEK 1970; CAUCHY 1971; NAZERIAN 1971). Cell culture studies have almost exclusively been with serotype-1 MDV or HVT infections.

In general, the morphology and morphogenesis of MDV and HVT are similar. Their structure is typical of herpesviruses. In negatively stained preparations the unenveloped virion measures about 100nm in diameter and has 162 hollow capsomeres (Fig. 1) (CHURCHILL and BIGGS 1967; EPSTEIN et al. 1968; NAZERIAN and BURMESTER 1968). The dimensions of the capsomeres are about 6 × 10nm with a centre-to-centre distance between adjacent capsomeres of 10nm (EPSTEIN et al. 1968; NAZERIAN 1973). The diameter of enveloped particles varies considerably because of the lack of rigidity of the envelope. For example, in negatively stained preparations of lysed feather follicle epithelium, enveloped particles were found to range in measurement from 273 to 400nm (CALNEK et al. 1970a).

Electron microscope studies of thin sections of infected cell cultures reveal naked particles measuring 85–100nm in the nucleus, which can appear hexagonal, and less frequently small ring-shaped structures about 35nm in diameter, which have long been associated with herpesviruses (Fig. 2) (MORGAN et al. 1959; EPSTEIN et al. 1968; NAZERIAN and BURMESTER 1968; OKADA et al. 1972; HAMDY et al.

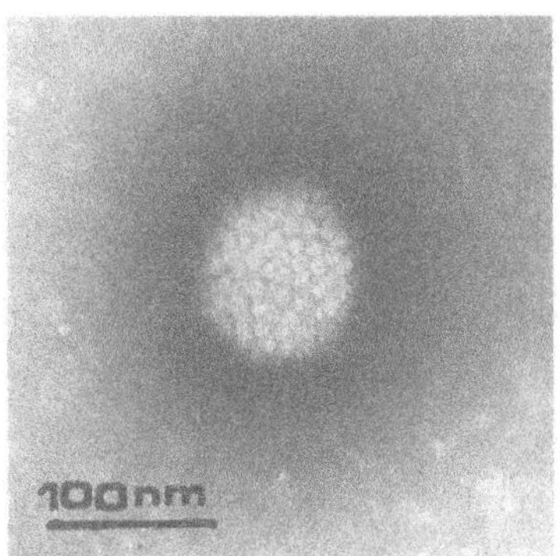

Fig. 1. Negative contrast preparation of serotype-1 HPRS-16 strain of Marek's disease herpesvirus unenveloped particle showing triangular surface facets and tubular capsomeres. (Kindly supplied by Dr. J.A. Frazier)

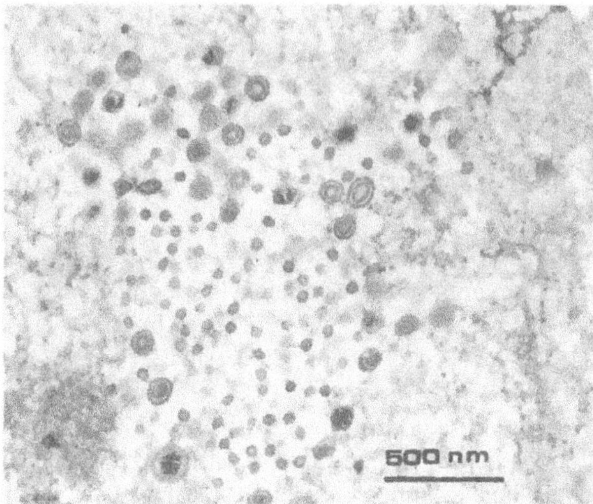

Fig. 2. Thin section of a cultured chick kidney cell infected with serotype-1 HPRS-16 strain of Marek's disease herpesvirus. Detail of the nucleus showing a number of unenveloped particles and small ring shaped structures. (Kindly supplied by Dr. J.A. Frazier)

1974). The nucleoid measures 50–60nm and is structured as a torus which is at right angles to and around a less electron-opaque cylindrical mass (NAZERIAN 1974). This accounts for the variable shape of the nucleoid seen in sections as it depends on, in addition to the stage of development of the particle, the angle of section. A characteristic of HVT is that in some particles the nucleoid has an electron-lucent cross appearance (NAZERIAN et al. 1971). More rarely enveloped particles are found measuring between 130 and 160nm in diameter in the perinuclear space or in nuclear vesicles (Fig. 3). Both naked and envelope particles are found in the cytoplasm but more rarely than in the nucleus.

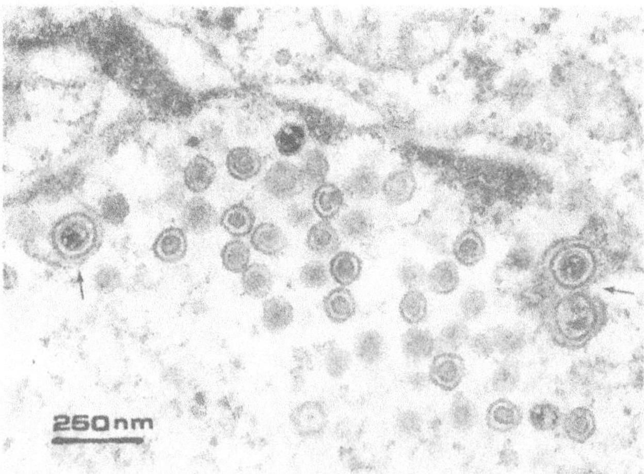

Fig. 3. Thin section of a cultured chick kidney cell infected with serotype-1 HPRS-16 strain of Marek's disease herpesvirus. Detail of the nucleus showing unenveloped particles and three enveloped particles in nuclear vesicles (*arrows*). (Kindly supplied by Dr. J.A. Frazier)

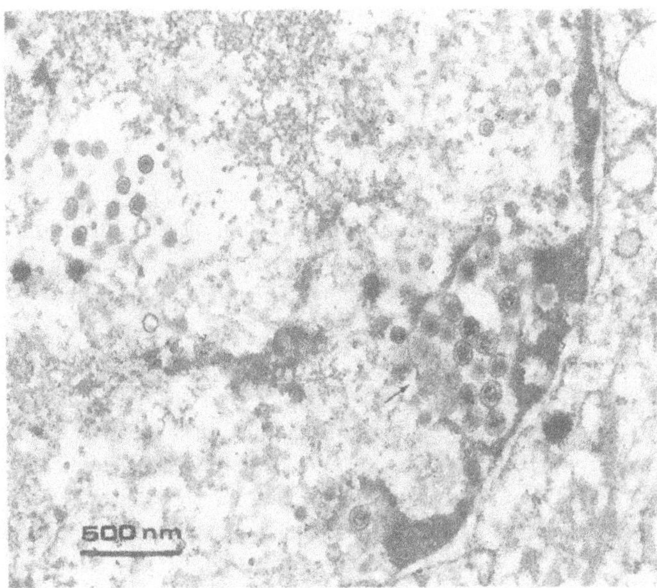

Fig. 4. Thin section of a cultured chick embryo fibroblast infected with serotype-2 HPRS-24 strain of Marek's disease herpesvirus. Detail of the nucleus showing unenveloped particles and enveloped particles in a nuclear vesicle (*arrow*). (From PAYNE et al. 1976)

The 35-nm structures appear first at about 8h post-infection followed about 2h later by the first appearance of nucleocapsids, and at 18h post-infection the first enveloped virions appear (HAMDY et al. 1974). The nucleocapsids mature to enveloped virions by budding through cellular membranes acquiring their envelope in the process. The enveloping of the nucleocapsid can occur at the inner nuclear membrane or membrane-bound cytoplasmic spaces. The former is more common.

Only limited studies have been made of the morphology of serotype-2 MDV, but what has been described is similar to serotype-1 MDV (Fig. 4) (PAYNE et al. 1976; SCHAT and CALNEK 1978).

3.5 Virus–Cell Interaction

There are four virus–cell interactions recognised in oncogenic MDV infection: productive, restrictively productive or abortive, and two forms of non-productive, latent and transforming.

3.5.1 Productive Infection

In productive infection, virus replication is complete and enveloped virus produced and released resulting in death of the cell. In the chicken, this only occurs in the feather follicle epithelium where large numbers of fully infectious virus particles are produced (CALNEK et al. 1970a,b). Full replication of the virus occurs variably in cell culture. In some virus–cell culture systems, virus is released into the supernatant and fully infectious virus can also be harvested by disruption of cultured cells (CALNEK et al. 1970c; BÜLOW et al. 1975; CHO 1978).

3.5.2 Restrictively Productive Infection (Abortive)

Restrictively productive or abortive infections are those where there is no fully infectious virus released from the cell or tissue, but expression of the viral genome may range from the production of virus-specific antigens to, in rare cases, enveloped intracellular virions. In the latter case it is a cytolytic infection. The presence of nucleocapsids, mainly in the cell nucleus, and fewer intracellular enveloped virions is seen in cytolytic cell culture infections. In infected chickens, enveloped virions are not seen except in the feather follicle epithelium or occasionally in other cells and tissues when young antibody-free chicks are infected. (FRAZIER and BIGGS 1972). Cytolytic infection also occurs in a number of tissues and organs in chickens infected with vv− and vv+MDV. Otherwise, nucleocapsids have only been seen in an occasional cell of infected chickens and then in small numbers and mainly in lymphoid cells. (SCHIDLOWSKY et al. 1969; CALNEK et al. 1970b; UBERTINI et al. 1970; NAZERIAN 1971). Most of the viral genome is transcribed in infected chick embryo fibroblasts in culture (ROSS 1985).

3.5.3 Latent Infection

Latent infection can only be detected by hybridization with viral DNA probes or methods that activate the viral genome. For example, in vitro cultivation of latently infected cells results in the expression of gene products and virus particles (CAMPBELL and WOODE 1969; CAUCHY and COURDET 1972; CALNEK et al. 1981). Latent infection has been reported in T lymphocytes and is present in chickens infected with either serotype-1, -2 or -3 viruses (SHEK et al. 1982; CALNEK et al. 1984). These cells contain less than five copies of viral DNA and do not express viral antigens (ROSS 1985).

3.5.4 Transforming Infection

Transforming infection differs from latent infection by the transcription and expression of a number of genes resulting in the presence of virus-associated non-structural proteins and transcripts in lymphoma cells and lymphoblastoid cell lines developed from MD lymphomas. A number of viral transcripts mapping in the inverted repeats flanking the long and short unique regions and to a lesser extent the adjacent unique sequences have been described (SUGAYA et al. 1990). Genes that have been described as expressed in lymphoma cells and lymphoblastoid cell lines are present in the same region of the genome (BIGGS 1997). For example, a gene found in the *Eco*R1 Q fragment called *meq*, an acronym for Marek's Eco Q, is highly expressed in lymphoblastoid tumours (JONES et al. 1992), and an MDV-specific phosphorylated protein, *pp38*, has been found in lymphoma cells and cells of lymphoblastoid cell lines (NAITO et al. 1986; NAKAJIMA et al. 1987). Transcripts which map antisense to the homologue of herpes simplex virus infected cell protein 4 (*ICP4*) gene have also been found in lymphomas and lymphoblastoid cell lines (Cantello et al. 1994; Li et al. 1994). These also map to the same region of the genome. Cells of lymphoma and of lymphoblastoid cell lines contain at least 10–20 copies of viral DNA per cell which is highly methylated (Ross et al. 1981; Kanamori et al. 1987).

3.6 Growth and Assay of Marek's Disease Virus

Marek's Disease Virus by definition will grow in chicks and chickens and this system has been used to identify, characterise and assay serotype-1 MDV. The virus will also grow in chick embryos, but the chick embryo is rarely used for its assay. MDV is usually assayed in cell culture. Both chickens and cell culture are used for the propagation of cell-free and cell-associated MDV to provide stocks of infectious material for experimental studies.

3.6.1 Chickens

For the propagation of cell-free and cell-associated virus, 1-day-old chicks, preferably of a susceptible line free of MDV antibodies, are inoculated with infectious material intra-abdominally. Harvest of infectious material can be made 6 weeks or more later. For cell-associated virus, blood or tumours can be collected. For the preparation of cell-free virus, the feather follicle epithelium is harvested. High passage virus grows less well in chickens than its low passage parent (Phillips and Biggs 1972; Witter et al. 1990).

Chicks of similar status are preferably used for assay of MDV serotype-1 infectious material usually inoculated intra-abdominally at 1-day-old. Histological lesions are present in peripheral nerves and some organs 2–4 weeks post-infection, and the presence or absence of these lesions can be used to record the proportion of chicks infected in a quantitative dilution assay (Biggs and Payne 1967; AAAP Report 1970). Other criteria of infection such as the presence of antigen or antibody to MDV can be used for detection and assay of non-pathogenic viruses. However, cell culture is the preferred system.

3.6.2 The Developing Chick Embryo

Inoculation of the chorioallantoic membrane (CAM) of developing chick embryos with MDV results in the development of pocks. Such a system can be used for a quantitative assay, providing the inoculum does not include immunologically competent cells, with a sensitivity equivalent to the chick kidney cell culture plaque assay. Inoculation of the yolk sac of chick embryos also results in pocks on the CAM with the number of pocks directly proportional to the dose. The assay using the yolk sac route is at least as sensitive as, but more variable than, the chick kidney cell culture plaque assay, but is detrimentally affected by the presence of antibody to MDV in the embryo. Infection by the yolk sac route results in a splenomegaly and infection of the liver, lung, kidney and heart (BÜLOW 1968; CHURCHILL 1968; BIGGS and MILNE 1971). All the studies with chick embryos have been with serotype-1 viruses.

3.6.3 Cell Cultures

MDV will grow in a number of types of cell in culture with a differing response for each serotype. Primary isolation and quantification of infected cells derived from chicken is best done using either chick kidney cells (CKC) or duck embryo fibroblasts (DEF). In both cell systems, infection results in a cytopathic effect in the form of characteristic plaques which can be used for quantitative assays (CHURCHILL and BIGGS 1967, 1968; CHURCHILL 1968; SOLOMON et al. 1968). Serotype-2 MDV, HVT and attenuated serotype-1 grow in CKC and DEF, but also propagate well in chick embryo fibroblasts (CEF). They also produce plaques that can be used for quantitative assays. A number of changes can be present in the plaques which include clusters of rounded degenerating cells, area of cell death, multinucleate cells and the presence of intranuclear type A inclusion bodies. The presence and combinations of these changes and plaque size vary with the combination of serotype and cell type. The plaques in most combinations are less than 1mm in diameter and are counted under magnification. The various responses of serotype–cell combination have been described and illustrated by a number of authors (CHURCHILL 1968; SOLOMON et al. 1968; WITTER et al. 1969b, 1990; BIGGS and MILNE 1972; SCHAT 1985; CALNEK and WITTER 1997).

Passage of oncogenic and/or pathogenic serotype-1 MDV in cell culture results in the loss of oncogenicity and pathogenicity (CHURCHILL et al. 1969a; EIDSON and ANDERSON 1970; NAZERIAN 1970; RISPENS et al. 1972a,b; WITTER 1982). Attenuated virus generally grows to higher titres and produces larger plaques in cell culture than its parent. Other changes that occur include what at first was believed to be the loss of an antigen known at the time as the "A" antigen (CHURCHILL et al. 1969a) and the expansion of 132bp direct repeat in the internal and terminal inverted repeat regions flanking the unique long region of the genome (ROSS et al. 1983; FUKUCHI et al. 1985; MAOTANI et al. 1986; ROSS et al. 1993). It is now known that the A antigen is the homologue of glycoprotein C of herpes simplex virus and that it is present in reduced amounts rather than absent or altered in attenuated viruses (ROSS et al. 1973; VAN ZAANE et al. 1982; BINNS and ROSS 1989; WILSON et al. 1994).

Some of these changes have been described for serotype-2 MDVs after passage in cell culture (WITTER et al. 1990).

4 Conclusions

Marek's disease is the response of the chicken to infection with an alpha herpesvirus. It forms a continual threat to the poultry industry requiring effective vaccines and strategies to keep it under control. The virus appears to be continuously evolving requiring new vaccine strategies and possibly in the future more effective vaccines. It is a disease of particular interest because infection with MDV can result in a cytolytic and/or proliferative response of lymphoid cells resulting in lymphomata, both of which can be fatal. The availability of well-studied and documented experimental animal systems, together with oncogenic and nononcogenic viruses in the MDV group, provide unparalleled opportunities for understanding the oncogenesis of this herpesvirus infection and for the increased understanding of the neoplastic process.

References

A.A.A.P. Report (1970) Methods in Marek's disease research. Avian Dis 14:820–828

Ahmed M, Schidlovsky G (1968) Electron microscopic localization of herpes-type particles in Marek's disease. J Virol 2:1443–1457

Anon (1941) Tentative pathologic nomenclature. Am J Vet Res 2:116

Bankowski RA, Moulton JE, Mikami T (1969) Characterization of Cal-1 strain of acute Marek's disease agent. Am J Vet Res 30:1667–1676

Beasley JN, Patterson LT, McWade DH (1970) Transmission of Marek's disease by poultry house dust and chicken dander. Am J Vet Res 31:339–344

Benton WJ, Cover MS (1957) The increased incidence of visceral lymphomatosis in broiler and replacement birds. Avian Dis 1:320–327

Biggs PM (1961) A discussion on the classification of the avian leucosis complex and fowl paralysis. Brit Vet J 117:326–334

Biggs PM (1962) Some observations on the properties of cells from the lesions of Marek's disease and lymphoid leukosis. Proc 13th Symp of the Colston Research Society pp 83–89. Butterworth London

Biggs PM (1965) Avian transmissible tumours. Proc 3rd Congress of the World Veterinary Poultry Association pp 61–67

Biggs PM (1966) Avian leukosis and Marek's disease. XIIIth World's Poultry Congr Symp Papers pp 91–118. Kiev USSR

Biggs PM (1997) Marek's disease herpesvirus: oncogenesis and prevention. Phil Trans R Soc Lond B 352:1951–1962

Biggs PM, Milne BS (1971) Use of the embryonating egg in studies on Marek's disease. Am J Vet Res 32:1795–1809

Biggs PM, Milne BS (1972) Biological properties of a number of Marek's disease isolates. In: Biggs PM, de Thé G, Payne LN (eds) Oncogenesis and herpesviruses pp 88–94. International Agency for Research on Cancer, Lyon

Biggs PM, Payne LN (1963) Transmission Experiments with Marek's disease. Vet Rec 75:177–179

Biggs PM, Payne LN (1964) Relationship of Marek's disease (neural lymphomatosis) to lymphoid leukosis. Natl Cancer Inst Monogr 17:83–97

Biggs PM, Payne LN (1967) Studies on Marek's disease. 1. Experimental transmission. J Natl Cancer Inst 39:267–280

Biggs PM, Purchase HG, Bee BR, Dalton PJ (1965) Preliminary report on acute Marek's disease (fowl paralysis) in Great Britain. Vet Rec 77:1339–1340

Biggs PM, Churchill AE, Rootes DG, Chubb RC (1968) The etiology of Marek's disease – an oncogenic herpes-type virus. In: Perspectives in Virology 6:211–237

Biggs PM, Payne LN, Milne BS, Churchill AE, Chubb RC, Powell DG, Harris AH (1970) Field trials with an attenuated cell associated vaccine for Marek's disease. Vet Rec 87:704–709

Biggs PM, Powell DG, Churchill AE, Chubb RC (1972) The epizootiology of Marek's disease. 1. Incidence of antibody, viraemia and Marek's disease in six flocks. Avian Path 1:5–25

Binns MM, Ross LJN (1989) Nucleotide sequence of the Marek's disease virus (MDV) RB1B A antigen gene and the identification of the MDV A antigen as the herpes simplex-1 glycoprotein C homologue. Virus Res 12:371–382

Blakemore F (1939) The nature of fowl paralysis (neurolymphomatosis). J comp Path 52:144–159
Bülow Vv (1968) Untersuchungen mit embryonierten Eiern bei der Marek'schen Hühnerlähmung. Berl Münch tierärztl Wchnschr 81:365–367
Bülow Vv (1977) Further characterisation of the CVI 988 strain of Marek's disease. Avian Path 6:395–403
Bülow Vv, Biggs PM (1975a) Differentiation between strains of Marek's disease virus and turkey herpesvirus by immunofluorescence assays. Avian Path 4:133–146
Bülow Vv, Biggs PM (1975b) Precipitating antigens associated with Marek's disease viruses and a herpesvirus of turkeys. Avian Path 4:147–162
Bülow Vv, Biggs PM, Frazier JA (1975) Characterization of a new serotype of Marek's disease herpesvirus. In: de Thé G, Epstein MA, Zur Hausen H (eds) Oncogenesis and herpesviruses II pp 329–336. International Agency for research in Cancer, Lyon
Calnek BW, Hitchner SB (1969) Localization of viral antigen in chickens infected with Marek's disease herpesvirus. J Natl Cancer Inst 43:935–949
Calnek BW, Madin SH (1969) Characteristics of in vitro infection of chicken kidney cell cultures with a herpesvirus from Marek's disease. Am J vet Res 30:1389–1402
Calnek BW, Witter RL (1997) Marek's disease. In: Calnek BW, Barnes HJ, Beard CW, McDougald LR, Saif YM (eds) Diseases of Poultry 10th Ed. Iowa State University Press, Ames Iowa USA
Calnek BW, Adldinger HK, Kahn DE (1970a) Feather follicle epithelium: a source of enveloped and infectious cell-free herpesvirus from Marek's disease. Avian Dis 14:219–233
Calnek BW, Ubertini T, Adldinger HK (1970b) Viral antigen, virus particles, and infectivity of tissues from chickens with Marek's disease. J Natl Cancer Inst 45:341–351
Calnek BW, Hitchner SB, Adldinger HK (1970c) Lyophilization of cell-free Marek's disease herpesvirus and a herpesvirus from turkeys. Appl Microbiol 20:723–726
Calnek BW, Shek WR, Schat KA (1981) Latent infections with Marek's disease virus and turkey herpesvirus. J Natl Cancer Inst 66:585–590
Calnek BW, Schat KA, Ross LJN, Shek WR, Chen C-LH (1984) Further characterization of Marek's disease virus-infected lymphocytes. I. In vivo infection. Int J Cancer 33:389–398
Campbell JG (1945) Neoplastic disease of the fowl with special reference to its history, incidence and seasonal variation. J Comp Path 55:398–321
Campbell JG (1956) Leucosis and fowl paralysis compared and contrasted. Vet Rec 68:527–529
Campbell JG (1961) A proposed classification of the leucosis complex and fowl paralysis. Brit Vet J 117:316–325
Campbell JG, Woode GN (1969) Demonstration of a herpes-type virus in short-term cultured blood lymphocytes associated with Marek's disease. J Med Microbiol 3:463–473
Cantello JL, Anderson AS, Morgan RW (1994) Identification of latency-associated transcripts that map antisense to the *ICP4* homolog gene of Marek's disease virus. J Virol 68:6280–6290
Carrozza JH, Fredrickson TN, Prince RP, Luginbuhl RE (1973) Role of desquamated epithelial cells. Avian Dis 17:767–781
Cauchy L (1971) La maladie de Marek. Histologie, ultrastructure des cellules et des particules virales. Annls Rech Vétér 2:5–32
Cauchy L, Courdet F (1972) Virologie – particules de type herpés de la maladie de Marek dans les lymphocytes infectes et maintenus in vitro. C R Acad Sci Paris 274:1864–1866
Cho BR (1977) Dual virus maturation of both pathogenic and apathogenic Marek's disease herpesvirus (MDHV) in the feather follicles of dually infected chickens. Avian Dis 21:501–507
Cho BR (1978) An improved method for extracting cell-free herpesvirus of Marek's disease and turkeys from infected cell cultures. Avian Dis 22:170–176
Cho BR, Kenzy SG (1972) Isolation and characterization of an isolate (HN) of Marek's disease virus with low pathogenicity. Appl Microbiol 24:299–306
Chomiak TW, Luginbuhl RE, Helmboldt CF, Kottaridis SD (1967) Marek's disease. I. Propagation of the Connecticut A (Conn A) isolate in chicks. Avian Dis 11:646–653
Churchill AE (1968) Herpes-type virus isolated in cell culture from tumors of chickens with Marek's disease. 1. Studies in cell culture. J Natl Cancer Inst 41:939–950
Churchill AE, Biggs PM (1967) Agent of Marek's disease in tissue culture. Nature 215:528–530
Churchill AE, Biggs PM (1968) Herpes-types virus isolated in cell culture from tumors of chickens with Marek's disease. II. Studies in vivo. J Natl Cancer Inst 41:951–956
Churchill AE, Chubb RC, Baxendale W (1969a) The attenuation, with loss of oncogenicity of the herpes-type virus of Marek's disease (strain HPRS-16) on passage in cell culture. J Gen Virol 4:557–564
Churchill AE, Payne LN, Chubb RC (1969b) Immunization against Marek's disease using a live attenuated virus. Nature 221:744–747

Cottral GE (1952) The enigma of avian leukosis. Proc 89th meeting Am vet Med Ass 89:285–293
DeOme KB (1943) Intraperitoneal injection of lymphomatous nerve tissue into resistant and susceptible chickens. Poult Sci 22:381–395
Durant AJ, McDougle HC (1939) Studies on the origin and transmission of fowl paralysis (neurolymphomatosis) by blood inoculation. Missouri Agric Exp Sta Res Bull 304:1–23
Durant AJ, McDougle HC (1945) Further investigations of the transmission of fowl paralysis (*neurolymphomatosis*) by direct transfusion. Missouri Agric Exp Sta Res Bull 393:1–18
Eidson CS, Anderson (1970) Immunisation against Marek's disease. Abstr 107th AVMA meeting. J Am Vet Med 156:1282
Eidson CS, Schmittle SC (1968) Studies on acute Marek's disease. I. Characteristics of isolate GA in chickens. Avian Dis 12:476–475
Eidson CS, Richey DJ, Schmittle SC (1969) Studies on acute Marek's disease. XI. Propagation of the GA isolate of Marek's disease in tissue culture. Avian Dis 13:636–653
Eidson CS, Ellis MN, Kleven SH (1981) Reduced vaccinal protection of turkey herpesvirus against field strains of Marek's disease herpesvirus. Poult Sci 60:317–322
Ellermann V (1922) The leucosis of fowls and leucaemia problems. Gyldendal London
Epstein MA, Achong BG, Churchill AE, Biggs PM (1968) Structure and development of the herpes type virus of Marek's disease. J Natl Cancer Inst 41:805–820
Ficken MD, Nasisse MP, Boggan GD, Guy JS, Wages DP, Witter RL, Rosenberger JK, Nordgren RM (1991) Marek's disease virus isolates with unusual tropism and virulence for ocular tissues: clinical findings, challenge studies and pathological features. Avian Path 20:461–474
Frazier JA, Biggs PM (1972) Marek's disease herpesvirus particles in tissues from chickens free of precipitating antibodies. J Natl Cancer Inst 48:1519–1523
Fritzsche K (1938) Versuche zur Erforschung und Bekämpfung der Marekschen Hühnerlähme. 1. Versuche über die Ausscheidung des Hühnerlähmevirus mit dem Kot und über die natürliche Infektionsweise. Z Infektkrankh Haustierre 52:51–69
Fritzsche K (1939) Ist die Mareksche Hühnerlähmung eine Form der Hühnerleukose? Z Infekt-Kr Haustiere 55:68–74
Fukuchi K, Tanaka A, Schierman LW, Witter RL, Nonoyama M (1985) The structure of Marek's disease virus DNA: the presence of unique expansion in nonpathogenic viral DNA. Proc Nat Acad Sci USA 82:751–754
Furth J (1935) Lymphomatosis in relation to fowl paralysis. Arch Path 20:379–428
Hamdy F, Sevoian M, Holt SC (1974) Biogenesis of Marek's disease (type II leukosis) virus in vitro: electron microscopy and immunological study. Infect Immun 9:740–749
Hirai K, Ikuta K, Kato S (1979) Comparative studies on Marek's disease virus and herpesvirus of turkey DNAs. J Gen Virol 45:119–131
Hirai K, Ikuta K, Kato S (1981) Structural changes of the DNA of Marek's disease virus during serial passage in cultured cells. Virology 115:385–389
Hlozánek I, Mach O, Jurajda V (1973) Cell-free preparations of Marek's disease virus from poultry dust. Folia Biol (Praha) 19:118–123
Ikuta K, Homma H, Maotani S, Ueda S, Kato S, Hirai K (1982) Monoclonal antibodies specific to and cross reactive with Marek's disease virus and herpesvirus of turkeys. Biken J 25:171–175
Imai K, Yuasa N, Iwakiri K, Nakamura K, Hihari H, Ishita T, Inamoto A, Okamoto I, Ohta K, Maeda M (1992) Characterization of very virulent Marek's disease viruses isolated in Japan. Avian Path 21:119–126
Jackson CAW, Biggs PM, Bell RA, Lancaster FM, Milne BS (1976) The epizootiology of Marek's disease 3. The inter-relationship of virus pathogenicity, antibody and the incidence of Marek's disease. Avian Path 5:105–123
Johnson EP (1934) The etiology and histogenesis of leucosis and lymphomatosis of fowls. Virginia Agric Exp Sta Tech Bull 56:1–32
Jones D, Lee L, Liu J-L, Kung H-J, Tillotson JK (1992) Marek's disease virus encodes a basic leucine zipper gene resembling the *fos/jun* oncogenes that is highly expressed in lymphoblastoid tumors. Proc Nat Acad Sci USA 89:4042–4046
Jungherr E (1937) Studies on fowl paralysis 2. Transmission experiments. Storrs Agric Exp Sta Bull 218:1–47
Jurajda V, Klimeš B (1970) Presence and survival of Marek's disease agent in dust, Avian Dis 14:188–190
Kanamori A, Ikuta S, Ueda S, Kato S, Hirai K (1987) Methylation of Marek's disease virus DNA in chicken T-lymphoblastoid cell lines. J Gen Virol 68:1485–1490
Kato S, Hirai K (1985) Marek's disease virus. Adv Virus Res 30:225–277

Kaup BF (1921) Paralysis of the domestic fowl. J Amer Ass Instructors and Invest in Poultry Husbandry 7:25–31

Kawamura H, King Jr DJ, Anderson DP (1969) A herpesvirus isolated from kidney cell culture of normal turkeys. Avian Dis 13:853–863

King D, Page D, Schat KA, Calnek BW (1981) Difference between influences of homologous and heterologous maternal antibodies on response to serotype-2 and serotype-3 Marek's disease vaccines. Avian Dis 25:74–81

Kross I (1996) Isolation of highly lytic serotype 1 Marek's disease viruses from recent field outbreaks in Europe. In: Silva RF, Cheng HH, Coussens PM, Lee LF, Velicer LF (eds) Current Research in Marek's disease. Proc 5th International Symp on Marek's disease 113–118

Lee LF, Nazerian K, Witter RL, Leinbach SS, Boezi JA (1978) A phosphonoacetate-resistant mutant of herpesvirus of turkeys. J Natl Cancer Inst 60:1141–1146

Lee LF, Liu X, Witter RL (1983) Monoclonal antibodies with specificity for three different serotypes of Marek's disease viruses in chicken. J Immun 130:1003–1006

Lerche F, Fritzsche K (1933) Die Aetiologie der Geflügellähme. Z Infektkrankh Haustiere 44:11–4

Li D-S, Pastorek J, Zelník J, Smith GD, Ross LJN (1994) Identification of novel transcripts complementary to the Marek's disease virus homologue of the *ICP4* gene of herpes simplex virus. J Gen Virol 75:1713–1722

Maotani K, Kanamori A, Ikuta K, Ueda S, Kato S, Hirai K (1986) Amplification of tandem direct repeat within inverted repeats of Marek's disease virus DNA during serial in vitro passage. J Virol 58: 657–660

Marek J (1907) Multiple Nervenentzündung (Polyneuritis) bei Hühnern. Dtsch Tierärztl Wschr 15: 417–421

Morgan C, Rose HM, Holden M, Jones EP (1959) Electron microscopic observations on the development of herpes simplex virus. J Exp Med 110:643–656

Naito M, Nakajima K, Iwa N, Ono K, Yoshida I, Konobe T, Ikuta K, Ueda S, Kato S, Hirai K (1986) Demonstration of Marek's disease virus-specific antigen in tumour lesions of chickens with Marek's disease using monoclonal antibody against a virus phosphorylated protein. Avian Path 15:503–510

Nakajima K, Ikuta K, Naito M, Ueda S, Kato S, Hirai K (1987) Analysis of Marek's disease virus serotype-1 specific phosphorylated polypeptides in virus-infected cells and Marek's disease lymphoblastoid cells. J Gen Virol 68:1379–1389

Nazerian K (1970) attenuation of Marek's disease virus and study of its properties in two different cell cultures. J Natl Cancer Inst 44:1256–1267

Nazerian K (1971) Further studies on the replication of Marek's disease virus in the chicken and in cell culture. J Natl Cancer Inst 47:207–217

Nazerian K (1973) Marek's disease: A neoplastic disease of chickens caused by a herpesvirus. Adv Cancer Res 17:279–315

Nazerian K (1974) DNA configuration in the core of Marek's disease virus. J Virol 13:1148–1150

Nazerian K, Burmester BR (1968) Electron microscopy of a herpesvirus associated with the agent of Marek's disease in cell culture. Cancer Res 28:2454–2462

Nazerian K, Solomon JJ, Witter RL, Burmester BR (1968) Studies on the etiology of Marek's disease. II. Finding of a herpesvirus in cell culture Proc Soc Exp Biol Med 127:177–182

Nazerian K, Lee LF, Witter RL, Burmester BR (1971) Ultrastructural studies of a herpesvirus of turkeys antigenically related to Marek's disease virus. Virology 43:442–452

Nii S, Yamada M, Yoshida M, Arao Y, Uno F, Ishikawa T, Hayashi M, Ono K, Hirai K (1988) Growth of MDV II in MDCC-MSB1-41C. In: Kato S, Horiuchi T, Mikami T, Hirai K (eds) Advances in Marek's disease research. Japanese Association on Marek's disease, Osaka pp 197–203

Okado K, Fujimota Y, Mikami T, Yonohara K (1972) The fine structure of Marek's disease virus and herpesvirus of turkey in cell culture. Japanese J Vet Res 20:57–68

Okazaki W, Purchase HG, Burmester BR (1970) Protection against Marek's disease by vaccination with a herpesvirus of turkeys. Avian Dis 14:413–429

Owen JJT, Moore MAS, Biggs PM (1966) Chromosome studies in Marek's disease. J Natl Cancer Inst 37:199–209

Pappenmeimer AM, Dunn LC, Cone V (1926) A study of fowl paralysis (neuro-lymphomatosis gallinarum). Storrs Agric Exp Sta Bull No.143:186–290

Pappenheimer AM, Dunn LC, Cone V (1929a) Studies on fowl paralysis (neurolymphomatosis gallinarum). I. Clinical features and pathology. J Exp Med 49:63–86

Pappenheimer AM, Dunn LC, Seidlin SM (1929b) Studies on fowl paralysis (neurolymphomatosis gallinarum). II. Transmission experiments. J Exp Med 49:87–102

Patterson FD, Wilcke HL, Murray C, Henderson EW (1932) So-called range paralysis of the chicken. J Amer Vet Med Ass 81:747–767

Payne LN, Frazier JA, Powell PC (1976) Pathogenesis of Marek's disease. International Review of Experimental Pathology 16:59–154

Phillips PA, Biggs PM (1972) Course of infection in tissues of susceptible chickens after exposure to strains of Marek's disease virus and turkey herpesvirus. J Natl Cancer Inst 49:1367–1373

Purchase HG, Biggs PM (1967) Characterization of five isolates of Marek's disease. Res vet Sci 8:440–449

Purchase HG, Okazaki W (1971) Effect of vaccination with herpesvirus of turkeys (HVT) on horizontal spread of Marek's disease herpesvirus. Avian Dis 15:391–397

Purchase HG, Burmester BR, Cunningham CH (1971) Response of cell cultures from various avian species to Marek's disease virus and herpesvirus of turkeys. Am J Vet Res 32:1811–1823

Rispens BH, Van Vloten H, Maas HJ, Hendrik JL (1972a) Control of Marek's disease in the Netherlands. I. Isolation of an avirulent Marek's disease virus strain (CVI 988) and its use in laboratory vaccination trials. Avian Dis 16:108–125

Rispens BH, Van Vloten H, Mastenbroek N, Maas HJ, Hendrik JL (1970b) Control of Marek's disease in the Netherlands. II. Field trials on vaccination with an avirulent strain (CVI 988) of Marek's disease virus. Avian Dis 16:126–138

Ross LJN (1985) Molecular biology of the virus. In: Payne LN (ed) Marek's disease. Scientific basis and methods of control. 113–150 Martinus Nijhoff Publishing, Boston Dordrecht Lancaster

Ross LJN, Biggs PM, Newton AA (1973) Purification and properties of the 'A' antigen associated with Marek's disease virus infections. J Gen Virol 18:291–304

Ross LJN, Delorbe W, Varmus HE, Bishop MJ, Brahie M, Haase A (1981) Persistence and expression of Marek's disease virus DNA in tumour cells and peripheral nerves studied by in situ hybridization. J Gen Virol 57:285–296

Ross LJN, Milne B, Biggs PM (1983) Restriction endonuclease analysis of Marek's disease virus DNA and homology between strains. J Gen Virol 64:2785–2790

Ross N, Binns MM, Sanderson M, Schat KA (1993) Alterations in DNA sequence and RNA transcription of the Bam H1-H fragment accompany attenuation of oncogenic Marek's disease herpesvirus. Virus Genes 7:33–51

Schat KA (1985) Characteristics of the virus. In: Payne LN (ed) Marek's disease. Scientific basis and methods of control, 77–112. Martinus Nijhoff Publishing, Boston Dordrecht Lancaster

Schat KA, Calnek BW (1978) Characterization of an apparently nononcogenic Marek's disease virus. J Natl Cancer Inst 60:1075–1082

Schat KA, Calnek BW, Fabricant J (1982) Characterisation of two highly oncogenic strains of Marek's disease virus. Avian Path 11:593–605

Schidlovsky G, Ahmed M, Jensen KE (1969) Herpesvirus in Marek's disease tumors. Science 164:959–961

Seagar EA (1933) The pathology of fowl paralysis with some aspects of its cause and control. Brit Vet J 89:454–473

Sevoian M, Chamberlain DM (1962) Avian lymphomatosis. Experimental reproduction of the ocular form. Vet Med 57:608–609

Sevoian M, Chamberlain DM, Counter F (1962) Avian lymphomatosis. Experimental reproduction of the neural and visceral forms. Vet Med 57:500–501

Sevoian M, Chamberlain DM, Larose RN (1963) Avian lymphomatosis. V. Air-borne transmission. Avian Dis 7:102–105

Sharma JM, Kenzy SG, Rissberger A (1969) Propagation and behavior in chicken kidney cultures of the agent associated with classical Marek's disease. J Natl Cancer Inst 43:907–916

Shek WR, Schat KA, Calnek BW (1982) Characterization of nononcogenic Marek's disease virus-infected and turkey herpesvirus infected lymphocytes. J Gen Virol 63:333–341

Silva RF, Barnett JC (1991) Restriction endonuclease analysis of Marek's disease virus DNA: differentiation of viral strains and determination of passage history. Avian Dis 35:487–495

Solomon JJ, Witter RL, Nazerian K, Burmester BR (1968) Studies on the etiology of Marek's disease. I. Propagation of the agent in cell culture. Proc Soc Exp Biol Med 127:173–177

Spencer JL, Calnek BW (1967) Storage of cells infected with Rous sarcoma virus or JM strain of avian lymphomatosis agent. Avian Dis 11:274–287

Spencer JL, Gilka F, Gavora JS, Hampson RJ, Caldwell DJ (1992) Studies with a Marek's disease virus that caused blindness and a high mortality in vaccinated flocks. Proc XIX World's Poultry Congress volume 1:199–201

Sugaya K, Bradley G, Nnonoyama M, Tanaka A (1990) Latent transcripts of Marek's disease virus are clustered in the short and long repeat regions. J Virol 64:5773–5782

Ubertini T, Calnek BW (1970) Marek's disease herpesvirus in peripheral nerve lesions. J Natl Cancer Inst 45:507–514

Van der Walle N, Winkler-Junius E (1924) De Neuritisepizootie bij Kippen te Barnveld in 1921. T. Vergelijk Geneesk Gezondhleer 10:34–50

Van Zaane D, Brinkhof JMA, Gielkins ALJ (1982) Molecular-biological characterization of Marek's disease virus. II. Differentiation of various MDV and HVT strains. Virology 121:133–146

Venugopal K, Bland AP, Ross LJN, Payne LN (1996) Pathogenicity of an unusual highly virulent Marek's disease virus isolated in the United Kingdom. In: Silva RF, Cheng HH, Coussens PM, Lee LF, Velicer LF (eds) Current Research in Marek's disease. Proc 5th International Symp on Marek's disease, pp 119–124

Warrack GH, Dalling T (1932) So-called "fowl paralysis". Also called neuritis in chickens, range paralysis, neurolymphomatosis gallinarum. Brit Vet J 88:28–43

Weiss RA, Biggs PM (1972) Leukosis and Marek's disease viruses of feral red jungle fowl and domestic fowl in Malaya. J Natl Cancer Inst 49:1713–1725

Wilson WR, Southwick RA, Pulaski JT, Tieber VL, Hong Y, Coussens PM (1994) Molecular analysis of glycoprotein C-negative phenotype of attenuated Marek's disease virus. Virology 199:393–402

Witter RL (1982) Protection by attenuated and polyvalent vaccines against highly virulent strains of Marek's disease virus. Avian Path 11:49–62

Witter RL (1983) Characteristics of Marek's disease viruses isolated from vaccinated commercial chicken flocks: Association of viral pathotype with lymphoma frequency. Avian Dis 27:113–132

Witter RL (1985) Principles of vaccination. In: Payne LN (ed) Marek's disease. Scientific basis of methods of control, 203–250. Martinus Nijhoff Publishing, Boston Dordrecht Lancaster

Witter RL (1996) Evolution of virulence of Marek's disease virus: evidence for a novel pathotype, In: Silva RF, Cheng HH, Coussens PM, Lee LF, Velicer LF (eds) Current Research in Marek's disease. Proc 5th International Symp on Marek's disease 86–91

Witter RL (1997) Increased virulence of Marek's disease virus field isolates. Avian Dis 41:149–163

Witter RL, Offenbecker L (1979) Nonprotective and temperature-sensitive variants of Marek's disease vaccine viruses. J Natl Cancer Inst 62:143–151

Witter RL, Burgoyne GH, Burmester BR (1968) Survival of Marek's disease agent in litter and droppings. Avian Dis 12:522–530

Witter RL, Burgoyne GH, Solomon JJ (1969a) Evidence for a herpesvirus as an etiologic agent of Marek's disease. Avian Dis 13:171–184

Witter RL, Solomon JJ, Burgoyne GH (1969b) Cell culture techniques for primary isolation of Marek's disease-associated herpesvirus. Avian Dis 13:101–118

Witter RL, Nazerian K, Purchase HG, Burgoyne GH (1970) Isolation from turkeys of a cell-associated herpesvirus antigenically related to Marek's disease virus. Am J Vet Res 31:525–538

Witter RL, Sharma JM, Offenbecker L (1976) Turkey herpesvirus infection in chickens: induction of lymphoproliferative lesions and characterization of vaccinal immunity against Marek's disease. Avian Dis 20:676–692

Witter RL, Sharma JM, Fadly AM (1980) Pathogenecity of variant Marek's disease virus isolants in vaccinated and unvaccinated chickens. Avian Dis 24:210–232

Witter RL, Lee LF, Sharma JM (1990) Biological diversity among serotype-2 Marek's disease viruses. Avian Dis 34:944–957

Pathogenesis of Marek's Disease Virus Infection

B.W. Calnek

1 Introduction

The term *pathogenesis* is most simply defined as the development of disease. Many uncomplicated infectious diseases proceed in a direct fashion from infection to a

Unit of Avian Health, Department of Microbiology and Immunology, College of Veterinary Medicine, Cornell University, Ithaca, NY 14853, USA specific pathologic expression. However, other diseases are relatively complex in terms of pathogenesis; Marek's disease (MD) most certainly belongs in this category. This chapter will concern itself with the spectrum of sequential events triggered by infection with MD virus (MDV), and the consequences of those events at both the cellular and host level. It will also consider the many factors that affect the potential manifestations of MD. Attention will be given to mechanisms and interactions associated with those factors. Major emphasis will be placed on the lymphomagenic nature of the disease.

2 Sequential Events

2.1 Pattern of Virus Infection in Susceptible Chickens Infected with Oncogenic (Serotype-1) Virus

Infection with MDV leading to lymphoma formation can be generally divided into four phases: (1) early cytolytic, (2) latent, (3) late cytolytic, and (4) transforming. Although essentially sequential, these are not necessarily discrete phases. A fairly sharp line demarcates the first two stages, and latent infection in certain cell types is a prerequisite to transformation, but both transforming and latent infections may exist intermixed with cytolytic infections in different cell populations as lymphomas are developing in the later stages. A permanent immunosuppression develops concurrent with the phase-3 cytolytic infection.

The various stages of infection with serotype-1 virus in genetically susceptible chickens have been described in detail in a number of comprehensive reviews (Biggs 1973; Calnek 1980, 1985b, 1986, 1998; Calnek and Witter 1997; Nazerian 1979; Payne 1972; Payne et al. 1976; Purchase 1972; Schat 1987; Settnes 1982). The earliest event, i.e., the initial infection of the chicken, occurs via the respiratory tract following inhalation of infectious cell-free MDV from a contaminated environment. The role of the lung as a site of primary infection is not entirely clear, since evidence of infection of parenchymal cells as an initial event has been meager at best (Adldinger and Calnek 1973; Phillips and Biggs 1972; Purchase 1970). On the other hand, the lung is the site from which phagocytic cells are presumed to pick up the virus and carry it to lymphoid organs such as the bursa of Fabricius (bursa), the thymus, and the spleen. At those sites, splenic ellipsoid-associated reticulum cells in blood vessel walls appear to be involved in the access to lymphocytes for MDV (Jeurissen et al. 1992). A productive, cytolytic, infection in those organs ensues and is particularly evident between 3 and 6 days postinfection (dpi). This necrotizing infection provokes an acute inflammation in which there is an influx of many cell types including macrophages, thymus-derived (T) and bursa-derived (B) lymphocytes, and various granulocytes. Both uncommitted and immunologically committed lymphocytes are present (Payne and Roszkowski 1973).

At 7–8 dpi, or sometimes slightly later, the infection in lymphoid organs switches from productive to latent and a "viremia", i.e., latent infection of peripheral blood lymphocytes (PBL), can be detected. There is a transient immunosuppression at about 7 dpi, apparently attributable to macrophage functions (Lee et al. 1978), and at the same time, a transient hyperplasia may be seen in the spleen. It seems likely that latency results from extrinsic factors associated with the immune responses that become evident at about 6–7 dpi (Higgins and Calnek 1975), although the lymphocyte subset might also be influential, given the fact that most cytolytically infected cells are B cells whereas most latently infected cells are T cells (see Sect. 3.2). The immune-response association with latency induction is compelling. Various cytokines, including interferon, are influential in the matter of latency (Buscaglia and Calnek 1988; Volpini et al. 1995, 1996), and immunocompetence is a requirement for latency to develop and to be maintained (Buscaglia et al. 1988). Latency develops in embryonally bursectomized chickens (Schat et al. 1981), so it can be presumed that cell-mediated immune functions are those of importance. Omar and Schat (1996) reported that cytotoxic T lymphocytes (CTLs) directed against early viral antigens are detectable by 6 dpi. It should be noted that the recrudescence of MDV cytolytic infection correlates with a concurrent permanent immunosuppression. In addition to specific immune responses, nonspecific immunity factors may also be involved in the switch to latency. Nitric oxide, produced by macrophages, and recombinant gamma interferon were able to inhibit the replication of MDV in vitro and in vivo (Xing and Schat 2000).

Infected PBL probably are the disseminators of virus to a large variety of tissues of epithelial origin where, by the end of the second week, a second phase of productive cytolytic infection is evident based on immunofluorescence tests and histopathology. No extracellular virus can be detected in the blood and so it must be concluded that virus spread is cell-to-cell. The question of virus receptors has not been resolved. Cell-free virus binds to, but does not easily penetrate, target cells in vitro (Adldinger and Calnek 1972); on the other hand, intracellular bridges appear to be an effective means for cell-to-cell spread of MDV (Hlozanek 1970; Kaleta and Neumann 1997).

Organs found to have infected non-lymphoid cells include the kidney, adrenal gland, proventriculus, uropygial gland, lung, liver, peripheral nerves, gonad, esophagus, crop, Harder's gland, and feather follicle (Adldinger and Calnek 1973; Calnek and Hitchner 1969; Phillips and Biggs 1972; Purchase 1970; Spencer and Calnek 1970; Ubertini and Calnek 1970). Probably there are few tissues truly spared. Organs which develop cytolytic infections in epithelial-derived tissues generally have infiltrating cells, including inflammatory cells and immunologically committed lymphocytes. Proliferating lymphoblasts are seen in some of the infected tissues, but the pleomorphic nature of the infiltrations suggests that neoplastically transformed lymphocytes are intermixed with nontransformed cells. A few of the infiltrating cells, including some that are neoplastically transformed, may show evidence of productive cytolytic infection; others are most certainly carrying viral genome based on the ability to rescue virus or otherwise demonstrate the presence of MDV genome.

Of all of the cytolytically infected tissues, only the feather follicle epithelium (FFE) produces virus that is infectious in the cell-free state (Calnek et al. 1970).

Although other tissues may contain a few enveloped virus particles detected by electron microscopy, no infectious cell-free virus can be obtained from disrupted cells. Thus, the feather follicle is unique in its epizootiologic role as the origin of virus that can spread from bird to bird. Virus is generally associated with desquamated, keratinized FFE cells which are shed along with molted feathers and as dander, thus contaminating the environment to complete the infection cycle (BEASLEY et al. 1970; CALNEK et al. 1970). Latent infection of PBL and splenic lymphocytes, and productive infection of the FFE, are lifelong events regardless of the consequences of infection, i.e., whether or not the bird develops MD (WITTER et al. 1971).

Beginning as early as 12–14 days, to as late as several weeks or even months after infection, microscopic and gross lesions may be seen in one or more of a large number of sites including lymphoid organs, visceral organs, muscle, skin, the eye, peripheral nerves, and brain. Even earlier, at 8–10 dpi, there may be an acute degenerative disease or a transient paralysis under the right conditions. Lymphoma development after 2 weeks or more is generally coupled with a permanent immunosuppression affecting both humoral and cell-mediated immunity (CMI) (see the chapter by K.A. Schat and C.J. Markowski, this volume). The degree of immunosuppression can vary and may be so severe as to permit lethal infections with other organisms in the absence of any lymphomatous lesions normally associated with MD. Immunosuppression generally accompanies the second wave of cytolytic infection but it is not known which is cause and which is effect, assuming they are related. Although lymphomatous lesions may regress (SHARMA et al. 1973), they often progress to massive size, thus compromising the function to the organ(s) in which they develop and resulting in paralysis (if nerves are involved), blindness (if eyes are affected), cachexia (particularly with visceral involvement), and death. In some cases, the development of disease can be very rapid and may cause death before clinical signs are prominent. Lymphomas can be diffuse or discrete in character.

2.2 Pattern of Virus Infection with Nononcogenic MDV and HVT

There are three types of nononcogenic virus that can be compared with oncogenic serotype-1 MDV: attenuated serotype-1 MDV, e.g., HPRS-16 att (CHURCHILL et al. 1969), naturally nononcogenic serotype-2 MDV, e.g., strains like SB-1 (SCHAT and CALNEK 1978), and serotype-3 HVT strains such as FC126 (WITTER et al. 1970). These all have a general pattern of tissue and organ infection somewhat similar to that of oncogenic serotype-1 MDV (CALNEK et al. 1979; FABRICANT et al. 1982; PAYNE et al. 1976; PHILLIPS and BIGGS 1972; SCHAT et al. 1985), but there are significant differences. The most prominent difference is a complete absence of productive, cytolytic infection in lymphoid organs or other tissues in chickens infected with HVT or attenuated MDV, even though virus can be isolated from latently infected cells in lymphoid organs or blood. In contrast, serotype-2 viruses cause at least a low level of cytolytic infection of lymphoid organs during the early infection period, but as with oncogenic viruses, the infection then becomes latent

(Calnek et al. 1979; Lin et al. 1991). Lin et al. (1991) found that productive infection with the ML-6 strain of serotype-2 virus was restricted to lymphoid organs with a single exception of a very minor involvement of the lung at 3 dpi and the FFE from the 14th day onward. Serotype-2 viruses spread horizontally (Lin et al. 1991; Schat and Calnek 1978), but neither attenuated MDV nor HVT do so, even though viral antigen can be detected for a limited period in the feather follicles of chickens infected with the latter (Fabricant et al. 1982).

3 Target Cells

Target cells for infection with oncogenic strains of MDV are of considerable interest because of their importance to the pathogenesis of the disease. After it was determined that the neoplastically transformed cells in lymphomas are T cells, it became important to learn the steps in pathogenesis that determine how and why they are singled out, and the mechanism(s) by which they are selectively transformed.

3.1 Targets for Cytolytic Infection

Shek et al. (1983) first determined that B lymphocytes are the primary targets for the early cytolytic infection in lymphoid organs. This finding was consistent with a report by Schat et al. (1981) showing that embryonal bursectomy obviated the early cytolytic infection phase following exposure to MDV and also with studies by Calnek et al. (1981, 1982) who found that B lymphocytes were the major target for in vitro infection of splenocytes. Baigent and Davison (1999) reported that most spleen cells expressing the viral gene product pp38 during the 4–6-dpi period were in aggregates of B cells surrounding sheathed capillaries. A few infected T cells could also be detected during the early infection period (Calnek et al. 1984). Those expressing pp38 were of both CD4$^+$ and CD8$^+$ phenotypes (Baigent and Davison 1999). Undoubtedly, these were activated T cells since resting T cells are refractory to infection with MDV (Calnek et al. 1985). The extent to which activated T cells support productive infection with MDV is not known, but they participate, to at least a minor extent, in the early cytolytic infection phase and probably to a greater extent as they emerge in larger numbers during the last part of this phase just before latency occurs. Aside from lymphocytes, a number of epithelial cells can be targets for cytolytic infection, as noted above.

3.2 Targets for Latent Infection

In contrast to the early cytolytic infection target cells, the predominant lymphocyte targets for latent infection are activated T cells, mainly of the CD4$^+$ phenotype, with only a minor population of latently infected B cells (Calnek et al. 1984; Hiramoto et al. 1996; Lee et al. 1999; Morimura et al. 1998; Shek et al. 1983). Probably, latently infected lymphocytes are those that become infected during the cytolytic phase when virus can easily spread from cell to cell, and then are prevented from completing the replicative cycle by intrinsic and/or extrinsic factors. The reason for the marked predominance of T cells rather than B cells during the latent phase is not known, but it could be that B cells are quickly lost because of their short life span compared to longer-lived T cells, or that the latent state in B cells is not as well induced and/or controlled by external influences as it is in T cells, or that T cells are intrinsically less likely to support a productive infection. There is some reason to suspect that the latter is at least of some importance; during the second wave of cytolytic infection that is responsible for the biphasic pattern of MDV infection. Latently infected, transformable T cells apparently coexist with productively infected lymphocytes in lymphoid organs in chickens that are severely immunosuppressed.

Although epithelial cells also are particularly prone to cytolytic infection, little work has been done to determine if they may also become latently infected. Pepose et al. (1981) reported that latent infections can be detected in satellite cells and non-myelinating Schwann cells of peripheral nerves and spinal ganglia.

3.3 Targets for Transformation

The heterogenous makeup of lymphomas made characterization of transformation targets difficult until continuous cell lines of MD tumor cells provided uniform populations for study. Akiyama et al. (1973) first reported the development of MD tumor cell lines and shortly thereafter, Powell et al. (1974) described two cell lines which they characterized as T cells. Other lines subsequently examined were invariably identified as T cells, including the original lines developed by Akiyama et al. (Matsuda et al. 1976; Nazerian and Sharma 1975; Nazerian et al. 1977; Ross et al. 1977; Schat et al. 1989, 1991). A variety of other antigens were subsequently found on cultured MD cells including so-called Marek's disease tumor-associated surface antigen (MATSA) (Matsuda et al. 1976; Powell and Rennie 1984; Witter et al. 1975), now known to be a marker associated with normal activated T cells (McColl et al. 1987), chicken fetal antigens (Murthy et al. 1979; Powell et al. 1983), and heterophile and Forssman antigens (Ikuta et al. 1981a,b). The significance of the latter antigens is unknown, but because MATSA seems to be a marker associated with activated T cells, it is reasonable to expect that transformed cells would express it.

Another antigen, AV37, recently described by Burgess et al. (1996), can also be found associated with MD tumor cells. The authors speculated that AV37 is linked with activation and may actually play a role in pathogenesis because it is thought to be associated with relatively high expression of MHC class I and II, thus affecting the immune recognition process. The antigen appears early (3–7 dpi) after MDV infection, and expression was found by Ross et al. (1997) to correlate with expression of *meq*, a putative MDV oncogene, and with small RNAs antisense to ICP4.

Through the use of monoclonal antibodies, SCHAT et al. (1982, 1991) were able to characterize a large number of MD cell lines and found that the vast majority were Ia-bearing (activated) CD4$^+$ cells that expressed either T-cell receptor (TCR) $\alpha\beta1$ (TCR2$^+$) or $\alpha\beta2$ (TCR3$^+$). Only a single line, of 45 tested, failed to express CD4 antigen and all were negative for CD8 expression. It could be deduced that the usual transformation targets in Marek's disease are a very specific subset, i.e., activated T-helper cells. However, it was also learned that although the usual transformation targets are CD4$^+$ T cells, this is apparently due more to the sequence of pathogenetic events than to a uniqueness of the subset as MDV-transformation target cells. CALNEK et al. (1989b) tested the hypothesis that activation of T cells is of crucial importance in lymphomagenesis because it provides an abundance of transformation targets. They inoculated alloantigens along with MDV in the wing-web or in the pectoral muscle of young chickens with the anticipation that the additional stimulation to the immune response would increase the number of activated T cells and thus enhance tumor development. Indeed, it did, because tumors developed at the site of inoculation in many of the test birds. Unexpectedly, MD lymphoblastoid tumor cell lines could be established from lymphocytes harvested from the early (4–6 dpi) inflammatory lesions and, surprisingly, these were found to represent a large number of combinations when tested for the presence of CD4, CD8, and TCR markers. Of 56 lines examined, only 12 (21%) were CD4$^+$, whereas 25 (45%) were CD8$^+$ and the remaining 19 (34%) were double negative, i.e., neither CD4$^+$ nor CD8$^+$. Both TCR2$^+$ and TCR3$^+$ markers were found, but with no obvious relationship to the CD4$^+$ or CD8$^+$ status of the lines. Probably, the alloantigens accelerated the immune response resulting in a more mixed population of T-cell subsets being present at the site of the cytolytic-phase infection before latency developed. The lack of uniqueness of CD4$^+$ cells as transformation targets was suggested by the SCHAT et al. (1991) study in which a single cell line was found to be double negative (neither CD4$^+$ nor CD8$^+$), a point confirmed by OKADA et al. (1997), who found a double-negative tumor in a chicken rendered deficient for CD4$^+$ and CD8$^+$ lymphocytes. Furthermore, tumor cells positive for the T-cell receptor (TCR1$^+$) were detected by BURGESS et al. (1996) in MD tumors associated with the superficial pectoral muscle and the jejunum where this phenotype is more prevalent than in other sites.

Three prerequisites for transformation of a given population of cells have been noted (CALNEK 1998): (1) they must be available at the time of, and at the site of, active infection; (2) they must be able to restrict viral replication, either intrinsically or because they are responsive to extrinsic inhibitors; and (3) they must be responsive to viral genes capable of effecting transformation either directly or indirectly. The foregoing observations suggest that these conditions are met by a variety of activated T cells (albeit most often a CD4$^+$ subset) but not B cells in chickens.

Although T cells are apparently the invariable transformation target in chickens, both B- and T-cell lymphomas have been observed in turkeys infected with MDV (NAZERIAN and SHARMA 1985; POWELL et al. 1984). Interestingly, infection of turkeys with oncogenic MDV strains produced little or no evidence of cytolytic infection in the various lymphoid organs during the period when acute necrotizing infection of these organs occurs in chickens (ELMUBARAK et al. 1981; POWELL et al. 1984), so the events leading to lymphoma formation are clearly different. The absence of productive infection in the bursa suggests that B cells are either not a favored early target, or they are intrinsically unable to support virus replication. The ability of MDV to transform both B and T cells, on the other hand, indicates that neither of these cell types is refractory to infection.

3.4 Targets for Infection by Nononcogenic Virus Strains

The target cells for attenuated MDV are not known. PAYNE et al. (1976) considered the possibility that attenuated virus is nononcogenic because of an inability to infect lymphocytes, a point supported in studies by SCHAT et al. (1985) who found that it was not possible to infect splenocytes with attenuated virus in vitro. Based on virus isolation attempts from splenocytes enriched or depleted of IgM-bearing B cells, SHEK et al. (1982) were unable to find evidence of B-cell infection in SB-1- or HVT-infected chickens at 5 dpi when there is little or no cytolytic infection of lymphoid tissues by these viruses (CALNEK et al. 1979; FABRICANT et al. 1982; SCHAT and CALNEK 1978). They did detect a low level of B-cell infection with SB-1 at 15 dpi, and CALNEK et al. (1981a) also concluded that a low level of B-cell infection occurred with SB-1. LIN et al. (1991), on the other hand, found substantial productive infection of B-cells in lymphoid organs (particularly spleen and bursa) from 5 to 14 dpi in chickens exposed intranasally to a different strain (ML-6) of serotype-2 MDV.

4 Consequences of Infection

The consequences of infection with MD herpesvirus depend on a number of single or interactive factors and these in turn affect not only the outcome of infection at the cellular level but, most importantly, the pathogenesis of the infection, and therefore the outcome of infection at the host level.

4.1 Cellular Level

Infection at the cellular level can result in at least four different responses: cytolysis, latency, proliferation, and transformation. The details of these virus–cell interactions are covered elsewhere in this book (particularly the chapters by P.M. Biggs, K.A. Schat and C.J. Markowski, R.W. Morgan et al., and H.-J. Kung et al., this volume). In general, it should be noted that the cell type, the immune-competency of the host coupled with its immune status, the virus strain, and vaccinal immunity all can influence the outcome in one way or another.

The importance of the cell type has already been discussed. For instance, B cells appear to be particularly prone to cytolytic infection whereas activated T cells are most often latently infected. With both B and T cells, this apparent difference is dictated partly, though perhaps not entirely, by the phase of infection (prior to or after immune responses) and the immunocompetence status and immune response(s) of the host. Proliferation, as a response to infection, has been observed in vitro. Two examples are a study by SPENCER (1970), who observed focal proliferation in cultured duck embryo fibroblasts following infection with MDV, and one by CALNEK and SCHAT (1991) in which in vitro MDV infection of activated T cells resulted in prolonged proliferation but not transformation. Finally, as already noted, only T cells are susceptible to transformation in chickens.

4.2 Host Level

The initial description, in 1907, of what we now call Marek's disease made no mention of many of the responses we now attribute to MDV infection; rather, it consisted of a detailed description of pathology classified as a "polyneuritis" (MAREK 1907). It was only later that it was learned that visceral lymphomas and eye lesions could be associated with the disease (PAPPENHEIMER et al. 1926), and that nerve lesions previously described as "polyneuritis" could be complex and involve both neoplastic and inflammatory components (PAPPENHEIMER et al. 1929; PAYNE and BIGGS 1967). The disease is now known to be very complex and to have many additional potential consequences, including apparently unrelated conditions that result from inflammatory/degenerative lesions, e.g., an early-mortality syndrome, a transient paralysis syndrome, and severe lymphoid organ atrophy with immunosuppression and the attendant consequences (CALNEK and WITTER 1997). Furthermore, MDV has been found to be a cause of atherosclerosis in chickens (FABRICANT et al. 1978).

4.2.1 Inflammatory and Necrotizing Disease

The inflammatory responses and necrotizing effects of cytolytic infection in lymphoid or other tissues are, by themselves, important aspects of the disease in addition to their obvious role in the cascade of events leading to lymphomas. Even though we now tend to consider the lymphomagenic nature of MD as being the primary concern, our present understanding of the disease recognizes that inflammation by itself may have serious consequences, and that the degenerative lesions that derive from productive infection are themselves sometimes the major effect of infection by MDV. This is particularly true with certain strains of virus, or under conditions that favor inflammatory or degenerative lesion development.

4.2.1.1 Transient Paralysis

Some strains of chickens infected with certain strains of serotype-1 MDV may develop lesions of the central nervous system (CNS) that cause a syndrome called transient paralysis (TP) (Kenzy et al. 1973). This condition is characterized by flaccid paralysis, often first involving the neck and then becoming more generalized. It occurs suddenly in young chickens at about 8–12 dpi and is generally followed by full recovery within 1–2 days. Wight (1968) described histopathological lesions including mild, diffuse cellular infiltrations, mild interfolial meningitis, swelling and proliferation of the capillary endothelium, and mild perivascular cuffing by mononuclear cells; these lesions were particularly evident in the cerebellum. Based on studies by Kornegay et al. (1983a,b), and by Swayne et al. (1988, 1989a,b,c), it is now evident that the clinical syndrome is the result of vasogenic brain edema. When edema is resolved, the clinical signs disappear. The edema is secondary to a vasculitis responsible for altered permeability of blood vessels. The pathogenesis of the vasculitis is not known, although it has been shown that there is B-cell involvement, given the need for a humoral immune response (Parker and Schierman 1983), and that the genotype is important since the condition is governed by genes within, or closely linked to, the major histocompatibility complex (Schierman and Fletcher 1980). Resistance to TP is dominant and appears to be independent of genetically controlled resistance to lymphomas or neural lesions characteristic of MD. However, TP-susceptible birds that recover from the transient disease often succumb a few weeks later from lymphomas and/or the more usual neural form of MD (Kenzy et al. 1973). Recent reports by Witter et al. (1999) and Gimeno et al. (1999) described a more acute form of TP in which chicks infected with a highly virulent strain of MDV failed to recover but instead succumbed within 1–3 days after the onset of clinical signs.

4.2.1.2 Ocular Disease

Ocular lesions consisting of pupil irregularities and loss of pigmentation of the iris ("gray eye") have long been associated with MD. Although proliferating lymphoblasts may be found in ocular lesions (Smith et al. 1974), suggesting a neoplastic element, often there is only evidence of inflammation in the form of edema and cellular infiltration of the iris, cornea, retina, pectin, uveal tract, and other structures (Ficken et al. 1991; Jungherr and Hughes 1965; Smith et al. 1974; Spencer et al. 1992). The result is blindness, a serious problem in its own right. As in the case of the vasculitis associated with transient paralysis, the pathogenesis of these inflammatory eye lesions is not known. Ficken et al. (1991) observed intranuclear inclusion bodies in both mononuclear cells and retinal cells. It could be that the degenerative changes associated with an active virus infection at the site are responsible for the inflammatory infiltrations, but it has not been determined if the viral infection of eye tissues precedes the infiltrations or if the infection is introduced by previously infected, infiltrating lymphocytes constituting part of the lesion. It appears that some cases, particularly those in which ocular lesions are the predominant problem, involve certain virus strains with a peculiar ocular tropism (Ficken et al. 1991; Spencer et al. 1992).

4.2.1.3 Degenerative Lesions, Immunosuppression, and Early-Mortality Syndrome

Severe systemic disease can result from the necrotizing infection of the two major immunity-related organs, the bursa and thymus, and other tissues including bone marrow (Jakowski et al. 1970; Witter et al. 1980). The absence of protective maternal antibody and infection with the more highly virulent strains of MDV enhances this non-neoplastic response which has been called "early-mortality syndrome" (EMS) (Witter et al. 1980). The consequences were reported to be aplasia of bone marrow (causing aplastic anemia) and marked atrophy of the bursa and thymus with an attendant depression of immunity (Jakowski et al. 1970), although it is important to remember that some of the MDV strains studied may have contained chicken anemia virus as a contaminant, which could cloud the interpretation of certain of the described consequences. An extravascular form of hemolytic anemia caused by high-virulence MDV isolates also has been described (Gilka and Spencer 1995). With certain virus isolates, high mortality may occur after infection at a young age (Buscaglia et al. 1995; Goodwin and Antillon 1995; Kross 1996; Venugopal et al. 1996; Witter 1996; Witter et al. 1980). Based on persistence of early cytolytic infection and the degree of atrophy of the bursa and thymus, Calnek et al. (1998) determined that there is a relationship between the immunosuppressive potential and the pathotype of MDV isolates, but the reason for this relationship is not known. It was suggested (Calnek et al. 1998) that it could be due to: (1) a particular propensity of high-virulence viruses to target lymphoid organs, above and beyond that of lower-virulence pathotypes; (2) infectivity for an expanded range of target lymphocyte subsets; (3) a greater likelihood that infection with the more virulent viruses results in cytolytic rather than latent infection of some lymphocyte subsets; or (4) differences in responsiveness to factors governing virus replication. Regardless, it is clear that certain isolates of MDV, when infecting genetically susceptible chickens in the absence of maternal antibody or vaccinal immunity, cause a virtual collapse of the immune system, which in turn is undoubtedly a major factor in EMS. Witter et al. (1999) noted that acute transient paralysis (see Sect. 4.2.1.1) also contributes to EMS.

The immunosuppression associated with MD could be directly related to loss of effector cell populations due to productive infection, but apoptosis and/or downregulation of CD8 (Morimura et al. 1995, 1996), or immunosuppression by certain MDV-encoded proteins (Cui and Lee 1997; Cui and Qin 1996) could also be involved (see the chapter by K.A. Schat and C.J. Markowski, this volume).

4.2.2 Nerve Lesions

The pathogenesis of neural lesions in MD is somewhat confusing since there is no published evidence of early virus infection that could serve as an incitant in peripheral nerves. Viral antigen and and/or virions have occasionally been observed in lymphocytes and Schwann cells of affected peripheral nerves (ADLDINGER and CALNEK 1973; UBERTINI and CALNEK 1970), but this was after gross lesions had developed. Most of the available information on the development of neural lesions in MD is derived from histopathological and ultrastructural examinations of peripheral nerves.

PAYNE and BIGGS (1967) described the pathologic changes in peripheral nerves as being of two main types. One, called A-type, is essentially neoplastic in character with proliferation of lymphoid cells, sometimes with demyelination and Schwann cell proliferation. B-type lesions, on the other hand, are inflammatory in nature being characterized by edema and infiltration with small lymphocytes and plasma cells.

A chronological study of ultrastructural changes in the peripheral nerves of MDV-infected chickens (LAWN and PAYNE 1979) provides valuable insight into the sequence of events. Cellular infiltration of nerves occurred as early as 5 days postinfection, with progression to proliferative A-type lesions by 3 weeks. Demyelination was evident at 4 weeks followed by the appearance of B-type lesions in some birds. It is interesting that the earliest nerve lesions coincided with the period of peak cytolytic infection of lymphoid organs, but the absence of virions or other evidence of MDV infection of peripheral nerves at that time raises the question of why the initial infiltrations occurred. No satisfactory explanation has been put forth, although LAWN and PAYNE (1979) pointed out that it is unlikely that autosensitization (as the result of demyelination) is involved because the damage follows rather than precedes the early infiltrations. LAWN and WATSON (1982) favored the hypothesis that lymphocytes infiltrate both brain and peripheral nerves since T lymphocytes stimulated as the result of the early MDV infection become highly motile, have an abnormal migration behavior, and are able to penetrate endothelial barriers. Because demyelination that occurs later in the evolution of nerve lesions was associated with the infiltration of the basement membrane of Schwann cells by lymphocytes and macrophages, it was concluded that the demyelination is likely the result of an allergic sensitization to normal nerve antigens (LAWN and PAYNE 1979). PAYNE (1979) suggested that at some point, probably during a later stage of nerve infiltration, an autosensitization to myelin occurs, provoking a progressive cell-mediated primary demyelination. This conclusion is consistent with observations of others who noted a similarity between MD nerve lesions and experimental allergic neuritis and the Landry-Guillain-Barré syndrome (FUGIMOTO and OKADA 1977; ICHIJO et al. 1981; LAMPERT et al. 1977; PEPOSE et al. 1981). The possibility of an autoimmune component for MD is further supported by studies in which immune complexes were found in the kidneys of MDV-infected chickens and quail (KAUL and PRADHAN 1991; PRADHAN et al. 1988). It is believed that the inflammatory B-type lesions develop following some repair and remyelination (PAYNE 1979). This presumes that either there is regression of neoplastic elements or that the earlier proliferative lesions did not contain transformed cells. It is not uncommon in MD for there to be mixtures of A- and B-type lesions in peripheral nerves supporting the idea that the pathologic scenario is complex.

4.2.3 Lymphomas

Lymphomagenesis is the consequence of MDV infection that is most commonly associated with Marek's disease. Lymphomatous tumors can appear as early as 12–14 dpi following infection of young, genetically susceptible chickens by a highly virulent strain of MDV, or they may not appear until several weeks or months after infection with less virulent viruses or under circumstances in which there are moderating factors such as older age at infection, vaccinal immunity, maternal antibodies, genetic resistance, etc. Lymphomas may occur in almost any visceral organ and also in various other sites including muscle, peripheral nerves, the eye, and skin.

Most MD lymphomas consist of a variable mixture of inflammatory cells, immunologically committed and uncommitted lymphocytes, macrophages, plasma cells, and neoplastically transformed T lymphocytes (HUDSON and PAYNE 1973; PAYNE et al. 1974; PAYNE and RENNIE 1976). Those in visceral organs may be mostly proliferative in character, but in other tissues, particularly, skin, eye, and nerves, inflammatory lesions may predominate in some cases. MDV infection in transformed tumor cells is largely nonproductive, although occasional cells may "turn on" and show evidence of viral antigen and virions. The selective involvement of various tissues is affected to some extent by both the virus strain and the genotype of the host, but the reasons for this are unknown. It could be that certain virus strains have an affinity for particular organs and that the ensuing infection there could attract transformed (or potentially transformable) T cells to the site with local lymphomas as the consequence. This appeared to be the case when lymphomas were induced at the site of inoculation in the local-lesion model (CALNEK et al. 1989b). Also, as reported by SHEARMAN and LONGENECKER (1981), transformed T cells may home to a particular site because of an organ-specific metastasis-associated antigen.

The pathogenesis of the lymphomatous lesions remains somewhat enigmatic. Much is known about some of the factors that affect the incidence of tumors (see following sections), but the exact mechanism(s) by which some of these operate is not well defined. Essentially, it has been determined that a specific cascade of events leading to infection of different populations of lymphocytes is critical to lymphomagenesis (see Fig. 1). It begins with the infection of B lymphocytes responsible for the early cytolytic infection phase, followed by a period of latency and ultimately transformation of T cells. The two latter stages are influenced by the first. One factor seemingly not involved in determining the probability of lymphoma development is host susceptibility to infection per se. Although the incidence of tumors is influenced by host genotype, age at infection, and virus strain, the intensity of the early cytolytic infection is largely unaffected by these variables (CALNEK et al. 1998; FABRICANT et al. 1977; LEE et al. 1981) (see Sect. 5). Of course the early infection phase involves mostly B cells and, as CALNEK (1998) pointed out, there could be differences in the rate of infection of activated T cells associated with genotype, virus strain, etc.

The importance of the early cytolytic infection cannot be overstated. It is undoubtedly this necrotizing infection and the ensuing inflammatory response that

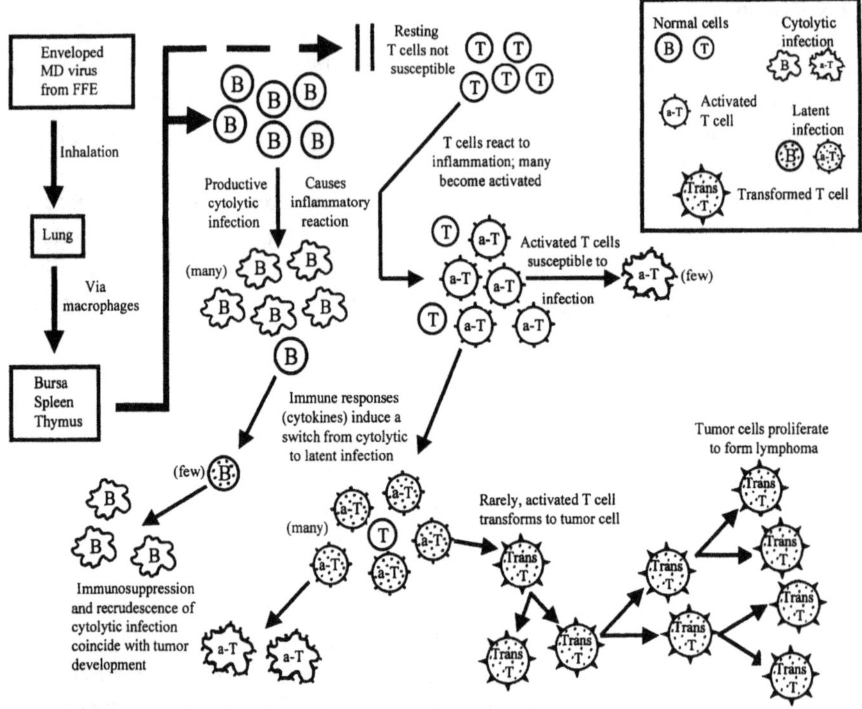

Fig. 1. Sequential events in the pathogenesis of Marek's disease lymphomas

stimulates the activation of T cells infiltrating the site of active productive infection. The greater the inflammation/immune response, the larger the number of activated T cells that might be expected to be present and therefore exposed to the productively infected B cells at the site. Indeed, enhancement of the inflammatory/immune response at a given site of MDV infection increases the probability of tumors (Calnek et al. 1989b; Petek and Dovadola 1970). Conversely, the reduction or absence of early cytolytic infection, whether due to maternal antibodies (Calnek 1972; Payne and Rennie 1973), vaccination (Calnek et al. 1980), virus serotype (Calnek et al. 1979), virus attenuation (Schat et al. 1985), or the absence of key early target organs (Schat et al. 1981) correlates with the absence or reduced incidence of lymphomas. Many of the modifying factors that affect tumor incidence are not absolute in their effects. Vaccinal immunity is a good example since vaccination usually does not totally eliminate MD. This fits very nicely with the concept that transformation is a relatively rare event governed by a set of probabilities. Thus, the more transformable targets that become infected, the greater the likelihood that a successful transformation will occur. Vaccination, as an example of a modifying factor, simply lowers the odds by reducing the level of active infection which, in turn, has two key effects: (1) it reduces the inflammatory response and thus keeps the number of activated T cells low, and (2) it eliminates foci of active infection where activated T cells that are already present could become infected.

Given the assumption that transformation is virus-directed, the most obvious conclusion is that one or more "oncogenes" provoke transformation either directly or indirectly. Approaches to identifying various genes that could contribute to oncogenicity of MDV have been described (Ross et al. 1997; Xie et al. 1996; see also the chapters by R.F. Silva et al., B. Lupiani et al., and H.-J. Kung et al., this volume) and a number of potentially important genes or latency-related RNAs have been described (Anderson et al. 1992; Bradley et al. 1989; Burgess et al. 1996; Cantello et al. 1994; Cui et al. 1990; Jones et al. 1992; Li et al. 1994; Maotani et al. 1986; McKie et al. 1995; Ohashi et al. 1994; Peng et al. 1992). Perhaps the most interesting of these is *meq*, which has transactivating activity and whose structure resembles that of the *jun/fos* oncogene family (see the chapter by H.-J. Kung et al.). Interference with the expression of several candidate oncogenes, including *meq*, ICP4, and a 1.8-kb gene family, by antisense sequences appears to alter the transformed phenotype of an MD cell line, suggesting that they are all involved in the maintenance of transformation (Kawamura et al. 1991; Xie et al. 1996).

In addition to the possible role of viral oncogenes in the transformation event, Venugopal and Payne (1995) noted that certain host cell genes, e.g., the tumor suppressor gene, *p53*, or protooncogenes such as *Bcl-2* could be responsible for interactions important in the molecular pathogenesis of MD.

Unfortunately, it is often difficult to take a direct approach in testing the significance of specific candidate oncogenes. For instance, the lack of oncogenicity of attenuated derivatives could be attributed to the inability of these viruses to infect lymphocytes efficiently (Schat et al. 1985), and the significance of the apparent loss of oncogenicity in an MDV deletion mutant lacking a functional *meq* coding region is tempered by the fact that the mutant virus induced infection of chickens less efficiently than the oncogenic parent virus.

A role for a chromosomal aberration in the pathogenesis of lymphomas was suggested by the discovery of an extra G-band on one homolog of the short arm of chromosome-1 in several MD lymphoblastoid cell lines (Bloom 1981; Moore et al. 1993). This extra band was due to the amplification of genomic DNA, but the potential of its significance was diminished by several subsequent observations: (1) it was infrequently seen in MD cell lines established from MD local lesions (Moore et al. 1994); (2) the site of the aberration was not a site of MDV integration (Delecluse and Hammerschmidt 1993); and (3) there was not a consistent correlation in MD cell lines between the occurrence of the aberration and features usually associated with neoplastic transformation such as colony formation in soft agar and transplantability (B.W. Calnek, unpublished data).

Regardless of the molecular basis, transformation is a relatively rare event in infected T cells. Assuming that latently infected, activated T cells all have the potential of becoming transformed, and given the knowledge that large numbers of activated T cells are infected during the period of latency, then it must be concluded that the transformation event itself is a very infrequent consequence of infection. This is supported when the question of clonality of tumors is considered. Delacluse and coworkers (Delecluse and Hammerschmidt 1993;

DELECLUSE et al. 1993) proposed that MD tumors are clonal in origin based on their finding that MDV DNA was randomly integrated into the genomes of lymphoma cells but the pattern of integration sites in the cells of a given lymphoma was consistent. On the other hand, multiple tumors in the same bird can be different, i.e., they may represent different T-cell phenotypes (SCHAT et al. 1991), and some cell lines established from individual local MD lesions (CALNEK et al. 1989b) contained transformed cells of more than one phenotype (B.W. Calnek, unpublished data).

4.2.4 Atherosclerosis

In 1973, FABRICANT et al. (1973) made the highly interesting observation that cell cultures infected with a feline herpesvirus contained both intracellular and extracellular cholesterol crystals, and they duly noted that such viral infections might have a role in degenerative vascular diseases. Based on various scattered reports suggesting the possible involvement of viruses, particularly herpesviruses, in humans (FABRICANT 1985), and especially a paper by PATERSON and COTTRAL (1950) in which there seemed to be an association between MD and coronary sclerosis in chickens, FABRICANT (1975) hypothesized a possible causal relationship between herpesvirus infection and human atherosclerosis. She and her coworkers subsequently tested the hypothesis using a low-virulence stain of MDV and genetically susceptible, specific-pathogen-free chickens (FABRICANT et al. 1978; MINICK et al. 1979). They found that the lesions induced by MDV included proliferative and fatty-proliferative changes in aortic, coronary, celiac, gastric, and mesenteric arteries. These changes were described as being strikingly similar to those seen in occlusive atherosclerosis in humans.

The pathogenesis of the lesions is of considerable interest from the viewpoint of comparative medicine. Although supplemental cholesterol in the diet appeared to enhance the severity of arterial lesions, the requirement for Marek's disease virus infection as the principal incitant was proven by the failure of a high cholesterol intake alone to produce atherosclerotic lesions, and more importantly, by the effectiveness of MD vaccines in preventing atherosclerosis in MDV-challenged chickens (FABRICANT 1985). Also, beginning about 1 month after infection, MD viral antigens were detected adjacent to lesions in smooth muscle cells in affected arteries, and cells comprising arterial lesions were found to be latently infected with MDV (FABRICANT 1985). Finally, it was shown that lipid metabolism was altered in arterial smooth muscle cells infected by MDV in vitro; there was an accumulation of phospholipids, free fatty acid, cholesterol, and cholesterol esters (FABRICANT et al. 1981; HAJJAR et al. 1985), and HAJJAR et al. (1986) detected similar alterations in cholesterol/cholesteryl ester metabolism in vivo during early stages of the disease. In spite of these highly interesting observations about the association between the MDV herpesvirus and atherosclerosis, the exact mechanism of the lesion evolution is presently unknown. In view of the known capacity of MDV to induce proliferation in certain other cells (CALNEK and SCHAT 1991; SPENCER 1970), it is interesting to speculate that a similar response of infected smooth muscle cells, coupled with the known alteration in lipid metabolism, might be a significant aspect of the pathogenesis of the lesion.

5 Factors Affecting Pathogenesis

The incidence of MD following exposure to MDV can be highly variable. Many factors are known to influence the likelihood that a given chicken will develop pathologic lesions and clinical disease after infection. Virus-related factors include both the serotype and the pathotype. Host factors include genotype, age, and sex. Modifying factors related to the host are those associated with its immune status, such as maternal antibody, immunosuppression from other conditions, and vaccinal immunity. Detailed discussions of several of these factors are found elsewhere: virus serotypes and pathotypes, and several other factors, are covered in the chapter by P.M. Biggs; the effect of genotype is addressed in the chapter by L.D. Bacon et al.; vaccinal immunity is considered in the chapters by R.L. Witter and by K.A. Schat and C.J. Markowski; and maternal antibody and immunocompetence are reviewed in the chapter by K.A. Schat and C.J. Markowski. It is important in the present chapter to identify how these differences might affect pathogenesis.

5.1 Virus Serotype and Pathotype

Neither serotype-2 nor serotype-3 viruses have been shown to have any oncogenic potential, but the reason for this is presently unknown. Determination of whether they lack genes required for transformation must await complete sequence information for various virus strains along with clarification regarding exactly which genes are involved in neoplastic transformation by known oncogenic viruses (see the chapters by R.F. Silva et al., B. Lupiani et al., Y. Izumiya et al., R.W. Morgan et al., and H.-J. Kung et al., this volume). The pattern of infection with non-oncogenic serotypes or with attenuated variants of serotype-1 viruses could also be instrumental, as previously noted.

The pathotype is of considerable importance in determining the outcome of MDV infection. Marked differences exist among serotype-1 virus strains in their pathogenic potential, with an apparent evolutionary shift toward higher virulence, broader host spectrum, and more varied lesion response in recent years, perhaps related to selective pressure associated with the use of MD vaccines (Witter 1998). In comparisons involving viruses that would now be classified (Witter 1997) as mild (mMDV), or virulent (vMDV) strains, few differences were found between pathotype and pattern of infection during the early cytolytic period (3–6 dpi) (Calnek et al. 1979; Fabricant et al. 1977; Smith and Calnek 1974), although higher virulence strains differed from lower virulence strains by showing a significant increase in virus titers in tissues and blood after the early infection period. In more recent studies involving vMDV, very virulent (vvMDV) and very virulent plus (vv + MDV) strains of virus, Calnek et al. (1998) determined that the two higher-virulence pathotypes (vvMDV and vv + MDV) resulted in higher virus-isolation rates from splenocytes at 4–8 dpi than were found with the lower-virulence pathotype (vMDV). Furthermore, although all three pathotypes induced similar levels of early (4–5 dpi) cytolytic infection, there were marked differences at 7–8 dpi when the vv + MDV strains had significantly higher levels of cytolytic infection in the bursa of Fabricius and thymus than were seen in chickens infected with vMDV or vvMDV strains. This indicated a persistence of cytolytic infection beyond the time when a switch to latency would normally occur, a feature attributed to a loss of immune competence. Also, a correlation was found to exist between virulence and atrophy of the bursa and thymus. Higher-virulence pathotypes caused more severe immunosuppression as measured by organ weights and histopathological evidence of necrosis and atrophy. Additionally, lymphoid organs of chickens infected with vv + MDVs showed little recovery between 8 and 14 dpi, whereas those of birds infected with vMDVs evidenced a significant return to normal by 14 dpi. Chickens given vvMDVs were intermediate in their recovery rate. All of these observations suggested a correlation between immunosuppressive potential and virulence, at least in the case of some of the more highly pathogenic strains of virus. This is in keeping with reports by others in which newer isolates of MDV were associated with unusually high virulence based on severe early cytolytic infection and exceptionally marked necrobiotic changes in lymphoid organs, high early mortality, severe and persistent atrophy of the bursa of Fabricius and thymus, and high mortality at 9–11 dpi (Buscaglia et al. 1995; Goodwin and Antillon 1995; Kross 1996; Venugopal et al. 1996; Witter 1997). The reasons for the exceptional virulence of some MDV isolates remains to be determined, although Barrow and Venugopal (1999) found some changes in amino acids within the *meq* and ICP4 proteins that were conserved in three very virulent European MDV isolates, suggesting that they may be involved in their high pathogenicity.

The possible association between high virulence and high immunosuppressive potential raises interesting questions (Calnek et al. 1998) including: (1) do vv + MDVs have a special propensity for lymphoid organs above and beyond that of vvMDVs or vMDVs? (2) do they infect a higher proportion of B and/or T cells?; and (3) do they more frequently induce a cytolytic vs latent infection in target lymphocytes, either because of inherent differences in viral replication events or because of differences in responsiveness to factors that govern virus replication, such as the cytokines shown to affect the establishment or maintenance of latency? Immunofluorescence data failed to support the idea that the vv + MDVs infect more lymphocytes during the early period (Calnek et al. 1998). Rather, it seems likely that the failure to enter latency with the accompanying continued destruction of lymphocytes accounts for the more severe atrophy of bursa and thymus in birds infected with the more virulent MDVs. This enhanced immunosuppressive activity could be because the vv + MDVs infect specific subsets of cells involved in effecting the switch to latency, thus reducing the required immune factors. Or, cells infected with the more virulent viruses may fail to display surface markers required for interaction with cytokines known to help modulate virus replication and thus latency. In any case, the severe destruction of the bursa and thymus results in overwhelming immunosuppression, a significant factor in the incidence of MD (see the chapter by K.A. Schat and C.J. Markowski, this volume). The importance of immune competence was apparent in early studies in which neonatal thymectomy enhanced the tumorigenicity of low-virulence strains of MDV (CALNEK et al. 1977) and increased the incidence of lymphomas in genetically resistant fowl (PAYNE 1972).

5.2 Genotype

Genetic resistance to MD is covered in detail in the chapter by L.D. Bacon et al. (this volume). However, several aspects of genotypic effects on the pathogenesis of MD should be briefly considered here: (1) influence on virus-related events such as infection, replication, latency or transformation; (2) effects on target cell availability; and (3) impact on immune responses against either virus infection or tumor progression that could affect the course of the disease. It is important to point out that both MHC-controlled and nonMHC-controlled types of resistance to MD are recognized (CALNEK 1985a).

With MHC-controlled resistance, an effect of genotype on virus infection, per se, has been ruled out based on both in vivo and in vitro studies. Initial cytolytic infections in resistant and susceptible strains of chickens were generally found to be equivalent (ABPLANALP et al. 1985; CALNEK 1973; FABRICANT et al. 1977) and MDV replicated equally well in cultured fibroblasts or kidney cells from genetically resistant and susceptible donors (SHARMA and PURCHASE 1974). The situation with nonMHC-controlled resistance may be somewhat different. When inbred line-6_2 (resistant) and line-7_2 (susceptible) strains were compared, the susceptible line had higher levels of early virus infection (BAIGENT et al. 1998; LEE et al. 1981; POWELL et al. 1982). These differences were thought to be related to target cell availability rather than to preferential susceptibility of individual target cells (BAIGENT et al. 1998; LEE et al. 1981). However, because splenocytes from the susceptible line adsorbed more cell-free HVT and MDV than did those from the resistant line, and because it was possible to transfer susceptibility by transplanting thymic cells from line-7_2 to line-6_2 chickens, POWELL et al. (1982) concluded that the susceptibility of line-7_2 is attributable to the greater susceptibility of their T cells to infection and transformation by MDV. Also, there is evidence that replication and spread of MDV is more efficient in line-7_2 than in line-6_2 chickens (BAIGENT and DAVISON 1999).

There is a clear genotype-based divergence in pathogenetic events beginning at about 7 dpi, when immune responses first become evident. Genetically susceptible chickens develop higher levels of latent infection of T-cells (FABRICANT et al. 1977; SHEK et al. 1983), and lower levels of virus-neutralizing antibodies (CALNEK 1973). For genetically resistant strains, the picture remains essentially unchanged after the first week postinfection unless they are infected by virus strains of very high pathogenicity or there are complicating factors such as immunosuppression

(Calnek 1985a). No secondary phase of cytolytic infection occurs and there is no permanent immunosuppression. Also, although latent infection in PBL and spleen cells can be detected and virus continues to be shed from the feather follicles, the level of these infections is low. Genetic strains that resist challenge with vMDV strains may fail to do so with vvMDV strains (Schat et al. 1981).

It seems probable that genetic resistance is due, at least in part, to immune responses. It is age-dependent (Calnek 1973; Sharma et al. 1975), and Sharma et al. (1975) found that neonatal thymectomy obviated age resistance. Higher virulence viruses, which can overcome genetic resistance, could do so by compromising the immune mechanisms that govern the switch from cytolytic to latent infection. With a prolonged cytolytic infection, there is an opportunity for more activated T cells to become infected. Because the cutoff of cytolytic infection is a critical event in the pathogenesis of MD, then it can be presumed that the important immune responses are those directed at viral antigens. This conclusion is supported by studies showing the development of CTLs directed against several MDV proteins (Omar and Schat 1996, 1997). An intriguing observation by Omar et al. (1998) was that CTLs which recognize the immediate-early protein, ICP4, developed in the genetically resistant N2a strain but not in the genetically susceptible P2a strain. They speculated that a response against this protein would eliminate virus-infected cells before they transferred infection to other cells, thus explaining why MDV infection in resistant-strain chickens remains latent rather than entering a second cytolytic infection phase, a key event in the pathogenesis of MD.

The effects of genetic resistance on pathogenesis are probably more complex than just affecting antiviral immune responses, or even antitumor responses should they exist. Contrary to the logic suggesting that a superior, genetically controlled, general immune response would correlate with resistance to MD, it was noted by a number of investigators that quite the opposite was true in certain comparisons of genetically susceptible and resistant strains of chickens. Blastogenic responses or other measures of T-cell immune functions of PBL or splenocytes from resistant-strain chickens often were poorer than those from susceptible-strain chickens (Calnek et al. 1985; Carpenter and Sevoian 1983; Fredericksen and Gilmour 1981; Lee and Bacon 1983; Schat et al. 1978). This helped in the development of the hypothesis that explains the sequential changes constituting the pathogenesis of MD (Fig. 1) (Calnek 1985b, 1986; Schat et al. 1982a). Part of the hypothesis states that the greater the number of activated T cells generated during periods of active MDV infection, the greater likelihood that lymphomas will develop. In other words, genetic strains with superior CMI responses could be expected to be more susceptible to the disease. This hypothesis was supported by the work noted above in which Calnek et al. (1989b) showed that enhancement of CMI at local sites of MDV infection increased the number of tumors. Furthermore, Baigent and Davison (1999) concluded from studies on the development and composition of splenic lesions that T-cell responses in early infection are greater in susceptible line-7_2 than in resistant line-6_2 chickens.

A role for natural killer (NK) cell activity to help explain genotypic differences in susceptibility to MD cannot be overlooked, given the observation that NK activity is increased after MDV infection or vaccination in genetically resistant strains, but is decreased in susceptible strains with MD tumors (HELLER and SCHAT 1987; SHARMA 1981). However, exactly how this affects the pathogenesis of the disease is unknown.

Little is known about the possibility that genetic resistance or susceptibility could reside at the level of virus genome expression through an effect on latency or transformation. It is instructive to note that when local lesions were induced by simultaneous injection of MDV and alloantigens in resistant (strain N2a) and susceptible (strain P2a or S13) chickens, the resistant-strain birds had significantly lower numbers of tumors, but transformed MD cell lines were easily established from cells collected from the lesions at 6 dpi regardless of the genotype of the donors (CALNEK et al. 1989a). This suggested that the resistance did not relate to transformability of the target T cells.

5.3 Age

So-called age resistance is apparently only an age-related expression of natural (genetic) resistance. This resistance, seen only in genetically resistant strains, is expressed to a variable degree at hatching, and is acquired gradually during several weeks paralleling the acquisition of immune competence (CALNEK 1973; SHARMA 1976). There is no effect of age on susceptibility to infection by MDV or on the intensity of the early cytolytic infection (ANDERSON et al. 1971; CALNEK 1973; WITTER et al. 1973). Rather, age appears to be influential only after latency develops. BUSCAGLIA et al. (1988) showed that young, immunologically incompetent chicks failed to show a switch from cytolytic to latent infection at the expected time. Also, lesion regression is reported to be a mediating factor (SHARMA et al. 1973), and an immunological basis is confirmed by the ability of neonatal thymectomy to abrogate age resistance (SHARMA et al. 1975).

5.4 Maternal Antibody

JAKOWSKI et al. (1970) reported that MDV-infected specific-pathogen-free chicks developed an acute disease characterized by severe hematopoietic destruction and lymphoid organ degeneration, apparently due to the absence of MDV maternal antibody; however, it is also possible that maternal antibody against chicken infectious anemia virus (CIAV) may have been absent as well and that CIAV could have contributed to the damage reported. A degree of protective efficacy against MD mortality was attributed to maternal antibody by several groups (BALL et al. 1971; CHUBB and CHURCHILL 1969; SPENCER and ROBERTSON 1972). Subsequently, CALNEK (1972) and PAYNE and RENNIE (1973) confirmed the ameliorative effects of passively acquired antibody, showing that its effect was to markedly suppress the initial productive virus infection and acute inflammation seen in lymphoid organs. The ameliorative effect of maternally derived protection is likely due to virus-neutralizing antibodies directed against gB (SCHAT 1996).

5.5 Vaccinal Immunity

Vaccinal immunity induced by any of the three vaccine serotypes has the effect of markedly reducing the level of early cytolytic infection (Calnek et al. 1980; Schat et al. 1982b; Smith and Calnek 1974), thus incriminating antiviral immunity as a very significant factor in the protective efficacy of MD vaccines. Based on cell-mediated cytotoxicity assays, it has been shown that there are common CMI epitopes on cells infected with oncogenic or vaccine viruses (Omar and Schat 1996; Omar et al. 1998).The significance of antiviral immunity in protection against lymphomas was confirmed by the efficacy of inactivated viral antigens (Murthy and Calnek 1979) and the ability of recombinant fowl pox virus expressing the glycoprotein B of MDV to induce protection against MD (Nazerian et al. 1992) probably through the stimulation of CTLs (Omar et al. 1998). Also, although immunization does not prevent infection, the level of latent infection and productive infection of the FFE is low. Lee et al. (1999) reported that the MDV re-isolation rates from vaccinated chickens were significantly lower than those of nonvaccinated chickens, and that $CD4^+$ T cells appeared to be the preferential subset for MDV infection in both groups. Omar and Schat (1997) identified a population of $CD8^+$ TCRαβ1, but not $CD8^+$ TCRαβ2 nor $CD4^+$, T cells that were involved in antiviral immunity in chickens vaccinated with serotype-2 SB-1 virus. Morimura et al. (1998) confirmed that $CD8^+$ T cells are largely responsible for the control of viral infection, and offered the hypothesis that these cells, in conjunction with other effector mechanisms including NK cells, NK T cells, double negative T cells, and T cells destroy the remaining infected cells and/or transformed cells. The possibility that there is a direct effect of immune responses on transformed cells is debatable, particularly since no unique tumor-associated antigens have been found (see the chapter by K.A. Schat and C.J. Markowski, this volume). On the other hand, it is very reasonable to expect that only the antiviral activity of vaccines would be enough to reduce the incidence of lymphomas by interrupting the normal cascade of pathogenetic events. The low level of latent T-cell infection in vaccinated birds is evidence supporting this concept since this is governed by the extent of T-cell activation and infection provoked by the cytolytic/inflammatory phase.

5.6 Immunocompetence

MDV is, itself, immunosuppressive, and as noted above, there may be a relationship between immunosuppressive potential and virulence of certain MDV isolates. It is not determined whether the correlation between permanent immunosuppression, late cytolytic infection, and lymphoma development is one of cause or effect. Regardless, it is important to note that immunosuppression from other conditions can affect the incidence of MD. Powell and Davison (1986) observed an increase in MD incidence in chickens experimentally immunosuppressed with betamethasone or corticosteroids. Subsequently, Buscaglia et al. (1988) found that either experimentally induced immunosuppression or natural incompetence based on young age resulted in prolonged and more widespread early cytolytic infection, thus interfering with the establishment of latency. Furthermore, immunosuppression after latency had developed caused the reappearance of cytolytic infection in the spleen suggesting that immunocompetence is also required for the maintenance of latency. So, it appears that immunocompetence is very significant in the pathogenesis of MD, and that a vicious cycle can be initiated in which incompetence and failure to enter latency causes a continuation of cytolytic infection, thus inducing even more destruction to the immune system and further immunosuppression.

6 Conclusions

Clearly, the pathogenesis of Marek's disease is complex with many influential factors and with many possible expressions of pathology. Probably, the most complex of the varied potential outcomes of MDV infection is lymphomagenesis in which a very complicated and specific cascade of events must occur. Viral pathotype and host genotype are of critical importance in determining the likelihood of lymphoma development and these affect the influence imparted by other modifying factors such as age, maternal antibody, and vaccinal immunity. The underlying mechanisms by which many of these various factors affect the events comprising the pathogenesis of the disease are still the subject of investigation. Some of the specific questions that continue to beg clarification are:

1. What are the exact relationships among permanent immunosuppression, recrudescence of cytolytic infection, and lymphomagenesis, i.e., which comes first and how does one influence the other?
2. Why is transformation such a rare event in latently infected T cells and exactly what is the mechanism by which transformation is effected?
3. Why are more virulent strains of virus more immunosuppressive and is this a key factor in their enhanced virulence?
4. How do transformed cells escape immunosuppression?

These, and other questions, will assure that the subject of pathogenesis will be of continuing interest.

References

Abplanalp H, Schat KA, Calnek BW (1985) Resistance to Marek's disease of congenic strains differing in major histocompatibility haplotypes to 3 virus strains. In: Calnek BW, Spencer JL (eds) Intl Symp on Marek's Disease. American Association of Avian Pathologists, Kennett Square, PA, pp 347–358

Adldinger HK, Calnek BW (1972) Effect of chelators on the in vitro infection with Marek's disease virus. In: Biggs PM, de Thé G, Payne LN (eds) Oncogenesis and Herpesviruses. IARC, Lyon, France, pp 99–105

Adldinger HK, Calnek BW (1973) Pathogenesis of Marek's disease: early distribution of virus and viral antigens in infected chickens. J Natl Cancer Inst 50:1287–1298

Akiyama Y, Kato S, Iwa N (1973) Continuous cell culture from lymphoma of Marek's disease. Biken J 16:177–179

Anderson AS, Francesconi A, Morgan RW (1992) Complete nucleotide sequence of the Marek's disease virus ICP4 gene. Virology 189:657–667

Anderson DP, Eidson CS, Richey DJ (1971) Age susceptibility of chickens to Marek's disease. Am J Vet Res 32:935–938

Baigent SJ, Davison TF (1999) Development and composition of lymphoid lesions in the spleens of Marek's disease virus-infected chickens: association with virus spread and the pathogenesis of Marek's disease. Avian Pathol 28:287–300

Baigent SJ, Ross LJN, Davison TF (1998) Differential susceptibility to Marek's disease is associated with differences in number, but not phenotype or location, of pp38+ lymphocytes. J Gen Virol 79:2795–2802

Ball RF, Hill JF, Lyman J, Wyatt A (1971) The resistance to Marek's disease of chicks from immunized breeders. Poult Sci 50:1084–1090

Barrow A, Venugopal K (1999) Molecular characteristics of very virulent European MDV isolates. Acta Virologica 43:90–93

Beasley JN, Patterson LT, McWade DH (1970) Transmission of Marek's disease by poultry house dust and chicken dander. Am J Vet Res 31:339–344

Biggs PM (1973) Marek's disease. In: Kaplan AS (ed) The Herpesviruses. Academic Press, New York, pp 557–594

Bloom SE (1981) Detection of normal and aberrant chromosomes in chicken embryos and in tumor cells. Poult Sci 60:1355–1361

Bradley G, Hayashi M, Lancz G, Tanaka A, Nonoyama M (1989) Structure of the Marek's disease virus BamHI-H gene family: genes of putative importance for tumor induction. J Virol 63:2534–2542

Burgess SC, Kaiser P, Davison TF (1996) A novel lymphoblastoid surface antigen and its role in Marek's disease (MD). In: Silva RF, Cheng HH, Coussens PM, Lee LF, Velicer LF (eds) Current Research on Marek's Disease. American Association of Avian Pathologists, Kennett Square, PA, pp 29–39

Buscaglia C, Calnek BW (1988) Maintenance of Marek's disease herpesvirus latency in vitro by a factor found in conditioned medium. J Gen Virol 69:2809–2818

Buscaglia C, Calnek BW, Schat KA (1988) Effect of immunocompetence on the establishment and maintenance of latency with Marek's disease herpesvirus. J Gen Virol 69:1067–1077

Buscaglia C, Nervi P, Garbi JL, Piscopo M (1995) Isolation of very virulent strains of Marek's disease virus from vaccinated chickens in Argentina. In: Proc 44th Western Poultry Disease Conference, Sacramento, CA, pp 53–57

Calnek BW (1972) Effects of passive antibody on early pathogenesis of Marek's disease. Infect Immun 6:193–198

Calnek BW (1973) Influence of age at exposure on the pathogenesis of Marek's disease. J Natl Cancer Inst 51:929–939

Calnek BW (1980) Marek's disease virus and lymphoma. In: Rapp F (ed) Oncogenic Herpesviruses. CRC Press, Boca Raton, FL, pp 103–143

Calnek BW (1985a) Genetic Resistance. In: Payne LN (ed) Marek's Disease. Martinus Nijhoff, Boston, MA, pp 293–328

Calnek BW (1985b) Pathogenesis of Marek's disease: a review. In: Calnek BW, Spencer JL (eds) Proc Int Symp Marek's Dis American Association of Avian Pathologists, Kennett Square, PA, pp 374–390

Calnek BW (1986) Marek's disease: a model for herpesvirus oncology. CRC Crit Rev Microbiol 12: 293–320

Calnek BW (1998) Lymphomagenesis in Marek's disease. Avian Pathol 27:s54–s64

Calnek BW, Hitchner SB (1969) Localization of viral antigen in chickens infected with Marek's disease herpesvirus. J Natl Cancer Inst 43:935–949

Calnek BW, Schat K (1991) Proliferation of chicken lymphoblastoid cells after in vitro infection with Marek's disease virus. Avian Dis 35:728–737

Calnek BW, Witter RL (1997) Marek's disease. In: Calnek BW, Barnes HJ, Beard CW, McDougald LR, Saif YM (eds) Diseases of Poultry. Iowa State University Press, Ames, IA, pp 369–413

Calnek BW, Adldinger HK, Kahn DE (1970) Feather follicle epithelium: a source of enveloped and infectious cell-free herpesvirus from Marek's disease. Avian Dis 14:219–233

Calnek BW, Fabricant J, Schat KA, Murthy KK (1977) Pathogenicity of low-virulence Marek's disease viruses in normal versus immunologically compromised chickens. Avian Dis 21:346–358

Calnek BW, Carlisle JC, Fabricant J, Murthy KK, Schat KA (1979) Comparative pathogenesis studies with oncogenic and nononcogenic Marek's disease viruses and turkey herpesvirus. Am J Vet Res 40:541–548

Calnek BW, Schat KA, Fabricant J (1980) Modification of Marek's disease pathogenesis by in ovo infection or prior vaccination. In: Essex M, Todaro G, zur Hausen H (eds) Viruses in Naturally Occurring Cancers. Cold Spring Harbor, New York, pp 185–197

Calnek BW, Shek WR, Schat KA (1981) Latent infections with Marek's disease virus and turkey herpesvirus. J Natl Cancer Inst 66:585–590

Calnek BW, Schat KA, Shek WR, Chen C-LH (1982) In vitro infection of lymphocytes with Marek's disease virus. J Natl Cancer Inst 69:709–713

Calnek BW, Schat KA, Ross LJN, Shek WR, Chen C-LH (1984) Further characterization of Marek's disease virus-infected lymphocytes. I. In vivo infection. Int J Cancer 33:389–398

Calnek BW, Schat KA, Heller ED, Buscaglia C (1985) In vitro infection of T-lymphoblasts with Marek's disease virus. In: Calnek BW, Spencer JL (eds) Proc Int Symp Marek's Dis. American Association of Avian Pathologists, Kennett Square, PA, pp 173–187

Calnek BW, Lucio B, Schat KA (1989a) Pathogenesis of Marek's disease virus-induced local lesions. 2. Influence of virus strain and host genotype. In: Kato S, Horiuchi T, Mikami T, Hirai K (eds) Advances in Marek's Disease Research. Japanese Association on Marek's Disease, Osaka, Japan, pp 324–330

Calnek BW, Lucio B, Schat KA, Lillehoj HS (1989b) Pathogenesis of Marek's disease virus-induced local lesions. 1. Lesion characterization and cell line establishment. Avian Dis 33:291–302

Calnek BW, Harris RW, Buscaglia C, Schat KA, Lucio B (1998) Relationship between the immunosuppressive potential and the pathotype of Marek's disease virus isolates. Avian Dis 42:124–132

Cantello JL, Anderson AS, Morgan RW (1994) Identification of latency-associated transcripts that map antisense to the ICP4 homolog gene of Marek's disease virus. J Virol 68:6280–6290

Carpenter SL, Sevoian M (1983) Cellular immune response to Marek's disease: listeriosis as a model of study. Avian Dis 27:344–356

Chubb RC, Churchill AE (1969) Effect of maternal antibody on Marek's disease. Vet Rec 85:303–305

Churchill AE, Chubb RC, Baxendale W (1969) The attenuation, with loss of oncogenicity of the herpes-type virus of Marek's disease (strain HPRS-16) on passage in cell culture. J Gen Virol 4:557–564

Cui Z, Lee LF (1997) Construction of a recombinant CV1988 strain expressing virulent epitopes on 38 KD phosphorylated protein of Marek's disease virus. In: Proc XIth Inter Cong World's Vet Poult Assoc, Budapest, Hungary, pp 32

Cui Z, Qin A (1996) Immunodepressive effects of the recombinant 38 KD phosphorylated protein of Marek's disease virus. In: Silva RF, Cheng HH, Coussens PM, Lee LF, Velicer LF (eds) Current Research on Marek's Disease. American Association of Avian Pathologists, Kennett Square, PA, pp 278–283

Cui Z-Z, Ding Y, Lee LF (1990) Marek's disease virus gene clones encoding virus-specific phosphorylated polypeptides and serological characterization of fusion proteins. Virus Genes 3:309–322

Delecluse H-J, Hammerschmidt W (1993) Status of Marek's disease virus in established lymphoma cell lines: herpesvirus integration is common. J Virol 67:82–92

Delecluse H-J, Schüller S, Hammerschmidt W (1993) Latent Marek's disease virus can be activated from its chromosomally integrated state in herpesvirus-transformed lymphoma cells. Eur Mol Biol Organ J 12:3277–3286

Elmubarak AK, Sharma JM, Witter RL, Nazerian K, Sanger VL (1981) Induction of lymphomas and tumor antigen by Marek's disease virus in turkeys. Avian Dis 25:911–926

Fabricant C (1975) Herpesvirus induced cholesterol: an added dimension in the pathogenesis, prophylaxis, or treatment of atherosclerosis. Artery 1:361

Fabricant CG (1985) Atherosclerosis: the consequence of infection with a herpesvirus. Adv Vet Sci Comp Med 30:39–66

Fabricant CG, Krook L, Gillespie JH (1973) Virus-induced cholesterol crystals. Science 181:566–567

Fabricant CG, Fabricant J, Litrenta MM, Minick CR (1978) Virus-induced atherosclerosis. J Exp Med 148:335–340

Fabricant CG, Hajjar DP, Minick CR, Fabricant J (1981) Herpesvirus infection enhances cholesterol and cholesteryl ester accumulation in cultured arterial smooth muscle cells. Am J Pathol 105:176–184

Fabricant J, Calnek BW, Schat KA (1982) The early pathogenesis of turkey herpesvirus infection in chickens and turkeys. Avian Dis 26:257–264

Fabricant J, Ianconescu M, Calnek BW (1977) Comparative effects of host and viral factors on early pathogenesis of Marek's disease. Infect Immun 16:136–144

Ficken MD, Nasisse MP, Boggan GD, Guy JS, Wages DP, Witter RL, Rosenberger JK, Nordgren RM (1991) Marek's disease virus isolates with unusual tropism and virulence for ocular tissues: clinical findings, challenge studies and pathological features. Avian Pathol 20:461–474

Fredericksen TL, Gilmour DG (1981) Chicken lymphocyte alloantigen genes and responsiveness of whole blood cells to Concanavalin A. Fed Proc 40:977

Fugimoto Y, Okada Y (1977) Pathological studies of Marek's disease. III. Electron microscopic observation on demyelination of the peripheral nerves. Jap J Vet Res 25:59–70

Gilka F, Spencer JL (1995) Extravascular haemolytic anaemia in chicks infected with highly pathogenic Marek's disease viruses. Avian Pathol 24:393–410

Gimeno IM, Witter RL (1999) Four distinct neurologic syndromes in Marek's disease: effect of viral strain and pathotype. Avian Dis 43:721–737

Goodwin MA, Antillon A (1995) Necrotizing herpesvirus bursitis, thymusitis, and splenitis in chickens. Avian Dis 39:444–447

Hajjar DP, Falcone DJ, Fabricant C, Fabricant J (1985) Altered cholesteryl ester cycle is associated with lipid accumulation in herpesvirus-infected arterial smooth muscle cells. J Biol Chem 260:6124–6128

Hajjar DP, Fabricant CG, Minick CR, Fabricant J (1986) Virus-induced atherosclerosis. Am J Pathol 122:62–70

Heller ED, Schat KA (1987) Enhancement of natural killer cell activity by Marek's disease vaccines. Avian Pathol 16:51–60

Higgins DA, Calnek BW (1975) Fowl immunoglobulins: quantitation and antibody activity during Marek's disease in genetically resistant and susceptible birds. Infect Immun 11:33–41

Hiramoto W, Takeda K, Anzai R, Ogasawara K, Sakihara H, Sugiura K, Seki S, Kumagai K (1996) Reisolation of Marek's disease virus from T cell subsets of vaccinated and non-vaccinated chickens. In: Silva RF, Cheng HH, Coussens PM, Lee LF, Velicer LF (eds) Current Research on Marek's Disease. American Association of Avian Pathologists, Kennett Square, pp 130–135

Hlozanek I (1970) The influence of ultraviolet-inactivated Sendai virus on Marek's disease virus infection in tissue culture. J Gen Virol 9:45–50

Hudson L, Payne LN (1973) An analysis of the T and B cells of Marek's disease lymphomas of the chicken. Nature (New Biol) 241:52–53

Ichijo K, Fujimoto Y, Okada K (1981) Ultrastructural study of experimental allergic neuritis in the chicken. Zbl Vet Med B 28:210–225

Ikuta K, Kitamoto N, Shoji H, Kato S (1981a) Hanganutziu and Deicher type heterophile antigen expressed on the cell surface of Marek's disease lymphoma-derived cell lines. Biken J 24:23–37

Ikuta K, Kitamoto N, Shoji H, Kato S, Naiki M (1981b) Expression of Forssman antigen of avian lymphoblastoid cell lines transformed by Marek's disease virus or avian leukosis virus. J Gen Virol 52:145–151

Jakowski RM, Fredrickson TN, Chomiak TW, Luginbuhl RE (1970) Hematopoietic destruction in Marek's disease. Avian Dis 14:374–385

Jeurissen SHM, Janse EM, Wagenaar F, deBoer GF (1992) The role of splenic ellipsoid-associated reticulum cells in the pathogenesis of Marek's disease. In: de Boer GF, Jeurissen SHM (eds) 4th Intl Symp Marek's Disease. Ponsen & Looijen, Wageningen, Amsterdam, The Netherlands, pp 211–215

Jones D, Lee L, Liu JL, Kung HJ, Tillotson JK (1992) Marek's disease virus encodes a basic-leucine zipper gene resembling the fos/jun oncogenes that is highly expressed in lymphoblastoid tumors. Proc Natl Acad Sci USA 89:4042–4046

Jungherr EL, Hughes WF (1965) The avian leukosis complex. In: Biester H, Schwarte L (eds) Diseases of Poultry. Iowa State University Press, Ames, IA, pp 512–567

Kaleta EF, Neumann U (1977) Untersuchungen zum Ubertragungsmechanismus des Puten-Herpesvirus in vitro. Avian Pathol 6:33–39

Kaul L, Pradhan HK (1991) Immunopathology of Marek's disease in quails: presence of antinuclear antibody and immune complex. Vet Immunol and Immunopathol 28:89–96

Kawamura M, Hayashi M, Furuichi T, Nonoyama M, Isogai E, Namioka S (1991) The inhibitory effects of oligonucleotides, complementary to Marek's Disease virus mRNA transcribed from the BamHI-H region, on the proliferation of transformed lymphoblastoid cells, MDCC-MSB1. J Gen Virol 72:1105–1111

Kenzy SG, Cho BR, Kim Y (1973) Oncogenic Marek's disease herpesvirus in avian encephalitis (temporary paralysis). J Natl Cancer Inst 51:977–982

Kornegay JN, Gorgacz EJ, Parker MA, Brown J, Schierman LW (1983a) Marek's disease virus-induced transient paralysis: clinical and electrophysiologic findings in susceptible and resistant lines of chickens. Am J Vet Res 44:1541–1544

Kornegay JN, Gorgacz EJ, Parker MA, Duncan JR, Schierman LW (1983b) Marek's disease virus-induced transient paralysis: a comparison of lesions in susceptible and resistant lines of chickens. Acta Neuropathol 61:263–269

Kross I (1996) Isolation of highly lytic serotype 1 Marek's disease virus from recent field outbreaks in Europe. In: Silva RF, Cheng HH, Coussens PM, Lee LF, Velicer LF (eds) Current Research on Marek's Disease. American Association of Avian Pathologists, Kennett Square, PA, pp 113–118

Lampert PW, Garrett R, Powell H (1977) Demyelination in allergic and Marek's disease virus induced neuritis. Comparative electron microscopic studies. Acta Neuropathol 40:103–110

Lawn AM, Payne LN (1979) Chronological study of ultrastructural changes in the peripheral nerves in Marek's disease. Neuropathol Appl Neurobiol 5:485–497

Lawn AM, Watson JS (1982) Ultrastructure of the central nervous system in Marek's disease and the effect of route of infection on lesion incidence in the central nervous system. Avian Pathol 11:213–225

Lee LF, Bacon LD (1983) Ontogeny and line differences in the mitogenic response of chicken lymphocytes. Poult Sci 62:579–584

Lee LF, Sharma JM, Nazerian K, Witter RL (1978) Suppression of mitogen-induced proliferation of normal spleen cells by macrophages from chickens inoculated with Marek's disease virus. J Immunol 120:1554–1559

Lee LF, Powell PC, Rennie M, Ross LJN, Payne LN (1981) Nature of genetic resistance to Marek's disease in chickens. J Natl Cancer Inst 66:789–796

Lee S-I, Ohashi K, Morimura T, Sugimoto C, Onuma M (1999) Re-isolation of Marek's disease virus from T cell subsets of vaccinated and non-vaccinated chickens. Arch Virol 144:45–54

Li DS, Pastorek J, Zelnek V, Smith GD, Ross LJN (1994) Identification of novel transcripts complementary to the Marek's disease virus homologue of the ICP4 gene of herpes simplex virus. J Gen Virol 75:1713–1722

Lin JA, Kodama H, Onuma M, Mikami T (1991) The early pathogenesis in chickens inoculated with non-pathogenic serotype 2 Marek's disease virus. J Vet Med Sci 53:269–273

Maotani K, Kanamori A, Ikuta K, Ueda S, Kato S, Hirai K (1986) Amplification of a tandem direct repeat within inverted repeats of Marek's disease virus DNA during serial in vitro passage. J Virol 58:657–660

Marek J (1907) Multiple Nervenentzuendung (Polyneuritis) bei Huehnern. Dtsch Tierarztl Wochenschr 15:417–421

Matsuda H, Ikuta K, Kato S (1976) Detection of T-cell surface determinants in three Marek's disease lymphoblastoid cell lines. Biken J 19:29–32

McColl K, Calnek BW, Harris WV, Schat KA, Lee LF (1987) Expression of a putative tumor-associated antigen on normal versus Marek's disease virus-transformed lymphocytes. J Natl Cancer Inst 79: 991–100

McKie EA, Ubukata E, Hasegawa S, Zhang S, Nonoyama M, Tanaka A (1995) The transcripts from the sequences flanking the short component of Marek's disease virus during latent infection form a unique family of 3′-coterminal RNAs. J Virol 69:1310–1314

Minick CR, Fabricant CG, Fabricant J, Litrenta MM (1979) Atheroarteriosclerosis induced by infection with a herpesvirus. Am J Pathol 96:673–706

Moore FR, Schat KA, Hutchison N, LeCiel C, Bloom SE (1993) Consistent chromosomal aberration in cell lines transformed with Marek's disease herpesvirus: evidence for genomic DNA amplification. Int J Cancer 54:685–692

Moore FR, Calnek BW, Bloom SE (1994) Cytogenetic studies of cell lines derived from Marek's disease virus-induced local lesions. Avian Dis 38:797–779

Morimura T, Hattori M, Ohashi K, Sugimoto C, Onuma M (1995) Immunomodulation of peripheral T cells in chickens infected with Marek's disease virus: involvement in immunosuppression. J Gen Virol 76:2979–2985

Morimura T, Ohashi K, Kon Y, Hattori M, Sugimoto C, Onuma M (1996) Apoptosis and CD8-down regulation in the thymus of chickens infected with Marek's disease virus. Archiv Virol 141:2243–2249

Morimura T, Ohashi K, Sugimoto C, Onuma M (1998) Pathogenesis of Marek's disease (MD) and possible mechanisms of immunity induced by MD vaccine. J Vet Med Sci 60:1–8

Murthy KK, Calnek BW (1979) Pathogenesis of Marek's disease: effect of immunization with inactivated viral and tumor-associated antigens. Infect Immun 26:547–555

Nazerian K (1979) Marek's disease lymphoma of chicken and its causative herpesvirus. Biochim Biophys Acta 560:375–395

Nazerian K, Sharma JM (1975) Brief communication: detection of T-cell surface antigens in a Marek's disease lymphoblastoid cell line. J Natl Cancer Inst 54:277–279

Nazerian K, Sharma JM (1985) Pathogenesis of Marek's disease in turkeys. In: Calnek BW, Spencer JL (eds) Proc Int Symp Marek's Dis American Association of Avian Pathologists, Kennett Square, PA, pp 262–267

Nazerian K, Stephens EA, Sharma JM, Lee LF, Gailitis M, Witter RL (1977) A nonproducer T lymphoblastoid cell line from Marek's disease transplantable tumor (JMV). Avian Dis 21:69–76

Nazerian K, Lee LF, Yanagida N, Ogawa R (1992) Protection against Marek's disease by a fowlpox virus recombinant expressing the glycoprotein B of Marek's disease virus. J Virol 66:1409–1413

Ohashi K, Zhou W, O'Connell PH, Schat KA (1994) Characterization. of a Marek's disease virus BamHI-L-specific cDNA clone obtained from a Marek's disease lymphoblastoid cell line. J Virol 68:1191–1195

Okada K, Tanaka Y, Murakami K, Chiba S, Morimura T, Hattori M, Goryo M, Onuma M (1997) Phenotype analysis of lymphoid cells in Marek's disease of CD4+ or CD8+ T-cell-deficient chickens: occurrence of double negative T-cell tumour. Avian Pathol 26:525–534

Omar AR, Schat KA (1996) Syngeneic Marek's disease virus (MDV)-specific cell-mediated immune responses against immediate early, late, and unique MDV proteins. Virology 222:87–99

Omar AR, Schat KA (1997) Characterization of Marek's disease herpesvirus (MDV)-specific cytotoxic T lymphocytes in chickens inoculated with a nononcogenic vaccine strain of MDV. Immunology 90:579–595

Omar AR, Schat KA, Lee LF, Hunt HD (1998) Cytotoxic T lymphocyte response in chickens immunized with a recombinant fowlpox virus expressing Marek's disease herpesvirus glycoprotein B. Vet Immunol Immunopathol 62:73–82

Pappenheimer AM, Dunn LC, Cone V (1926) A study of fowl paralysis (neuro-lymphomatosis gallinarum). Storrs Agric Exp Stn Bull 143:187–290

Pappenheimer AM, Dunn LC, Cone VA (1929) Studies on fowl paralysis (neuro-lymphomatosis gallinarum). I. Clinical features and pathology. J Exp Med 49:63–86

Parker MA, Schierman LW (1983) Suppression of humoral immunity in chickens prevents transient paralysis caused by a herpesvirus. J Immunol 130:2000–2001

Paterson JC, Cottral GE (1950) Experimental coronary sclerosis. III: Lymphomatosis as a cause of coronary sclerosis in chickens. Acta Pathol 49:699–709

Payne LN (1972) Pathogenesis of Marek's disease A review. In: Biggs PM, de Thé G, Payne LN (eds) Oncogenesis and Herpesviruses. IARC, Lyon, France, pp 21–37

Payne LN (1979) Marek's disease in comparative medicine. J Royal Soc Med 72:635–638

Payne LN, Biggs PM (1967) Studies on Marek's disease. II. Pathogenesis. J Natl Cancer Inst 39:281–302

Payne LN, Rennie M (1973) Pathogenesis of Marek's disease in chicks with and without maternal antibody. J Natl Cancer Inst 51:1559–1573

Payne LN, Rennie M (1976) Sequential changes in the numbers of B and T lymphocytes and other leukocytes in the blood in Marek's disease. Int J Cancer 18:510–520

Payne LN, Roszkowski J (1973) The presence of immunologically uncommitted bursa and thymus dependent lymphoid cells in the lymphomas of Marek's disease. Avian Pathol 1:27–34

Payne LN, Powell PC, Rennie M (1974) Response of B and T lymphocytes and other blood leukocytes in chickens with Marek's disease. Cold Spring Harb Symp Quant Biol 39:817–826

Payne LN, Frazier JA, Powell PC (1976) Pathogenesis of Marek's disease. Int Rev Exp Pathol 16:59–154

Peng Q, Bradley G, Tanaka A, Lancz G, Nonoyama M (1992) Isolation and characterization of cDNAs from Bam:HI-H family RNAs associated with the tumorigenicity of Marek's disease virus. J Virol 66:7389–7396

Pepose JS, Stevens JG, Cook ML, Lampert PW (1981) Marek's disease as a model for the Landry-Guillain-Barré Syndrome: latent viral infection in nonneuronal cells is accompanied by specific immune responses to peripheral nerve and myelin. Am J Pathol 103:309–332

Petek M, Dovadola E (1970) Local tumours produced by NDV-inactivated adjuvant vaccine in Marek's disease affected birds. In: Avian Leukosis and Marek's Disease Symposium. Bulgarian Academy of Sciences, Sofia, Bulgaria, pp 243–245

Phillips PA, Biggs PM (1972) Course of infection in tissues of susceptible chickens after exposure to strains of Marek's disease virus and turkey herpesvirus. J Natl Cancer Inst 49:1367–1137

Powell PC, Davison TF (1986) Induction of Marek's disease in vaccinated chickens by treatment with betamethasone or corticosterone. Isr J Vet Med 42:73–78

Powell PC, Rennie M (1984) The expression of Marek's disease tumor-associated surface antigen in various avian species. Avian Pathol 13:345–349

Powell PC, Payne LN, Frazier JA, Rennie M (1974) T lymphoblastoid cell lines from Marek's disease lymphomas. Nature 251:79–80

Powell PC, Lee LF, Mustill BM, Rennie M (1982) The mechanism of genetic resistance to Marek's disease in chickens. Int J Cancer 29:169–174

Powell PC, Hartley KJ, Mustill BM, Rennie M (1983) The occurrence of chicken foetal antigen after infection with Marek's disease virus in three strains of chicken. Oncodev Biol Med 4: 261–271

Powell PC, Howes K, Lawn AM, Mustill BM, Payne LN, Rennie M, Thompson MA (1984) Marek's disease in turkeys: the induction of lesions and the establishment of lymphoid cell lines. Avian Pathol 13:201–214

Pradhan HK, Mohanty GC, Lee WY, Kaul L, Kataria JM (1988) Immune complex-mediated glomerulopathy in Marek's disease. Vet Immunol Immunopathol 19:165–171

Purchase HG (1970) Virus-specific immunofluorescent and precipitin antigens and cell-free virus in tissues of birds infected with Marek's disease. Cancer Res 30:1898–1908

Purchase HG (1972) Recent advances in the knowledge of Marek's disease. Adv Vet Sci Comp Med 16:223–258

Ross LJN, Powell PC, Walker DJ, Rennie M, Payne LN (1977) Expression of virus-specific, thymus-specific and tumor-specific antigens in lymphoblastoid cell lines derived from Marek's disease lymphomas. J Gen Virol 35:219–235

Ross N, O'Sullivan G, Rothwell C, Smith G, Burgess SC, Rennnie M, Lee LF, Davison TF (1997) Marek's disease virus EcoRI-Q gene (meq) and a small RNA antisense to ICP4 are abundantly expressed in CD4+ cells and cells carrying a novel lymphoid marker, AV37, in Marek's disease lymphomas. J Gen Virol 78:2191–2198

Schat KA (1987) Marek's disease: a model for protection against herpesvirus-induced tumours. Cancer Surveys 6:1–37

Schat KA (1996) Immunity to Marek's disease, lymphoid leukosis and reticuloendotheliosis. In: Davison F, Payne LN, Morris TR (eds) Poultry Immunology. Carfax Publishing Company, Abingdon, England, pp 209–223

Schat KA, Calnek BW (1978) Characterization of an apparently nononcogenic Marek's disease virus. J Natl Cancer Inst 60:1075–1082

Schat KA, Schultz RD, Calnek BW (1978) Marek's disease: effect of virus pathogenicity and genetic susceptibility on responses of peripheral blood lymphocytes to concanavalin-A. In: Bentvelzen P, Hilgers J, Yohn DS (eds) Advances in Comparative Leukemia Research. Elsevier, Amsterdam, pp 183–185

Schat KA, Calnek BW, Fabricant J (1981a) Influence of oncogenicity of Marek's disease virus on evaluation of genetic resistance. Poult Sci 60:2559–2566

Schat KA, Calnek BW, Fabricant J (1981b) Influence of the bursa of Fabricius on the pathogenesis of Marek's disease. Infect Immun 31:199–207

Schat KA, Chen C-LH, Shek WR, Calnek BW (1982a) Surface antigens on Marek's disease lymphoblastoid tumor cell lines. J Natl Cancer Inst 69:715–720

Schat KA, Calnek BW, Fabricant J (1982b) Characterisation of two highly oncogenic strains of Marek's disease virus. Avian Pathol 11:593–605

Schat KA, Calnek BW, Fabricant J, Graham DL (1985) Pathogenesis of infection with attenuated Marek's disease virus strains. Avian Pathol 14:127–146

Schat KA, Chen C-LH, Lillehoj H, Calnek BW, Weinstock D (1989) Characterization of Marek's disease cell lines with monoclonal antibodies specific for cytotoxic and helper T cells. In: Kato S, Horiuchi T, Mikami T, Hirai K (eds) Advances in Marek's Disease Research. Japanese Association on Marek's Disease, Osaka, Japan, pp 220–226

Schat KA, Chen C-LH, Calnek BW, Char D (1991) Transformation of T-lymphocyte subsets by Marek's disease herpesvirus. J Virol 65:1408–1413

Schierman LW, Fletcher OJ (1980) Genetic control of Marek's disease virus-induced transient paralysis: association with the major histocompatibility complex. In: Biggs PM (ed) Resistance and Immunity to Marek's Disease. Commission European Communities, Luxembourg, pp 429–442

Settnes OP (1982) Marek's disease A common naturally herpesvirus-induced lymphoma of the chicken. Nord Veterinaermed Suppl:11–132

Sharma JM (1976) Natural resistance to Marek's disease at hatching in chickens lacking maternal antibody, and relationship between this early resistance and resistance acquired with age. Avian Dis 20:311–323

Sharma JM (1981) Natural killer cell activity in chickens exposed to Marek's disease virus: inhibition of activity in susceptible chickens and enhancement of activity in resistant and susceptible chickens. Avian Dis 25:882–893

Sharma JM, Purchase HG (1974) Replication of Marek's disease virus in cell cultures derived from genetically resistant chickens. Infect Immun 9:1092–1097

Sharma JM, Witter RL, Burmester BR (1973) Pathogenesis of Marek's disease in old chickens: lesion regression as the basis for age-related resistance. Infect Immun 81:715–724

Sharma JM, Witter RL, Purchase HG (1975) Absence of age-resistance in neonatally thymectomised chickens as evidence for cell-mediated immune surveillance in Marek's disease. Nature 253:477–479

Shearman PJ, Longenecker BM (1981) Clonal variation and functional correlation of organ-specific metastasis and an organ-specific metastasis-associated antigen. Int J Cancer 27:387–395

Shek WR, Schat KA, Calnek BW (1982) Characterization of nononcogenic Marek's disease virus-infected and turkey herpesvirus infected lymphocytes. J Gen Virol 63:333–341

Shek WR, Calnek BW, Schat KA, Chen C-LH (1983) Characterization of Marek's disease virus-infected lymphocytes: discrimination between cytolytically and latently infected cells. J Natl Cancer Inst 70:485–491

Smith MW, Calnek BW (1974) Comparative features of low-virulence and high-virulence Marek's disease virus infections. Avian Pathol 4:229–246

Smith TW, Albert DM, Robinson N, Calnek BW, Schwabe O (1974) Ocular manifestations of Marek's disease. Invest Ophthalmol 13:586–592

Spencer JL (1970) Marek's disease herpesvirus: comparison of foci (macro) in infected duck embryo fibroblasts under agar medium with foci (micro) in chicken cells. Avian Dis 14:565–578

Spencer JL, Calnek BW (1970) Marek's disease: application of immunofluorescence for detection of antigen and antibody. Am J Vet Res 31:345–358

Spencer JL, Robertson A (1972) Influence of maternal antibody on infection with virulent or attenuated Marek's disease herpesvirus. Am J Vet Res 33:393–400

Spencer JL, Gilka F, Gavora JS, Hampson RJ, Caldwell DJ (1992) Studies with a Marek's disease virus that caused blindness and high mortality in vaccinated flocks. In: de Boer G, Jeurissen SHM (eds) Proc 4th Int Symp Marek's Dis Ponsen & Looijen, Wageningen, Amsterdam, The Netherlands, pp 199–201

Swayne DE, Fletcher OJ, Schierman LW (1988) Marek's disease virus-induced transient paralysis in chickens: alterations in brain density. Acta Neuropathol 76:287–291

Swayne DE, Fletcher OJ, Schierman LW (1989a) Marek's disease virus-induced transient paralysis in chickens: demonstration of vasogenic brain oedema by an immunohistochemical method. J Comp Path 101:451–462

Swayne DE, Fletcher OJ, Schierman LW (1989b) Marek's disease virus-induced transient paralysis in chickens. 1. Time course association between clinical signs and histological brain lesions. Avian Pathol 18:385–396

Swayne DE, Fletcher OJ, Schierman LW (1989c) Marek's disease virus-induced transient paralysis in chickens. 2. Ultrastructure of central nervous system. Avian Pathol 18:397–412

Ubertini T, Calnek BW (1970) Marek's disease herpesvirus in peripheral nerve lesions. J Natl Cancer Inst 45:507–514

Venugopal K, Payne LN (1995) Molecular pathogenesis of Marek's disease Recent developments. Avian Pathol 24:597–609

Venugopal K, Bland AP, Ross LJN, Payne LN (1996) Pathogenicity of an unusual highly virulent Marek's disease virus isolated in the United Kingdom. In: Silva RF, Cheng HH, Coussens PM, Lee LF, Velicer LF (eds) Current Research on Marek's Disease. American Association of Avian Pathologists, Kennett Square, PA, pp 119–124

Volpini LM, Calnek BW, Sekellick MJ, Marcus PI (1995) Stages of Marek's disease virus latency defined by variable sensitivity to interferon modulation of viral antigen expression. Vet Microbiol 47:99–109

Volpini LM, Calnek BW, Sneath B, Sekellick MJ, Marcus PI (1996) Interferon modulation of Marek's disease virus genome expression in chicken cell lines. Avian Dis 40:78–87

Wight PAL (1968) The histopathology of transient paralysis of the domestic fowl. Vet Rec 82:749–755

Witter RL (1996) Evolution of virulence of Marek's disease virus: evidence for a novel pathotype. In: Silva RF, Cheng HH, Coussens PM, Lee LF, Velicer LF (eds) Current Research on Marek's Disease. American Association of Avian Pathologists, Kennett Square, PA, pp 86–91

Witter RL (1997) Increased virulence of Marek's disease virus field isolates. Avian Dis 41:149–163

Witter RL (1998) The changing landscape of Marek's disease. Avian Pathol 27:s46–s53

Witter RL, Nazerian K, Purchase HG, Burgoyne GH (1970) Isolation from turkeys of a cell-associated herpesvirus antigenically related to Marek's disease virus. Am J Vet Res 31:525–538

Witter RL, Solomon JJ, Champion LR, Nazerian K (1971) Long term studies of Marek's disease infection in individual chickens. Avian Dis 15:346–365

Witter RL, Sharma JM, Solomon JJ, Champion LR (1973) An age-related resistance of chickens to Marek's disease: some preliminary observations. Avian Pathol 2:43–54

Witter RL, Stephens EA, Sharma JM, Nazerian K (1975) Demonstration of a tumor-associated surface antigen in Marek's disease. J Immunol 115:177–183

Witter RL, Sharma JM, Fadly AM (1980) Pathogenicity of variant Marek's disease virus isolants in vaccinated and unvaccinated chickens. Avian Dis 24:210–232

Witter RL, Gimeno IM, Reed WM, Bacon LD (1999) An acute form of transient paralysis induced by highly virulent strains of Marek's disease virus. Avian Dis 43:704–720

Xie Q, Anderson AS, Morgan RW (1996) Marek's disease virus (MDV) ICP4, pp38, and meq genes are involved in the maintenance of transformation of MDCC-MSB1 MDV-transformed lymphoblastoid cells. J Virol 70:1125–1131

Xing Z, Schat KA (2000) Inhibitory effects of nitric oxide and interferon-gamma on the in vitro and in vivo replication of Marek's disease virus. J Virol (in press)

Protective Efficacy of Marek's Disease Vaccines

R.L. Witter

USDA Agricultural Research Service, Avian Disease and Oncology Laboratory, 3606 East Mount Hope Road, East Lansing, MI 48823, USA

1 Introduction

For the past 30 years, the poultry industry has relied on a series of avirulent or attenuated live virus vaccines to provide protection to young chickens against natural challenge with field strains of Marek's disease (MD). This strategy has been unusually effective. In the United States, losses from the condemnation at slaughter of young broiler chickens with visible lesions of MD have decreased from 1.5% in 1970 to 0.0121% in 1999, a reduction of over 99% (Fig. 1). Even more dramatic reductions are evident in Georgia and Delaware, states with intensive broiler production. Similar benefits have been realized in commercial layer and breeder flocks. Vaccination for the control of MD in the field has been one of the great successes in veterinary medicine.

Epidemiological characteristics of infection with MD virus (MDV) include shedding of enveloped, infectious viral particles from the feather follicle, rapid horizontal transmission by aerosol routes, and survival of infectivity in the environment for weeks to years, depending on conditions. Infection is ubiquitous and can be devastating in unvaccinated chickens, producing up to 60% mortality within a few weeks. Vaccination protects against tumor induction and mortality but does not prevent viral replication or shedding. Therefore, most poultry environments are heavily seeded with fully infectious MDV, and exposure at relative early ages is typical. If there was no vaccination against MD, the poultry industry would not exist in its present form. The presence of this universal lethal challenge and continued evolution of the virus towards greater virulence warrant continued vigilance and concern.

Several MD vaccines have been used with success in the field and can be loosely grouped into four generations, each succeeding generation providing additional benefits (Fig. 2). The first effective vaccines against MD were

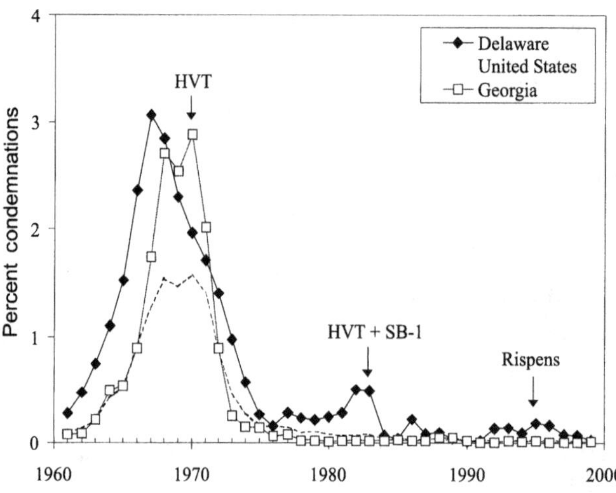

Fig. 1. Forty years of MD control in the United States measured by condemnation of young broiler chickens. Disease incidence varies between different states, and has been markedly reduced through a succession of increasingly effective vaccines (*arrows*)

attenuated serotype 1 viruses (CHURCHILL et al. 1969b; BIGGS et al. 1970; BAXENDALE 1997), using technology developed at the Houghton Poultry Research Station in the United Kingdom. The HPRS-16att strain (CHURCHILL et al. 1969b) was used extensively in Europe and other countries for several years. However, it was soon largely superceded by the naturally nononcogenic turkey herpesvirus (serotype 3), also known as HVT (WITTER et al. 1970a; OKAZAKI et al. 1970). The FC126 strain of HVT was licensed in the United States in 1971 and has been in continuous use since that time. A bivalent vaccine (CALNEK et al. 1983), introduced in 1983, was based on a unique synergistic activity (WITTER and LEE 1984) between the two component viruses, HVT and SB-1, the latter a nononcogenic chicken herpesvirus (serotype 2) (SCHAT and CALNEK 1978). The bivalent vaccine offered superior protection against challenge with the newly discovered very virulent (vv) pathotypes of MDV. Strain CVI988 (also called Rispens), is an attenuated serotype 1 vaccine virus introduced in 1971 (RISPENS et al. 1972a,b). This strain was used initially in The Netherlands, later gained favor in many European and Asian countries, and has been used extensively in the United States since the mid-1990s. It offers superior protection against challenge with vv or vv+ pathotypes compared to either HVT alone or bivalent serotype 2+3 vaccines (WITTER 1992a).

The foregoing succession of MD vaccines paralleled, and indeed was prompted by, the continued evolution of field strains of serotype 1 MDV to greater virulence (WITTER 1997). The v pathotype emerged in the late 1950s and was responsible for the catastrophic losses from MD in the 1960s. The vv pathotype emerged in the late 1970s, after HVT vaccines had been in use about 8–10 years. The vv+ pathotype emerged in the early 1990s after bivalent vaccine had been in use about 10 years. It appears that MDV has the ability to mutate in response to the selection pressure created by intensive vaccination. If so, this does not bode well for the future of MD control. Herein lies the dilemma and challenge that make the study of MD vaccines so intriguing and important – can the cycle of mutation around vaccinal immunity be broken and, if so, how?

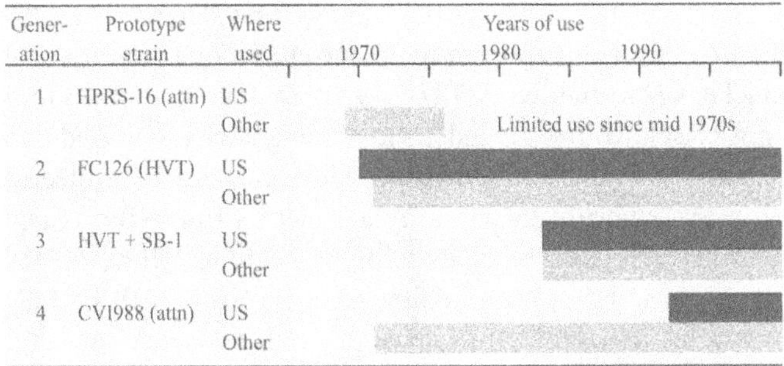

Fig. 2. Usage patterns of MD vaccines in the United States and other countries

2 Types of Vaccines and Vaccine Viruses

There are several types of vaccines and vaccine viruses in use or proposed for use to protect chickens against challenge with virulent MDV. Some of the more important of these are discussed below.

2.1 Serotype 3

HVT (serotype 3) viruses are nononcogenic viruses of turkeys, also known as turkey herpesviruses, that were identified in turkey kidney cell cultures (Kawamura et al. 1969) or cultures inoculated with turkey blood (Witter et al. 1970a). Antigenically, HVT is closely related to MDV (Witter et al. 1970a). The FC126 strain (Witter et al. 1970a) has been widely used in the United States and elsewhere although several other strains are in commercial production. Licensed HVT strains appear to offer good protection, but inferior protection by some strains has been described (Witter 1992b). A unique feature of HVT is the ability to recover enough cell-free virus by sonication of infected cell cultures to permit production of cell-free vaccine. Although cell-free HVT vaccines have been used in many countries, especially where the logistics of handling cell-associated vaccines have been difficult or expensive, the cell-associated form of this vaccine is now recognized as more protective and is generally preferred. The basis probably lies in the greater susceptibility of cell-free HVT to neutralization by maternal antibody (Prasad 1978; Witter and Burmester 1979). This virus replicates very well in chicken embryo fibroblast (CEF) cultures and is probably the most economical of all the MD vaccine strains to produce. This vaccine was originally propagated in duck embryo cells, but CEF cultures are preferred because more is known about chicken-embryo-transmitted pathogens and methods for their detection (Payne 1992).

2.2 Serotype 2

The protective ability of serotype 2 viruses was probably first demonstrated by immunization of chickens by natural exposure to seeder chicks or inoculation of blood from infected chickens (Zander et al. 1971, 1972; Zander and Raymond 1985). The HN-1 strain was isolated from Zander's material and characterized as a low-virulence isolate (Cho and Kenzy 1972). Also, chickens could be protected against MD lesions by prior exposure to naturally avirulent viruses (Biggs and Milne 1972), some of which were later typed as serotype 2 strains. However, the SB-1 strain (Schat and Calnek 1978) was the first serotype 2 strain to be fully characterized as nononcogenic and shown to induce protection against challenge with the GA strain. Subsequently, additional serotype 2 strains were isolated from field flocks and characterized (Witter et al. 1987, 1990). One of these strains,

301B/1, was shown to provide high levels of protection in comparative trials with other serotype 2 viruses (WITTER et al. 1987). The SB-1 and 301B/1 strains are both licensed for use as vaccines in the United States and are generally combined with HVT or other vaccine strains as components of bivalent or trivalent products. The serotype 2 viruses grow moderately well in CEF cultures and are exclusively used as cell-associated vaccines.

2.3 Attenuated Serotype 1

Serotype 1 vaccine strains have been traditionally attenuated by serial passage in cell culture (CHURCHILL et al. 1969a). Many serotype 1 strains have been attenuated (CHURCHILL et al. 1969a; NAZERIAN 1970; ZANELLA et al. 1970; EIDSON et al. 1971; MEULEMANS et al. 1971; VIELITZ and LANDGRAF 1971; SPENCER et al. 1972; WITTER 1982; RISPENS et al. 1972a), but relatively few have found sustained popularity as commercial vaccines. As previously mentioned, the attenuated HPRS-16 strain was a successful vaccine that induced immunity against challenge with v pathotype viruses. Although this strain is still commercially available in some markets, it has been largely replaced by other strains that can provide equal or better protection. The CVI988 strain had low levels of oncogenicity upon initial isolation but was attenuated by serial passage in cell culture. Interestingly, the exceptional protective ability of this strain was not recognized, at least in the United States, until the 1990s when highly virulent pathotypes were available as challenge viruses (WITTER 1992a). An attenuated preparation of the vv pathotype Md11 strain, known as Md11/75C/R2/23, also provided high levels of protection (WITTER 1991) and is licensed for commercial use, although it now is considered less protective than CVI988. Because the attenuated CVI988 vaccine strain was reported to have some residual oncogenicity (BÜLOW 1977b; POL et al. 1986), DE BOER and colleagues developed two higher passaged, cloned variants identified as clone C (DE BOER et al. 1986) and clone C/R6 (DE BOER et al. 1988). These have also been used as commercial vaccines but are now considered less protective than the original CVI988. In contrast to earlier reports, the CVI988 preparations presently available appear to lack oncogenicity and spread poorly by contact (WITTER et al. 1995). As a group, the attenuated serotype 1 vaccine strains grow well in CEF cultures and are exclusively used as cell-associated vaccines.

2.4 Polyvalent Vaccines

Combinations of the SB-1 strain (serotype 2) and the FC126 strain (serotype 3, HVT) provide better protection than either vaccine alone (SCHAT et al. 1982; WITTER 1982). This important finding provided the stimulus for the commercial development of polyvalent vaccines, which are highly effective and widely used (CALNEK et al. 1983; WITTER et al. 1984, 1987). Interestingly, synergism proved to be most pronounced with combinations of serotype 2 and 3 viruses (Witter 1988). Some characteristics of synergism are discussed in a later section.

2.5 Recombinant Vaccines

The prospect of recombinant DNA (rDNA) vaccines has been viewed with considerable optimism (Finkelstein and Silva 1989; Sondermeijer et al. 1993; Ross 1998; Witter and Hunt 1994). Many research laboratories and vaccine companies have made major commitments to rDNA-based vaccination for MD. A parallel objective was the production of rDNA vaccines for other poultry diseases based on HVT or MDV vector systems. However, the results from these endeavors have not thus far been commensurate with expectations. Several rDNA vaccines for MD have been developed but none have yet been licensed for commercial use in the United States. There are several problems. The technology needed to create recombinants of MDV or HVT has been problematic at best and is only now starting to be routinely successful. There is only limited knowledge on which viral genes are involved with immunity or virulence and what combination of genes must be expressed (or deleted) to produce an effective vaccine. Also, for reasons not well understood, genetic manipulation appears likely to reduce replication ability in vivo.

Several strategies have been considered. Thus far, most attention has been directed to live virus vectors based on HVT (Morgan et al. 1992, 1993; Darteil et al. 1995; Heckert et al. 1996; Cronenberg et al. 1999), serotype 1 MDV (Nakamura et al. 1992; Sonoda et al. 1996; Tsukamoto et al. 1999), or fowlpox virus (Nazerian et al. 1992, 1996; Schat et al. 1996) which express inserted genes obtained from other MDV serotypes or from other avian viruses. Some level of protection has been reported against MD with such vaccines (Reddy et al. 1996; Nazerian et al. 1996). Gene-deleted serotype 1 MDVs have been produced and although some of these appear to be attenuated (Zelnik et al. 1995; Morgan et al. 1998), protective efficacy appears to be limited. Some protection has been reported with a DNA vaccine based on serotype 1 MDV (Li et al. 1996). These approaches remain important because future vaccines for MD and other diseases will likely be based on rDNA technology. Recombinant vaccines are considered in detail in the chapter by K. Hirai and M. Sakaguchi, and will not be discussed further here.

2.6 Other Vaccines

Early studies provided evidence of some protection following vaccination with inactivated cells or subunit vaccines (see review, Witter 1985a). Partially purified subunit preparations of individual glycoproteins, such as gB, provide at least partial protection against MD challenge (Ono et al. 1985). However, there seems to be little interest at present in inactivated or subunit vaccines for MD.

3 Biological Principles

Two biological phenomena have played central roles in the development of MD vaccines. Attenuation is not unique to MD but is complex and the key to development of serotype 1 vaccines. Synergism appears to be unique to MD; it is also complex and is the key to development of bivalent vaccines of serotypes 2 and 3. Because of the importance of these phenomena to MD vaccines, each will be reviewed briefly.

3.1 Biology of Attenuation

The recognition that oncogenicity of the HPRS-16 strain of serotype 1 MDV was reduced upon serial passage in chicken kidney cell culture was a seminal discovery (Churchill et al. 1969a), although hardly unexpected considering the large number of infectious agents that have been similarly attenuated for virulence by serial passage.

Attenuation means to weaken or lose virulence. For serotype 1 MDV strains, attenuation is functionally defined as the loss of oncogenicity, as measured by gross or microscopic tumor induction, following the inoculation of genetically susceptible, antibody-free chicks at hatch. Complete attenuation has been achieved in as few as 27 (Schat et al. 1985) or 33 (Churchill et al. 1969a) passages in chicken kidney cells but often requires longer. For example, the Md11 strain was still virulent after 55 passages (R.L. Witter, unpublished data) but was attenuated after 75 passages (5 in duck embryo cells and 70 in chicken embryo cells) (Witter 1982). The 648A strain required 100 passages for full attenuation (R.L. Witter, unpublished data). The process of attenuation is undoubtedly influenced by various factors including the pathotype (or virulence) of the starting virus, the days between passage, the number of cells transferred, and the type of cell substrate. Attenuation speed may also be a matter of chance – although no reports exist where the same virus has been attenuated in replicate passage series, variability would be expected.

Propagation of MD lymphoblastoid cell lines where infection is mostly nonproductive does not appear to result in attenuation even after up to 70 (Lee et al. 1975) or several hundred (K. Nazerian, personal communication) subcultures. Therefore, productive infection is probably required since it is most likely to generate the highest numbers of viral DNA copies at each replication.

In addition to the loss of oncogenicity, attenuated viruses also exhibit other changes. Compared to wild-type viruses, attenuated strains replicate faster in vitro and produce larger plaques which have a different morphology and must be enumerated earlier in assays in liquid medium (Churchill et al. 1969a). A limited amount of cell-free virus can be extracted from lysed cells (Nazerian 1970; Witter et al. 1987). Production of the so-called "A antigen" (otherwise known as glycoprotein C) is diminished or absent (Churchill et al. 1969a) as is the so-called membrane antigen (Nazerian 1973). Compared to wild-type viruses, attenuated strains replicate more slowly in vivo and exhibit limited if any cytolytic infection of lymphoid organs (Schat et al. 1985). They also have decreased ability to induce pocks on the chorioallantoic membrane of embryos (Sharma et al. 1976). Following inoculation of chickens, viremia titers are variable but often low (Phillips and Biggs 1972; Witter 1991) causing speculation that attenuated strains may lose their tropism for lymphoid cells (Schat et al. 1985). Also, attenuated viruses are shed poorly from the feather follicle and horizontal infection is limited or absent (Phillips and Biggs 1972). Attenuated viruses may lack the ability to cause splenomegaly after infection, to induce early mortality syndrome or transient paralysis, or to cause lymphoid organ atrophy or immunodepression (Witter 1991) (R.L. Witter and I.M. Gimeno, unpublished data). However, there are many exceptions.

The attenuation process is undoubtedly the result of mutations in the viral genome during productive infection and replication. Under conditions of serial passage, mutant clones that replicate more efficiently than the wild type will eventually predominate. This can be an extended process, requiring many months of culture. During this time, additional mutations may be accumulated so that the resulting virus population becomes increasingly heterogeneous, differing from not only the wild-type strain but also other attenuated preparations. Thus, attenuation is most likely a step-wise process involving multiple mutations, only some of which directly influence oncogenicity. Different biological properties are affected at different passage levels, and the interval between reduction and total elimination of a characteristic may take many passages. Also, the rate of attenuation of a given characteristic may differ between viral strains (Fig. 3).

Excessive numbers of serial passages can lead to overattenuation, a condition where a virus cannot replicate in vivo or replicates so poorly in vivo that it becomes ineffective as a vaccine (Konobe et al. 1979; Witter and Offenbecker 1979). Partial overattenuation can also reduce but not eliminate the protective ability of vaccines. Baxendale has suggested that overattenuation of the HPRS-16att strain

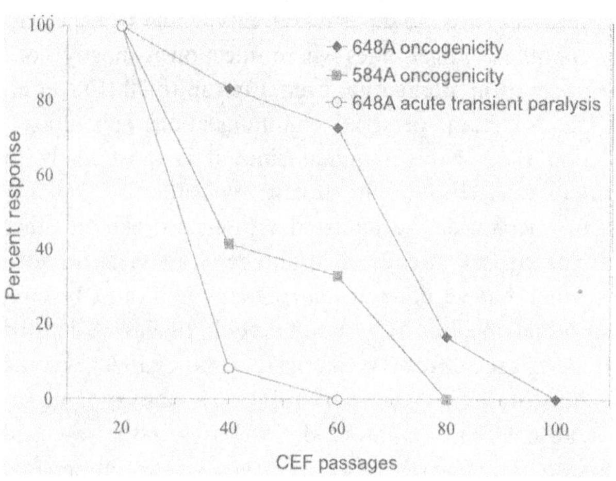

Fig. 3. Attenuation of two vv+MDV strains by serial passage in CEF. MD tumor induction is attenuated earlier for 584A than for 648A. For strain 648A, the ability to induce acute transient paralysis (TP) is attenuated earlier than the ability to induce MD tumors. (R.L. Witter and I.M. Gimeno, unpublished data)

may have reduced its efficacy (BAXENDALE 1997). Differences in protective efficacy of the same attenuated MD vaccine produced in two laboratories have been reported (WITTER et al. 1996), presumably due to overattenuation in the second laboratory (R.L. Witter, unpublished data). In contrast, backpassage of attenuated viruses can increase in vivo replication and oncogenicity, although not normally to the levels exhibited by the wild-type virus. These principles are illustrated by the following example. The Md11 strain was fully attenuated but poorly protective after 75 passages in CEF. After 2–8 backpassages in chickens, it became highly protective but was also mildly virulent (WITTER 1987). A further set of 30 cell culture passages reduced virulence to acceptable levels and the resulting virus was still highly protective (WITTER 1991). Backpassage has also been used (DE BOER et al. 1988) to increase the protective ability of highly passaged clones of strain CVI988.

The immunizing potential of attenuated serotype 1 vaccines may, at least in certain cases, be significantly compromised in chickens with homologous or heterologous maternal antibodies (WITTER 1982). For example, the influence of antibody on Md11/75C/R2 virus was greater than on CVI988 clone C (WITTER et al. 1987) indicating that the susceptibility of attenuated strains to in vivo neutralization may vary.

The specific mutations responsible for attenuation of oncogenicity or alterations in other biological properties are not known. It seems likely that oncogenicity is a complex trait influenced by multiple genes. Complete attenuation, therefore, may require an accumulation of multiple mutations in appropriate genes, thus explaining the stepwise progression of this phenomenon during serial passage.

Attenuation is not usually important with naturally nononcogenic viruses such as those of serotypes 2 and 3. However, other undesirable properties of such viruses are amenable to attenuation. For example, the ability of a serotype 2 virus to enhance the oncogenicity of avian leukosis virus (BACON et al. 1989), as measured by the induction of bursal lymphomas in genetically susceptible chickens, could be attenuated by serial passage in culture (WITTER 1995), resulting in a serotype 2 vaccine that still provides good protection against challenge with virulent MDV.

There are several practical considerations of attenuation. Not all attenuated vaccines protect equally (WITTER 1992a; DE BOER et al. 1988). The protective ability of attenuated vaccines can be manipulated by additional passage in vitro or by backpassage in chickens. Thus, a highly protective vaccine may be adversely affected by the additional passages required for production serials. Viruses that are quickly attenuated may be superior vaccines because they do not accumulate as many extraneous mutations (WITTER et al. 1997). Site-directed mutagenesis may ultimately be the best approach, but is not yet feasible.

3.2 Biology of Protective Synergism

Synergism is the action of two or more substances to achieve an effect of which each alone is incapable. The discovery that serotype 2 and serotype 3 viruses provided better protection when used as a combination than when used separately revealed a phenomenon, later termed protective synergism (WITTER and LEE 1984), that is of basic interest and practical importance. Synergism is the basis for polyvalent vaccination, a concept that has provided one of the principal strategies for MD control since its introduction in the early 1980s. The anticipation of improved protection through synergism undoubtedly drives the continued use of various combinations of vaccine viruses worldwide.

The synergistic effects of multiple viruses on protective efficacy were observed independently by two groups. SCHAT and colleagues (SCHAT et al. 1982) found that combinations of the FC126 strain of HVT (serotype 3) and the SB-1 strain (serotype 2) protected up to 78% of chickens compared to 43–56% for either virus alone, but only in line P chickens. Witter (WITTER 1982) found a trivalent vaccine composed of serotypes 1, 2, and 3 to provide better protection (91%) than any single component virus (57–74%). These findings prompted field trials by both groups that confirmed a reduction of up to 78% in broiler condemnations for MD with bivalent (SB-1 plus FC126) vaccine compared to FC126 alone (CALNEK et al. 1983; WITTER et al. 1984).

The additional protection is not due to an additive dose effect (SCHAT et al. 1982), and can be achieved with as little as 80 PFU of SB-1 combined with normal field doses of HVT (WITTER and LEE 1984). Synergism is strongly serotype specific and is especially evident between combinations of serotypes 2 and 3 (WITTER 1988). No synergism occurs between two virus strains of the same serotype. Synergism between serotypes 1 and 3, or 1 and 2 was inconsistent and often undetectable (WITTER 1988). Although combinations of HVT and CVI988 are commonly used, there is little evidence that these provide more protection than CVI988 alone (VIELITZ and LANDGRAF 1985; ZANELLA et al. 1985; WITTER et al. 1995). Viral combinations that produce (or fail to produce) synergism are illustrated in Fig. 4. Interestingly, enhancement of protection of over 300% can be achieved with combinations of poorly protective serotype 2 and 3 strains (WITTER 1992b), thus

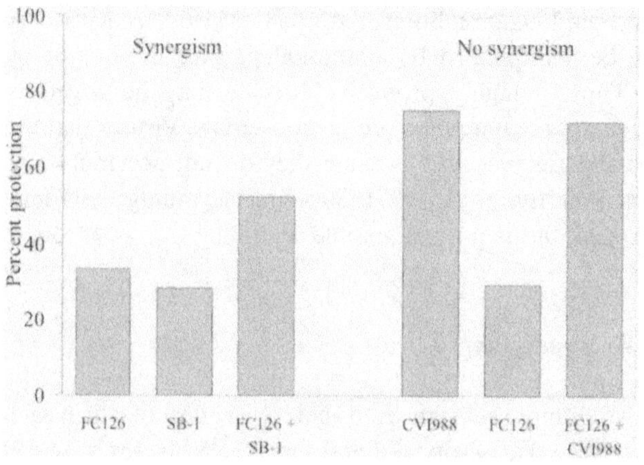

Fig. 4. Protective synergism between FC126 and SB-1 (from WITTER 1987); lack of synergism between CVI988 and FC126 (from WITTER et al. 1995)

providing model systems for the study of synergistic mechanisms. The mechanism of protective synergism is poorly understood and deserves additional study.

Augmentation of protection for MD vaccines has been observed with two additional systems. Combinations of HVT and glutaraldehyde-inactivated serotype 1 MDV obtained from feather follicles was reported to improve protection (PRUTHI et al. 1987). Combinations of HVT and a recombinant fowlpox virus expressing gB and other glycoproteins of serotype 1 MDV also provide improved protection (NAZERIAN et al. 1996). This latter finding suggests that strategic combinations between conventional and recombinant strains may be worthy of consideration.

There are several practical implications of synergism. Experience with serotype 2 and 3 combinations has validated this approach for increasing the efficacy of vaccination. Synergism has been utilized to design an experimental bivalent vaccine composed of recombinant FPV/gB and cell-free HVT that is, arguably, the most effective cell-free vaccine for MD (NAZERIAN et al. 1996). With sufficient knowledge, therefore, synergism can be a powerful tool for vaccine improvement with application to both conventional and genetically engineered products. On the other hand, synergism (or the prospect of synergism) may be over-utilized. Vaccine strains have been used in virtually all possible combinations (SARMA et al. 1992), but it now seems clear that only a few combinations show a clear benefit.

4 Vaccine Production and Licensing

4.1 Production Techniques

Commercial vaccine production is a sophisticated and proprietary process. The considerable knowledge developed by vaccine companies on production efficiency has resulted in low production costs, which in turn contribute to the establishment of favorable market prices to consumers. This topic has been reviewed by others (CHURCHILL 1985; PAYNE 1992; NATHAN and LUSTIG 1990) and will not be discussed in depth here. However, there are several important considerations.

Substrates are typically primary cultures of chicken embryo cells from specific-pathogen-free (SPF) flocks. This creates a significant demand for SPF eggs and constitutes one of the major costs of production. In recent years, various continuous cell lines have been evaluated as substrates for MD vaccines but to this author's knowledge, no commercial MD vaccine is currently produced with such substrates. Techniques of propagation between master seeds and production serials involve a minimum number of passages, each usually of very short duration (<48h) and where cultures are infected at high multiplicity. Most MD vaccines are cell-associated and the final product consists of a suspension of viable cells from infected cultures; the ratio of infected to total viable cells probably varies with the virus type but may approach 1:2 for very rapidly replicating strains such as FC126.

The number of total (or infected) cells per ampule is estimated by experience as the potency of vaccines cannot be determined in advance. The cells are cryopreserved by slow freezing with a suitable protectant, e.g., DMSO. Cell-free preparations of HVT, produced in limited quantities for specialized markets, require steps to lyse the infected cells by sonication prior to the filling of ampules. Cell-associated vaccines are stored at −196°C and are transported to customers in liquid nitrogen containers.

Polyvalent vaccines are prepared in two ways. Most are sold as monovalent vaccines that are combined after thawing just prior to use. Some are grown individually, but are mixed prior to the filling of ampules. To establish the titer of two or more constituent viruses representing different serotypes in a mixed product, it is necessary to stain plaques with serotype-specific monoclonal antibodies.

4.2 Quality Control

MD vaccines are normally regulated by the legal statutes of the country in which they are sold. Such regulation is an important safeguard and helps assure consumers that vaccines are pure, safe, and effective. A discussion of specific regulations is beyond the scope of this chapter and has been reviewed elsewhere (Thornton 1985; Payne 1992).

MD vaccines must be pure, i.e., be free of contamination with extraneous viruses or other agents. Although many viruses are potential contaminants, reticuloendotheliosis, avian leukosis, and chicken infectious anemia viruses warrant special attention. Contamination of MD vaccines with reticuloendotheliosis virus was reported in several countries (Yuasa et al. 1975; Bagust et al. 1979) and continues be a threat. Purity must be assured by tests on the master seed, the SPF flock supplying the eggs for cell culture, and on the final product.

It is a requirement that MD vaccines be safe, i.e., not cause disease in susceptible chickens. The criterion normally used to certify vaccine strains is the absence of gross MD lesions or clinical signs of disease, including reduced body weights, in normal, susceptible chickens receiving up to ten times of a field dose and studied for an extended observation period. All serotype 2 and 3 strains are naturally nononcogenic and routinely pass safety evaluations. Attenuated serotype 1 strains must be rigorously tested for oncogenicity, as attenuation is a gradual process (see previous section). In addition, attenuated strains are usually tested for reversion to virulence upon serial backpassage in susceptible chickens. Attenuated serotype 1 viruses should be used at the earliest passage where virulence is lost, in order to minimize overattenuation during subsequent multiplication passages.

MD vaccines must be efficacious, i.e., provide significant protection against challenge with virulent MDV. Standardized protection tests are designed to show evidence of at least minimal protection (Payne 1992; Nathan and Lustig 1990; Biggs et al. 1979). However, these tests do not usually compare the protection provided by a vaccine with that of other vaccines, and the ability to meet regulatory requirements does not necessarily guarantee a high level of effectiveness in the field.

The design of efficacy tests may vary and can have a profound effect on the final result (see Sect. 6.2).

MD vaccines must meet criteria for potency, i.e., contain a quantity of virus sufficient to provide good protection with a margin of safety. In the United States, each serotype contained in MD vaccines must each contain a minimum of 1000 PFU per chick dose. In practice, most vaccines exceed this dose by a substantial margin. Higher doses reduce the risk of production lots that fail potency tests and tend to provide a competitive advantage in marketing. Many customers purchase MD vaccines on the basis of the PFU content, as determined by an independent laboratory.

MD vaccines must be compatible with all additives such as antibiotics, preservatives, and other viruses. Some commercial MD vaccines are mixed with vaccines against infectious bursal disease, reovirus (tenosynovitis), or fowl pox in the same ampule. Clearly, the efficacy and potency of MD vaccines must be carefully controlled to insure that added materials do not compromise the integrity of the product. Licensed mixtures are required to meet strict standards and are safe, but empirical mixtures created in the hatchery may be risky and should be discouraged.

4.3 Regulatory Oversight

In most countries, vaccine manufacture is regulated on the basis of licenses granted by the respective government. Licenses will be granted to products that meet established criteria. Standards for MD vaccines for the United States are contained in the Code of Federal Regulations, volume 9, and for the European Union in the European Pharmacopoeia. Because licenses may be withdrawn if manufacturers produce a defective product, regular surveillance of vaccine serials by regulatory authorities is required.

5 Vaccine Handling and Administration

5.1 Vaccine Handling

Since MD vaccines must normally be administered to chicks at hatch or prior to hatch, the hatchery becomes the focal point for vaccine handling and administration. This has simplified vaccine distribution and allowed for a greater level of control compared to individual farm use. However, cell-associated vaccines are fragile and improper handling or storage may reduce potency so that chicks receive an insufficient dose. Basically, cell-associated vaccines must be transported from manufacturer to hatchery at −196°C, maintained at −196°C until use, thawed rapidly, diluted appropriately, and used quickly. Some important considerations for vaccine handling in the hatchery have been identified (Jackson 1999). If the vaccine is handled inappropriately, chicks may not receive sufficient virus to induce protective immunity and problems with excessive MD losses may result.

5.2 Day-Old Vaccination

Until recently, most chicks have been vaccinated for MD immediately after hatching and before transport from the hatchery to the rearing house. In the United States, chicks typically receive vaccine by the subcutaneous route, usually in the neck region. Intramuscular vaccination, usually in the leg, may be superior to subcutaneous vaccination, at least for CVI988 (Oei and de Boer 1986; van Eck 1997). In some situations, the individual component bivalent vaccines have been administered separately by different routes to the same chick, but this is probably unnecessary. By either intramuscular or subcutaneous routes, the vaccine is injected with a needle. Various semiautomated devices have been developed to speed the process so that one operator may deliver vaccines to 2000–3000 chicks per hour. However, poor technique will result in improperly vaccinated chickens. Some companies mix dye with the vaccine so that the process may be monitored. Needles are not typically disinfected between chicks so that pathogens contained in one chick may be mechanically transferred to another; such needle transmission has been demonstrated for avian leukosis virus (de Boer et al. 1980). The intra-abdominal route is sometimes used to administer vaccines in laboratory experiments and appears to produce satisfactory results.

5.3 In Ovo Vaccination

Following the demonstration that administration of HVT vaccine to 18-day-old embryos provided protection against challenge after hatch (Sharma and Burmester 1982), the course of MD vaccinology changed. Not only was immunity induced by in ovo administration but, more importantly, the onset of protective immunity was speeded up thus providing a practical advantage to chickens in the field. Hatchability was not reduced, and in some cases appeared to increase. The system could be used for vaccines of all three serotypes (Sharma and Witter 1983), and was successful even in embryos with maternal antibodies (Sharma and Graham 1982). The technology has been developed commercially and is now in widespread use. However, it was 10 years before a suitable automated delivery system was available, making in ovo vaccination a testament to innovation in both biology and engineering (Johnston et al. 1997; Ricks et al. 1999). Although the initial impetus was the promise of more effective immunity to MD, the ultimate driving force behind the technology is the promise of reduced costs of vaccine administration. Presently, more than 90% of broiler chicks in the United States receive in ovo vaccination. The technology is being adopted in many other countries, mostly for broilers. In ovo vaccination in layer and breeder chickens is being contemplated, but cost advantages are offset by the need to deliver vaccines to male chicks, which are usually discarded at hatch.

The advent of in ovo vaccination has practical implications to MD vaccinology. Vaccines may need to be specially formulated for in ovo use. Whereas serotype 3 vaccines replicate well in ovo and induce rapid immunity post-hatch, serotype 1 and 2 vaccines replicate less vigorously in ovo (Sharma 1987). It now appears that vaccines of all serotypes can be used successfully, but differences between the biology of MD vaccine viruses in embryos and day-old chicks exist (Sharma et al. 1984) and require additional study. In ovo vaccination is already well established in commercial practice for delivery of MD vaccines and may soon be used for delivery of a variety of other vaccines or biologics (Johnston et al. 1997; Karaca et al. 1998; Gagic et al. 1999; Rautenschlein et al. 1999).

6 Biology and Efficacy of Vaccines

6.1 Biology of Vaccine Viruses

Chickens are typically vaccinated at or before hatch with one to three different herpesviruses, each of which persist for long periods (probably permanently) in the host. Some are shed and can be transmitted to other chickens by contact. Vaccinated chickens are invariably exposed in the field to one or several strains of virulent MDV, each capable of initiating infection (since vaccination does not prevent superinfection) and persisting in the host. Under field conditions, therefore, chickens typically are concurrently infected with multiple herpesviruses. The vaccine strains thus become an important and integral part of the viral ecosystem of the host chicken and deserve attention from an epidemiological perspective.

Since biological characteristics of all viral serotypes are discussed in other chapters, only shedding and horizontal spread of vaccine strains will be discussed here. Serotype 2 strains are shed from the feather follicle and spread readily by contact among chickens (Calnek et al. 1979; Witter et al. 1987, 1990), although perhaps less efficiently than virulent serotype 1 strains (Cho 1977; Rangga-Tabbu and Cho 1982). Serotype 2 and mild serotype 1 strains are common in nature where they have been associated with protection against natural MD viral challenge (Biggs et al. 1972; Zander 1973; Jackson et al. 1974; Witter 1985b). In contrast, contact spread of attenuated serotype 1 and serotype 3 viruses is limited or absent (Witter et al. 1970a, 1987). However, serotype 3 spreads well in its natural host, the turkey (Witter and Solomon 1972). The CVI988 strain initially spread well by contact, prompting Rispens to investigate the concept of partial flock vaccination (Rispens et al. 1972b), but this did not work in practice as the spread of the vaccine strain was not rapid enough to precede challenge with field strains. More recent studies (Witter et al. 1995) indicated that the CVI988 virus spreads poorly, perhaps indicating a change in the virus due to additional passage in cell culture. Other attenuated serotype 1 vaccine strains, such as Md11/75C/R2 (Witter et al. 1987) and R2/23 (Witter et al. 1995), failed to spread at all. Even though serotype 2 strains spread horizontally, there appears to be very little carryover between flocks in the same house, even when old litter is not removed (Witter et al. 1984).

6.2 Efficacy Tests

The comparative efficacy of MD vaccines is traditionally evaluated by challenge experiments in susceptible chickens, using tumor formation as the response criterion. However, many factors influence the degree of protection obtained and standardization of test systems between laboratories has not been achieved. The protocol used at the Avian Disease and Oncology Laboratory involves vaccination of genetically susceptible, maternal antibody positive chicks at 1-day-old with 2000 PFU by the intra-abdominal route. Vaccinated and unvaccinated groups are challenged at 5 days post-vaccination (6 days of age) with 500 PFU of the selected challenge strain by the intra-abdominal route. All groups are held in separate isolators until 8 weeks post-challenge, killed, and evaluated for gross lesions of MD. Critical features of this model system are the use of chickens with maternal antibodies to all three serotypes of MDV (this mimics the field condition and depresses protection), early challenge (to emphasize the need for rapid onset of protection), and careful evaluation of gross lesions in nerves and visceral organs. Group size is 15–17 birds and experiments are replicated 2–3 times. It is usually appropriate to select vv or vv+ strains for challenge as vaccines need to protect against the highly virulent strains in the field. Protection by new (test) vaccines is contrasted with that of prototype vaccines in the same experiment. Models that produce 90–100% lesions in unvaccinated chickens and 30–70% lesions in control chickens that receive prototype vaccines seem to be most useful for detection of differences between prototype and test vaccines. If, however, one desires simply to test whether a vaccine can induce protective immunity, other protocols may be more appropriate. For validation of vaccines for official licensure, procedures approved by regulatory authorities must be followed.

Protection tests such as that described above are time-consuming and costly but are considered to be the gold standard. Alternate tests based on other criteria, e.g., cellular immunity, viral replication, early mortality, transient paralysis, etc., could have value but would require careful validation.

6.3 Immune Responses

Vaccination stimulates a variety of cellular and humoral immune responses primarily directed against viral antigens. Details of these responses are discussed in other chapters. Although vaccination protects against lymphoma induction and reduces virus load following field challenge, it does not prevent chickens from becoming infected and shedding virulent virus. The lack of a sterilizing immunity is critical because it permits the wild-type strains to persist and to mutate, gradually gaining virulence (WITTER 1997).

The mechanism by which immunity is maintained for long periods is not known. Vaccine viruses induce mainly latent infections although surely some cells may occasionally "turn on" to productive infection, develop antigens, and die (CALNEK et al. 1981). Some but not all vaccine viruses may also produce productively infected cells in the feather follicle epithelium, a uniquely permissive site for MDV replication (CHO 1977). The antigenic mass represented by cells in lymphoid organs and feather follicles productively infected with vaccine viruses may provide continuous stimulation to the immune system to maintain (or boost) immunity at high levels for long periods, but this hypothesis requires confirmation. The wild-type strains, which have greater ability than vaccine viruses to induce productive infection, may also present antigens to prime and boost the immune system. Better understanding of factors involved in maintenance of protective immunity may be important to the development of improved vaccines.

6.4 Safety Concerns

The safety record of MD vaccines, except for occasional batches contaminated with other agents, has been exceptional. Historically, MD vaccines have experienced critical scrutiny as the first effective vaccine against a viral-induced cancer. Even though no linkage between MD viruses and public health was or is recognized (PURCHASE and WITTER 1986), inoculation of animals in the human food chain with "cancer" viruses raised issues not experienced with other poultry vaccines. From the beginning, MD vaccines were required to be fully attenuated, i.e., completely nononcogenic. This criterion was met by serotype 2 and 3 viruses with little difficulty, as these viruses did not cause tumors in chickens; the minor histologic lymphoproliferative lesions induced in nerves by HVT (WITTER et al. 1976) and in immunodepressed or in ovo-inoculated chickens by SB-1 (SCHAT and CALNEK 1978) were not of sufficient concern to prevent their use.

The safety of attenuated serotype 1 viruses is a more problematic issue, and has focused on whether vaccines produce any gross MD lesions or whether vaccines will revert to virulence upon serial backpassage in chickens. Licensing requirements specify that serotype 1 vaccines satisfy both types of criteria.

Immunodepression by vaccine viruses of all 3 serotypes has been reported (LEE et al. 1978; FRIEDMAN et al. 1992) but these effects have been minor and transient, and are generally insufficient to cause clinical problems in the field.

At a time when options for creating yet more effective vaccines are becoming limited, it would be interesting to reexamine the rationale for complete attenuation of serotype 1 vaccines. WITTER showed that the R2 strain, which was mildly oncogenic in susceptible chicks lacking maternal antibodies, provided better protection than the completely attenuated R2/23 strain (WITTER 1991). The CVI988 strain was cloned and further passaged to achieve more complete attenuation and safety (de Boer et al. 1988), but the derivative clone C proved to be less efficacious than the original parent strain (Witter 1992a). The RM1 strain, an attenuated mutant of JM containing a retroviral LTR sequence, resembles its parent strain in replication and immunodepressive effects in vivo and induces mild histological nerve lesions, but provides surprisingly high levels of protection (Witter et al. 1997). Thus, it seems that the most effective vaccines are characterized by robust replication in vivo and may possess a low level of oncogenicity. However, more information is required before regulatory authorities will be willing to reconsider standards for MD vaccine safety.

6.5 Comparative Efficacy

Based on controlled laboratory trials and by extensive experience under field conditions, it is now well accepted that the relative protective efficacy of different types of MD vaccines varies significantly. The comparison of similar products produced by different companies will not be addressed here. However, it is appropriate to recognize general types or classes of MD vaccines that provide different degrees of protection. A large number of reports contain comparative data on vaccine efficacy (Vielitz and Landgraf 1972; Willemart 1972; Zanella and Granelli 1974; Blaxland et al. 1975; Eidson et al. 1976; Schat et al. 1982; Witter 1982, 1992a; Werner et al. 1992; Witter et al. 1995) and should be consulted for specific information. A report by Witter (Witter 1992a) contrasts many of the commercial vaccine types and provides much of the basis for the following discussion.

The commercially available MD vaccines can generally be placed in one of three groups that are discussed in order of ascending efficacy. The first group, represented by the FC126 strain of HVT, provides good protection against challenge with vMDV isolates. Other members of this group probably include CVI988/C and HPRS-16att, as well as serotype 2 strains when used alone. The second group, represented by the bivalent HVT + SB-1 vaccine, provides superior protection against challenge with vvMDV isolates. Other members of this group include CVI988/C/R6, R2/23, and another bivalent vaccine, HVT + 301B/1. The third group is represented by the CVI988 strain and is generally considered the most protective of the commercially available products, especially against challenge with vv + MDVs.

It is not always prudent to choose the most efficacious vaccine. Because of cost, many production managers select the least effective vaccine that provides sufficient protection under a given set of conditions or circumstances. As circumstances change, then a different vaccine may be used. For example, some broiler producers will use HVT alone during summer and fall months, but switch to HVT + SB-1 vaccine in winter months when challenge is more severe. Some producers have considered it prudent to avoid the use of CVI988, except in conditions where other vaccines are not sufficient, in order to reduce the selection pressure on field strains that may result in viruses of higher virulence or CVI988-resistant mutants.

7 Factors Affecting the Efficacy of Vaccines

Many factors have the potential to influence the efficacy of MD vaccination in the field. Some of these are of special interest as potential causes of vaccine failures or as strategies for improving vaccine efficacy.

7.1 Dose

Relatively low doses of vaccine strains may result in protection under laboratory conditions, but higher doses are used in practice in order to provide for a margin of safety. Commercial vaccines in the United States often exceed minimum PFU per dose titers by up to tenfold or more. It is common, although not necessarily recommended, for producers to dilute vaccines for use in commercial broilers to achieve the desired PFU concentration for the particular circumstances of the flock. On the other hand, producers of layer and breeder flocks, desiring higher levels of protection, rarely give less than a full dose. The rational for this practice lies in the expectation that a full dose provides better protection than a partial dose, an expectation that appears to be validated by field experience. Although protection (measured by quantitative protective dose 50% assays) varies with dose if doses are 690 PFU or less (DE BOER et al. 1981), it does not necessarily hold that protection is increased with very high doses of MD vaccines (SCHMIDT et al. 1984). Yoshida et al. (YOSHIDA et al. 1973) found doses of 173 and 17,800 PFU were equally effective. HVT doses up to 105,000 PFU offered no added benefit against challenge with vvMDV strains compared to doses of 1375 PFU or less (EIDSON et al. 1978; WITTER et al. 1980). Landgraf et al. (LANDGRAF et al. 1981) found doses of 1000 and 2000 PFU of HVT equally effective in providing protection against challenge and induction of viremias at 19 days. The principal function of dose is to establish infection. Higher doses may establish infection earlier (EIDSON et al. 1978; YOSHIDA et al. 1973) and may overcome the effect of maternal antibodies more effectively (WITTER and BURMESTER 1979). Thus, vaccine doses in excess of that needed to establish infection are not likely to be beneficial.

7.2 Maternal Antibody

Under normal commercial conditions, breeder chickens invariably have antibodies induced by natural exposure to field strains of MDV and by vaccination with various MD vaccine viruses representing several serotypes. These antibodies when passed to progeny chicks tend to reduce, but not eliminate, the effectiveness of MD vaccination (CALNEK and SMITH 1972; SPENCER and ROBERTSON 1972; ZYGRAICH and HUYGELEN 1972; CHURCHILL and BAXENDALE 1973; EIDSON et al. 1973). Maternal antibodies also slow the progression of MD lesions induced by early exposure to virulent serotype 1 strains in the absence of vaccination (CHUBB and CHURCHILL 1969; CALNEK 1972; BURGOYNE and WITTER 1973). However, effects on vaccine viruses are probably more important. MD vaccines of all serotypes provide lower levels of protection in chicks with maternal antibodies (Witter and Lee 1984; Witter 1982). In one study, protection by FC126 and Md11/75C strains was reduced by 39 and 64%, respectively, in the presence of maternal antibodies; reductions of vaccine viremia titers after vaccination were even greater (Witter and Lee 1984). The inhibitory effect of maternal antibody is greater for cell-free HVT vaccine than for cell-associated vaccines (Saijo et al. 1972; Witter and Burmester 1979; Sharma and Graham 1982). Adequate levels of protection can, nonetheless, be obtained with cell-free vaccines in antibody-positive chickens by increasing the vaccine dose (Witter and Burmester 1979).

Titers of neutralizing antibody may vary among chicken lines (Choi et al. 1978) and individual chickens (Biggs et al. 1980; von dem Hagen et al. 1980), which may result in variable responses of progeny to vaccination. Some vaccines, especially attenuated serotype 1, may be more susceptible to interference by maternal antibodies than others (Witter and Lee 1984). Although there is substantial cross-neutralization between serotypes, antibodies probably interfere most efficiently with vaccines of the same serotype (King et al. 1981). Alternate generation vaccination, where breeders are vaccinated with serotype 1 or 2 vaccines in order to increase the effectiveness of serotype 3 vaccines in progeny, has been recommended for some situations (Jackson et al. 1977; de Boer et al. 1988; King et al. 1981). This strategy is used currently by at least one major breeding company in the United States (K. Kreager, personal communication). Since serotype 1 and 2 viruses circulate naturally in commercial poultry populations, only serotype 3 can be intentionally avoided.

7.3 Revaccination

The practice of revaccinating chickens at 1–3 weeks of age to improve protection has become common in many countries. This circumstance is surprising, considering that published reports uniformly fail to show any greater protection from two vaccinations than from a single vaccination (Ball and Lyman 1977; Cho et al. 1976; Riddell et al. 1978; Spencer et al. 1972; Zanella et al. 1975). If chickens are "missed" during vaccination at hatch, then revaccination would be beneficial, especially if done immediately. However, vaccination of chickens after they are exposed to virulent strain in the field is not likely to be beneficial, since vaccination after challenge was ineffective in laboratory trials (Okazaki and Burmester 1971). Nonetheless, empirical evidence from field experience provides increasing support for revaccination. The discordance between laboratory findings and field experience on this issue is significant, and warrants further study.

7.4 Age at Challenge

Susceptibility of vaccinated chickens to MDV challenge at different ages is a function of the speed of onset and duration of immunity. The onset of protective immunity after administration of MD vaccines is a progressive process. Although following vaccination at hatch some protection can be measured by 2 days, protection is usually not complete until 5–8 days (OKAZAKI and BURMESTER 1971; WITTER and LEE 1984). The acquisition of immunity can be speeded by in ovo vaccination, at least for HVT vaccine (SHARMA and BURMESTER 1982). Although the specific timing of immunity acquisition may vary with circumstances, vaccinated chicks will be susceptible to challenge for the first few days of life even under optimal environmental conditions. Unfortunately, early exposure to virulent MDV strains is common (WITTER et al. 1970b) due to management practices such as use of built-up litter for broilers and multiple age rearing complexes for layers, and is one of the most important causes of excessive MD losses in vaccinated flocks.

Once established, immunity to challenge appears relatively long lasting. There are few laboratory experiments to test this point, but HVT immunity persisted through at least 40 weeks of age in one study (WITTER and OFFENBECKER 1978). Since vaccine viruses establish latent infection that undoubtedly persists for the life of the chicken, immunity should also persist providing that enough viral antigens are produced to maintain the population of immune cells and that immunity is not compromised by immunodepressive factors. Therefore, vaccines may usually be expected to protect well against challenge at older ages. Some chickens exhibit an age-related resistance to MD lymphoma induction (ANDERSON et al. 1971; CALNEK 1973; WITTER et al. 1973) which should work in concert with vaccinal immunity to produce high levels of resistance. So-called "late" vaccine breaks may occasionally occur and are discussed in a later section.

7.5 Challenge Strain

MDVs have evolved, beginning in the late 1950s, toward increasingly greater virulence. This is probably the direct result of selection pressures brought about by modern husbandry and the use of increasingly effective vaccines. Field strains of MDV are classified in one of four pathotypes: mild (m), virulent (v), very virulent (vv), and very virulent+ (vv+) according to their relative pathogenicity for chickens receiving different vaccines and thus representing a gradient of resistance (WITTER 1997). The more virulent viruses are defined by their ability to overcome the immunity induced by weaker vaccines. Under field conditions, excessive MD losses in vaccinated chickens are often associated with the presence of more virulent MDV pathotypes. Thus, challenge with vv or vv+ strains of MDV represents a major factor limiting the efficacy of vaccines. Not only have MDVs evolved to greater virulence, but these highly virulent strains have also increased in prevalence. For example, of MDV isolates obtained between 1997 and 1998 from farms in the United States where MD was excessive, 5 of 7 typed as vv or vv+ pathotypes (R.L. Witter, unpublished data). Strains with excessive virulence have also been reported in many other countries (BUSCAGLIA et al. 1999; KROSS 1996; LIN and CHEN 1996; LIU et al. 1996; VENUGOPAL et al. 1996). Associated syndromes related to infection with highly virulent strains, such as acute transient paralysis and early mortality syndrome (Witter et al. 1980, 1999), are also seen. It is interesting that the superior protective ability of CVI988 vaccine was not fully recognized until the 1990s when vv and vv+ strains first became available for challenge trials (Witter 1992a). Continued isolation and characterization of highly virulent strains may be essential to assess the protective efficacy of current and future vaccines.

7.6 Stress and Immunodepression

Establishment and maintenance of immunity to MD depends on an intact, fully functional immune system that is capable of producing a robust response. It is well known that stress and various immunodepressive viral infections may compromise immune responses in the chicken. Vaccinal immunity in MD can be compromised by infectious bursal disease virus infection (Jen and Cho 1980), although only in certain circumstances (Sharma 1984). Infections with reticuloendotheliosis virus (Bülow 1977a; Witter et al. 1979), reovirus (Rosenberger 1983), and chicken infectious anemia virus (Bülow et al. 1983; Otaki et al. 1988; Jeurissen and de Boer 1993) also may reduce immune responses to MD vaccines. Although Landgraf et al. (Landgraf et al. 1981) found cold stress or deprivation of food and water during the first few days after hatching had no effect on HVT-induced vaccinal immunity, the potential of physiological and environmental stresses to compromise MD immunity deserves consideration, especially at older ages.

7.7 Genetic Resistance

Genetic factors associated with the *B*-locus significantly influence the susceptibility of chickens to MD (Hansen et al. 1967; Biggs et al. 1968; Cole 1968) (see also the chapter by L.D. Bacon et al., this volume). However, genetic resistance does not totally prevent MD and can be overcome by challenge with highly virulent MDV strains (Bacon and Witter 1992; Witter 1983), and appeared to be best utilized as an adjunct to vaccination. Early studies showed that vaccination provided protection to both resistant and susceptible chicken strains and, moreover, the rank order of strains based on MD susceptibility did not change following vaccination (Spencer et al. 1972; Gavora 1975). Thus, genetic resistance and vaccinal immunity appeared to be additive or at least complementary.

In contrast to earlier findings, a subsequent report showed that HVT vaccination of 15.B congenic lines of chickens altered relative susceptibility rankings to challenge (Bacon and Witter 1992). This suggested that HVT vaccination could be utilized in selection programs as a means to identify resistant haplotypes that might otherwise be missed. The interaction between host genes and vaccines varied with the vaccine serotype; whereas attenuated serotype 1 vaccines gave better protection in certain B-haplotypes, serotype 2 vaccine was superior to others in B^5 chickens (Bacon and Witter 1994). The differential susceptibility of different genotypes to MD vaccines offers an opportunity for the development of genotype-specific vaccines. This problem can also be overcome, at least in part, by the use of polyvalent vaccines.

The genetic effects mentioned above have all been related to the major histocompatibility complex (MHC) or *B*-locus. However, non-MHC effects may also be important (Groot and Albers 1992). Up to 14 non-MHC quantitative trait loci for MD have been recently identified in recombinant congenic lines developed from line 6 (resistant) and line 7 (susceptible) chickens, both with similar MHC alleles (Vallejo et al. 1998; Yonash et al. 1999). Relationships between non-MHC resistance and vaccination have not been investigated.

8 The Anatomy of Vaccination Failures

A vaccination failure is loosely defined in practice as the occurrence of excessive MD losses in a vaccinated flock. However, not all cases of excessive losses result because of inadequate vaccine or vaccination procedures. The major variables are: (1) the vaccine, including its handling and administration; (2) the challenge virus; (3) the chicken; and (4) the rearing environment including various stress factors. A breakdown in any aspect of the process can adversely affect the development or maintenance of protective immunity. Descriptions of vaccination failures (breaks) and a discussion of possible causes have been previously reported (Witter 1985c; Schat 1997). Clinically, most MD breaks occur during the rearing or early egg production periods. However, breaks in mature layers or breeders, sometimes commencing after molt prior to a second laying cycle, have been recently reported (Kreager 1997a). Mortality patterns representing early and late MD breaks adapted from actual flock data are shown in Fig. 5.

8.1 Early Breaks

Historically, most vaccination failures in layers have usually occurred early, with an onset of between 8 and 24 weeks, a duration of 10–15 weeks, and a total mortality of 20–50% in the worst cases (Kreager 1997a,b). There are many possible causes of early breaks, and each must be carefully excluded. If vaccine is mishandled during shipment, stored incorrectly, diluted inaccurately, mixed with inappropriate additives, or stored too long after dilution, the titer may be insufficient to induce a robust infection. If the vaccination technique is poor, so that some chickens fail to receive a full dose of vaccine, then immunity will be uneven in a flock and some chicks may be highly susceptible. If chicks become immunodepressed due to the stress of poor management or intercurrent infectious diseases, then immunity may fail to develop or, if developed, may decline. If chickens are exposed too quickly to virulent field strains, i.e., early exposure, vaccination will be ineffective. If chickens are exposed to highly virulent pathotypes of MDV and are vaccinated with less

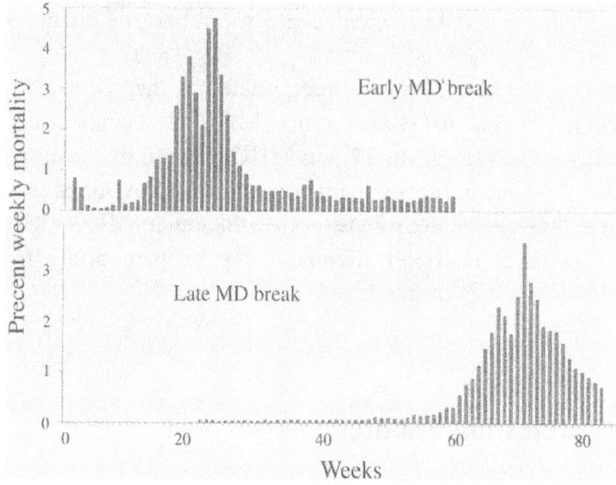

Fig. 5. Mortality pattern characteristics of MD vaccine breaks derived from flock data provided by K. Kreager

efficacious vaccines, then protection may be inadequate. More than one factor may operate concurrently.

Determination of the cause of an early break by retrospective analysis is very difficult. Usually, one starts by examining the vaccination process and management factors (SCHAT 1997; JACKSON 1998). Many problems have been resolved by improvement of procedures. Early exposure on multiple age pullet farms has been a problem in the past, and has in a few cases been resolved by a switch to an all-in-all-out rearing program. Breaks are frequently blamed on highly virulent pathotypes, but this is very difficult to confirm. The process of virus isolation and pathotyping by chicken inoculation is very slow and costly, and is currently available only as a research tool. Solutions to reduce the frequency of breaks start with establishment of high standards of management and biosecurity. Many times, improved management of vaccines or chicken rearing is beneficial. It may also be helpful to use a more effective vaccine formulation, if one is available. Vaccines provide differential levels of protection, as discussed earlier.

8.2 Late Breaks

Excessive losses from MD have occasionally been noted in older laying flocks, i.e., over the age of 40 weeks. Some breaks have occurred subsequent to molting (at 60–70 weeks). A suitable explanation of late breaks is elusive and alternate theories have been advanced. One theory (new infection) suggests that flocks became exposed to a new, highly virulent strain of MDV at an older age which induced infection, broke through existing levels of immunity, and caused lymphomas and mortality (KREAGER 1997a; LUCIO-MARTINEZ 1999). An alternate theory (old infection) suggests that flocks become exposed to an immunodepressive stress that compromises the delicate balance between immunity and disease and allows the progression of MD, caused by preexisting virulent strains. Both theories fit the typical epidemiological picture of late breaks but neither has been confirmed as the cause. The "new infection" theory implies that chickens vaccinated at hatch and exposed to field viruses for many months will be susceptible to challenge with a more virulent strain late in life. Adult SPF chickens are susceptible to challenge with highly virulent MDV strains (Rosenberger et al. 1997). In other studies, Witter (R.L. Witter, unpublished data) confirmed the susceptibility of adult SPF chickens but failed to induce disease by contact challenge with highly virulent MDVs of adult SPF birds previously immunized at hatch. On the other hand, the "old infection" theory predicts the presence of some factor, acting like an infectious agent, that compromises the immune system, but such a factor has not been identified. More work on this important issue is needed.

9 Integrated Control Strategies

The control of MD depends on a successful combination of vaccination, management and genetic resistance (Witter 1998). By themselves, good vaccines are not enough and can be defeated by many factors including early exposure and viral evolution. Accordingly, it is important to view the control of MD as having several components, each critical to the success of the whole.

Vaccines are the cornerstone of MD control, and will continue to be so. Present vaccines provide exceptionally high levels of protection, as judged by the difference between predicted and actual condemnation losses in young broiler chickens (Fig. 6). Surprisingly, the trend in MD losses in the United States continues downward, despite the increase in virulence of field strains of MDV, and in 1999 reached historic low levels (Fig. 7). In other countries, however, significant

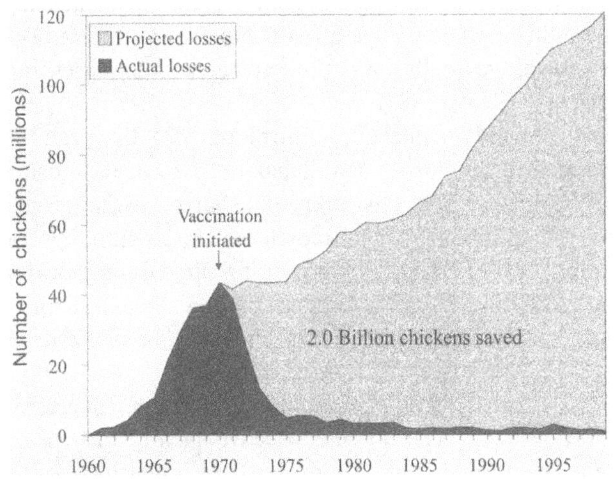

Fig. 6. Performance of MD vaccines in the United States by comparison of actual condemnations of young broiler chickens to condemnations projected at 1970 (pre-vaccine) rates

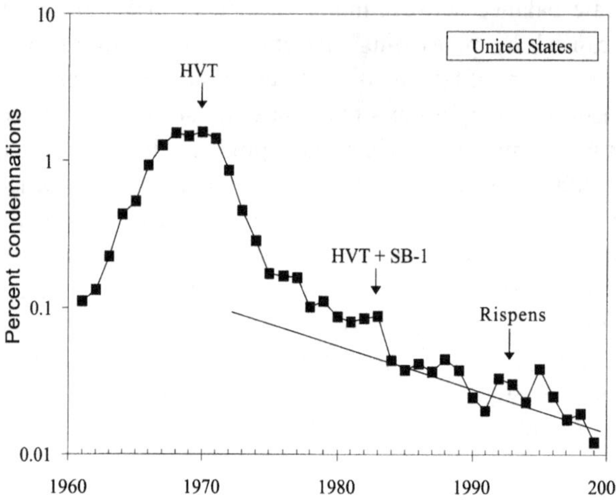

Fig. 7. MD condemnations in young broiler chickens continue to decrease. Note trend of recent condemnations on the $\log_{10}$ scale

losses from MD continue to be observed, and in some countries may be increasing (Kreager 1997b). The available vaccines should be used judiciously, as the time cannot be predicted when more protective products will be needed due to further evolution of viral strains. The useful life of past MD vaccines has approximated 10 years (Kreager 1997a). More importantly, the technology needed to produce vaccines with significantly better protective ability is not yet in hand and it is not clear whether or not this goal is even feasible (Baxendale 1997).

Increasingly, the poultry industry makes decisions on disease control based on short-term economic expectations (Kreager 1996). For example, broiler houses are not cleaned after every brood, even though this would surely reduce MD losses, because the cost to do so exceeds the benefit. Similarly, broilers in the United States do not usually receive a full dose of vaccine, even though this might achieve better levels of protection against MD, because the cost to do so exceeds the benefit. Multiple-age pullet farms may be converted to single-age operations when MD losses are high but later, when losses are less of a problem, multiple-age housing may be reestablished because this method of operation is more cost-efficient. The economic pressures of the competitive agribusiness marketplace, in some cases, run counter to efforts to control MD (Rudd 1996).

Early exposure is one obstacle to effective control of MD that can be addressed. The knowledge and technology is available to minimize exposure of chickens to virulent MDV challenge from environmental sources. Although not currently practical except for specific-pathogen-free flocks, large numbers of chickens can be raised entirely free of MDV exposure using filtered air, positive pressure environments (Anderson et al. 1972; Mitchell et al. 1989). Similar principles of biosecurity are applicable to field flocks and have been useful in individual operations.

Interventions designed to slow or stop the evolution of MDV to greater virulence are even more problematic. The specific selective factors that favor emergence of virulent mutants are not well understood. However, it may help to use full doses of vaccine, reduce exposure and develop genetic resistance based on multiple gene systems. This multiple barrier approach (Witter 1998) should make it more difficult for the virus to overcome host resistance by a single mutation. Additional options that restrict or prevent viral replication or shedding, perhaps by the transfer and expression in the chicken germ line of viral transgenes, have potential to be a more permanent solution, but technological, financial, public policy, and regulatory obstacles will be daunting.

References

Anderson DP, King DD, Eidson CS, Kleven SH (1972) Filtered air positive pressure (FAPP) brooding of broiler chickens. Avian Dis 16:20–26

Anderson K, Eidson CS, Richey DJ (1971) Age susceptibility of chickens to Marek's disease. Amer J Vet Res 32:935–938

Bacon LD, Witter RL (1992) Influence of turkey herpesvirus vaccination on the B-haplotype effect on Marek's disease resistance in 15. B-congenic chickens. Avian Dis 36:378–385

Bacon LD, Witter RL (1994) Serotype specificity of B-haplotype influence on the relative efficacy of Marek's disease vaccines. Avian Dis 38:65–71

Bacon LD, Witter RL, Fadly AM (1989) Augmentation of retrovirus-induced lymphoid leukosis by Marek's disease herpesviruses in white leghorn chickens. J Virol 63:504–512

Bagust TJ, Grimes TM, Dennett DP (1979) Infection studies on a reticuloendotheliosis virus contaminant of a commercial Marek's disease vaccine. Aust Vet J 55:153–157

Ball RF, Lyman J (1977) Revaccination of chicks for Marek's disease at twenty-one days old. Avian Dis 21:440–444

Baxendale W (1997) Marek's disease vaccines have to become more efficacious. World Poultry August: 20–21

Biggs PM, de Boer GF, Burmester BR, von Bülow V, Kaleta EF (1979) Recommendations for control in the production of Marek's disease vaccine. J Biol Stand March: 29–42

Biggs PM, Milne BS (1972) Biological properties of a number of Marek's disease virus isolates. In: Biggs PM, de-The G, Payne LN (eds) Oncogenesis and Herpesviruses. International Agency for Research on Cancer, Lyon, France, pp 88–94

Biggs PM, Payne LN, Milne BS, Churchill AE, Chubb RC, Powell DG, Harris AH (1970) Field trials with an attenuated cell associated vaccine for Marek's disease. Vet Rec 87:704–709

Biggs PM, Powell DG, Churchill AE, Chubb RC (1972) The epizootiology of Marek's disease. I. Incidence of antibody, viraemia and Marek's disease in six flocks. Avian Pathol 1:5–25

Biggs PM, Shilleto RFW, Milne BS (1980) A quantitative sequential study of viraemia and neutralising antibody to HVT and MDV in a commercial flock vaccinated with HVT. Avian Pathol 9: 511–523

Biggs PM, Thorpe RJ, Payne LN (1968) Studies on genetic resistance to Marek's disease in the domestic chicken. Brit Poult Sci 9:37–52

Blaxland JD, MacLeod AJ, Hall T (1975) Trials with Marek's disease vaccines prepared from a turkey herpes virus and an attenuated Marek's disease virus. Vet Rec 97:48–49

Burgoyne GH, Witter RL (1973) Effect of passively transferred immunoglobulins on Marek's disease. Avian Dis 17:824–837

Buscaglia C, Risso M, Antonini E, Prio MV, Prada MG, Villat MC, del Barrio E (1999) Behaviour of a very virulent Marek's disease virus in two vaccinated commercial chicken flocks in Argentina. Proc 48th West Poult Dis Conf 105–108

Bülow VV (1977b) Further characteristics of the CVI988 strain of Marek's disease virus. Avian Pathol 6:395–403

Bülow VV (1977a) Immunological effects of reticuloendotheliosis virus as potential contaminant of Marek's disease vaccines. Avian Pathol 6:383–393

Bülow VV, Fuchs B, Vielitz E, Landgraf H (1983) Early mortality syndrome of chickens after dual infection with Marek's disease virus (MDV) and chicken anaemia agent (CAA). Zbl Vet Med 30:742–750

Calnek BW (1972) Effects of passive antibody on early pathogenesis of Marek's disease. Infect Immun 6:193–198

Calnek BW (1973) Influence of age at exposure on the pathogenesis of Marek's disease. J Natl Cancer Inst 51:929–939

Calnek BW, Carlisle JC, Fabricant J, Murthy KK, Schat KA (1979) Comparative pathogenesis studies with oncogenic and nononcogenic Marek's disease viruses and turkey herpesvirus. Amer J Vet Res 40:541–548

Calnek BW, Schat KA, Peckham MC, Fabricant J (1983) Field trials with a bivalent vaccine (HVT and SB-1) against Marek's disease. Avian Dis 27:844–849

Calnek BW, Shek WR, Schat KA (1981) Latent infections with Marek's disease virus and turkey herpesvirus. J Natl Cancer Inst 66:585–590

Calnek BW, Smith MW (1972) Vaccination against Marek's disease with cell-free turkey herpesvirus: interference by maternal antibody. Avian Dis 16:954–957

Cho BR (1977) Dual virus maturation of both pathogenic and apathogenic Marek's disease herpesvirus (MDHV) in the feather follicles of dually infected chickens. Avian Dis 21:501–507

Cho BR, Balch RK, Hill RW (1976) Marek's disease vaccine breaks: differences in viremia of vaccinated chickens between those with and without Marek's disease. Avian Dis 20:496–503

Cho BR, Kenzy SG (1972) Isolation and characterization of an isolate (HN) of Marek's disease virus with low pathogenicity. Appl Microbiol 24:299–306

Choi CO, Sinkovic VP, Jackson CAW (1978) Development of neuralising antibody to turkey herpesvirus (HVT) in chickens with and without homologous maternal antibody. Vet Bull 48:5409

Chubb RC, Churchill AE (1969) The effect of maternal antibody on Marek's disease. Vet Rec 85: 303–305

Churchill AE (1985) Production of vaccines. In: Payne LN (ed) Marek's disease. Martinus Nijhoff, Boston, pp 251–266

Churchill AE, Baxendale W (1973) Virus effect of lyophilized turkey herpesvirus – Marek's vaccines in chicks with maternal antibodies against turkey herpesvirus. Wien Tierarztl Monatsschr 60:248–254

Churchill AE, Chubb RC, Baxendale W (1969a) The attenuation with loss of oncogenicity of the herpes-type virus of Marek's disease (Strain HPRS-16) on passage in cell culture. J Gen Virol 4:557–564

Churchill AE, Payne LN, Chubb RC (1969b) Immunization against Marek's disease using a live attenuated virus. Nature 221:744–747

Cole RK (1968) Studies on genetic resistance to Marek's disease. Avian Dis 12:9–28

Cronenberg AM, Van Geffen CEH, Dorrestein J, Vermeulen AN, Sondermeijer PJA (1999) Vaccination of broilers with HVT expressing an *Eimeria acervulina* antigen improves performance after challenge with *Eimeria*. Acta Virol 43:192–197

Darteil R, Bublot M, Laplace E, Bouquet JF, Audonnet JC, Riviere M (1995) Herpesvirus of turkey recombinant viruses expressing infectious bursal disease virus (IBDV) VP2 immunogen induce protection against an IBDV virulent challenge in chickens. Virology 211:481–490

de Boer GF, Groenendal JE, Boerrigter HM, Kok GL, Pol JMA (1986) Protective efficacy of Marek's disease virus (MDV) CVI-988 CEF65 clone C against challenge infection with three very virulent MDV strains. Avian Dis 30:276–283

de Boer GF, Orthel FW, Krasselt M, Oei HL, Pereboom WJ, Barendregt LG (1981) Comparative 50% protective dose assays (PD 50) of Marek's disease virus strain CVI988. J Biol Stand 9:15–22

de Boer GF, Pol JMA, Jeurissen SHM (1988) Marek's disease vaccination strategies using vaccines made from three avian herpesvirus serotypes. In: Kato S, Horiuchi T, Mikami T, Hirai K (eds) Advances in Marek's disease research. Japanese Association on Marek's Disease, Osaka, Japan, pp 405–413

de Boer GF, VanVloten J, Van Zaane D (1980) Possible horizontal spread of lymphoid leukosis virus during vaccination against Marek's disease. In: Biggs PM (ed) Resistance and Immunity to Marek's Disease. Commission of the European Communities, Luxembourg, pp 552–565

Eidson CS, Anderson DP, King DD (1971) Resistance of progeny from parental stock of chickens immunized against Marek's disease. Amer J Vet Res 32:2071–2076

Eidson CS, Kleven SH, Anderson DP (1973) Efficacy of cell-free and cell-associated herpesvirus of turkey vaccines in progeny from vaccinated parental flocks. Amer J Vet Res 34:869–872

Eidson CS, Page RK, Giambrone JJ, Kleven SH (1976) Long term studies comparing the efficacy of cell-free versus cell-associated HVT vaccines against Marek's disease. Poult Sci 55:1857–1863

Eidson CS, Page RK, Kleven SH (1978) Effectiveness of cell-free or cell-associated turkey herpesvirus vaccine against Marek's disease in chickens as influenced by maternal antibody, vaccine dose, and time of exposure to Marek's disease virus. Avian Dis 22:583–597

Finkelstein A, Silva RF (1989) Live recombinant vaccines for poultry. Trends Biotechnol 7:273–277

Friedman A, Shalem-Meilin E, Heller ED (1992) Marek's disease vaccines cause temporary B-lymphocyte dysfunction and reduced resistance to infection in chicks. Avian Pathol 21:621–631

Gagic M, St Hill CA, Sharma JM (1999) In ovo vaccination of specific-pathogen-free chickens with vaccines containing multiple agents. Avian Dis 43:293–301

Gavora JS (1975) Vaccination against Marek's disease: effectiveness of cell associated and lyophilized herpesvirus of turkeys in 9 strains of leghorns. Poult Sci 54:1765

Groot AJC, Albers GAA (1992) The effect of MHC on resistance to Marek's disease in White Leghorn crosses. In: 4th International Symposium on Marek's Disease, 19th World's Poultry Congress, Vol. 1. World's Poultry Science Assn., Amsterdam, pp 185–188

Hansen MP, Van Zandt JN, Law GRJ (1967) Differences in susceptibility to Marek's disease in chickens carrying two different B locus blood group alleles. Poult Sci 46:1268

Heckert RA, Riva J, Cook S, McMillen JK, Schwartz RD (1996) Onset of protective immunity in chicks after vaccination with a recombinant herpesvirus of turkeys vaccine expressing Newcastle disease virus fusion and hemagglutinin-neuraminidase antigens. Avian Dis 40:770–777

Jackson CAW (1998) Multiple causes of Marek's disease vaccination failure in Australian poultry flocks. Proc 47th West Poult Dis Conf 49–51

Jackson CAW (1999) Quality assurance of Marek's disease vaccine use in hatcheries. Proc 48th West Poult Dis Conf 34–38

Jackson CAW, Biggs PM, Bell RA, Lancaster FM, Milne BS (1974) A study of vaccination against Marek's disease with an attenuated Marek's disease virus. Avian Pathol 3:123–144

Jackson CAW, Sinkovic B, Choi CO (1977) Infectivity and immunogenicity of apathogenic and attenuated Marek's disease virus vaccines. Proc 54th Ann Conf Austral Vet Assn 149–152

Jen LW, Cho BR (1980) Effects of infectious bursal disease on Marek's disease vaccination: suppression of antiviral immune response. Avian Dis 24:896–907

Jeurissen SHM, de Boer GF (1993) Chicken anaemia virus influences the pathogenesis of Marek's disease in experimental infections, depending on the dose of Marek's disease virus. Vet Quart 15:81–84

Johnston PA, Liu H, O'Connell T, Phelps P, Bland M, Tyczkowski J, Kemper A, Harding T, Avakian A, Haddad E, Whitfill C, Gildersleeve R, Ricks CA (1997) Applications in ovo technology. Poult Sci 76:165–178

Karaca K, Sharma JM, Winslow BJ, Junker DE, Reddy S, Cochran M, McMillen JK (1998) Recombinant fowlpox viruses coexpressing chicken type I IFN and Newcastle disease virus HN and F genes: influence of IFN on protective efficacy and humoral responses of chickens following in ovo or post-hatch administration of recombinant viruses. Vaccine 16:1496–1503

Kawamura H, King DJ Jr, Anderson DP (1969) A herpesvirus isolated from kidney cell culture of normal turkeys. Avian Dis 13:853–863

King DD, Page D, Schat KA, Calnek BW (1981) Difference between influences of homologous and heterologous maternal antibodies on response to serotype-2 and serotype-3 Marek's disease vaccines. Avian Dis 25:74–81

Konobe T, Ishikawa T, Takaku K, Ikuta K, Kitamoto N, Kato S (1979) Marek's disease virus and herpesvirus of turkey noninfective to chickens, obtained by repeated in vitro passages. Biken J 22:103–107

Kreager K (1996) Industry concerns workshop. In: Silva RF, Cheng HH, Coussens PM, Lee LF, Velicer LF (eds) Current Research on Marek's Disease. American Association of Avian Pathologists, Inc., Kennett Square, Pennsylvania, pp 509–511

Kreager K (1997b) A global perspective on Marek's disease control in layers and layer breeders. World Poultry August:14–15

Kreager K (1997a) Marek's disease: clinical aspects and current field problems in layer chickens. In: Fadly AM, Schat KA, Spencer JL (eds) Diagnosis and control of neoplastic diseases of poultry. American Association of Avian Pathologists, Kennett Square, pp 23–26

Kross I (1996) Isolation of highly lytic serotype 1 Marek's disease viruses from recent field outbreaks in Europe. In: Silva RF, Cheng HH, Coussens PM, Lee LF, Velicer LF (eds) Current Research on Marek's Disease. American Association of Avian Pathologists, Inc., Kennett Square, Pennsylvania, pp 113–118

Landgraf H, Vielitz E, Huttner B (1981) HVT-viraemia and protection against Marek's disease following intramuscular or subcutaneous vaccination with different viral doses as well as following thermal stress broiler chicks. Deut Tierarztl Woch 88:524–526

Lee LF, Nazerian K, Boezi JA (1975) Marek's disease virus DNA in a chicken lymphoblastoid cell line (MSB-1) and in virus-induced tumours. In: Biggs PM, de-The G, Payne LN (eds) Oncogenesis and Herpesviruses II. IARC Publications, Lyon, pp 199–204

Lee LF, Sharma JM, Nazerian K, Witter RL (1978) Suppression and enhancement of mitogen response in chickens infected with Marek's disease virus and the herpesvirus of turkeys. Infect Immun 21: 474–479

Li J, Zhao X, Yang J, Liu C, Li X, Tian Z, Xu Y (1996) Primary studies on immunoprotectivity of Marek's disease virus DNA vaccine. In: Silva RF, Cheng HH, Coussens PM, Lee LF, Velicer LF (eds) Current Research on Marek's Disease. American Association of Avian Pathologists, Inc., Kennett Square, Pennsylvania, pp 367–371

Lin JA, Chen CP (1996) First isolation and characterization of very virulent Marek's disease virus in Taiwan. J Vet Med Sci 58:1011–1015

Liu X, Guan E, Wu C, Zhang R (1996) Characterization of very virulent strains of Marek's disease virus from vaccinated chickens in Eastern China. In: Silva RF, Cheng HH, Coussens PM, Lee LF, Velicer LF (eds) Current Research on Marek's Disease. American Association of Avian Pathologists, Inc., Kennett Square, Pennsylvania, pp 142–147

Lucio-Martinez B (1999) Impact of vv Marek's disease on mortality and production in a multiple-age farm. Proc 48th West Poult Dis Conf pp 55–56

Meulemans G, Halen P, Schyns P, Bruynooshe D (1971) Field trials with an attenuated Marek's disease vaccine. Vet Rec 89:325–329

Mitchell BW, Beard CW, Yoder HW (1989) Recent advances in filtered-air positive-pressure (FAPP) housing for the production of disease-free chickens. Avian Dis 33:792–800

Morgan RW, Gelb JJ, Pope CR, Sondermeijer PJA (1993) Efficacy in chickens of a herpesvirus of turkeys recombinant vaccine containing the fusion gene of Newcastle disease virus: onset of protection and effect of maternal antibodies. Avian Dis 37:1032–1040

Morgan RW, Gelb JJ, Schreurs CS, Lutticken D, Rosenberger JK, Sondermeijer PJA (1992) Protection of chickens from Newcastle and Marek's diseases with a recombinant herpesvirus of turkeys vaccine expressing the Newcastle disease virus fusion protein. Avian Dis 36:858–870

Morgan RW, Kent J, Anderson AS (1998) Marek's disease virus gC and *meq* mutants with dramatic reductions in tumour incidences and horizontal transmission. Avian Pathol 27:S89

Nakamura H, Sakaguchi M, Hirayama Y, Miki N, Yamamota M, Hirai K (1992) Protection against Newcastle Disease by recombinant Marek's disease virus serotype-1 expressing the fusion protein of Newcastle Disease virus. In: 4th International Symposium on Marek's Disease, 19th World's Poultry Congress, Vol. 1. World's Poultry Science Assn, Amsterdam, pp 332–335

Nathan DB, Lustig S (1990) Production of Marek's Disease vaccine. Viral Vacc 14:347–365

Nazerian K (1970) Attenuation of Marek's disease virus and study of its properties in two different cell cultures. J Natl Cancer Inst 44:1257–1267

Nazerian K (1973) Studies on intracellular and membrane antigens induced by Marek's disease virus. J Gen Virol 21:193–195

Nazerian K, Lee LF, Yanagida N, Ogawa R (1992) Protection against Marek's Disease by a fowlpox virus recombinant expressing the glycoprotein B of Marek's Disease virus. J Virol 66:1409–1413

Nazerian K, Witter RL, Lee LF, Yanagida N (1996) Protection and synergism by recombinant fowl pox vaccines expressing genes from Marek's disease virus. Avian Dis 40:368–376

Oei HL, de Boer GF (1986) Comparison of intramuscular and subcutaneous administration of Marek's disease vaccine. Avian Pathol 15:569–579

Okazaki W, Burmester BR (1971) The temporal relationship between vaccination with the herpesvirus of turkeys and challenge with virulent Marek's disease virus. Avian Dis 15:753–761

Okazaki W, Purchase HG, Burmester BR (1970) Protection against Marek's disease by vaccination with a herpesvirus of turkeys. Avian Dis 14:413–429

Ono K, Takashima M, Ishikawa T, Hayashi M, Konobe T, Ikuta K, Nakajima K, Ueda S, Kato S, Hirai K, Yoshida I (1985) Partial protection against Marek's disease in chickens immunized with glycoproteins gB purified from turkey-herpesvirus-infected cells by affinity chromatography coupled with monoclonal antibodies. Avian Dis 29:533–539

Otaki Y, Nunoya T, Tajima M, Kato S, Nomura Y (1988) Depression of vaccinal immunity to Marek's disease by infection with chicken anaemia agent. Avian Pathol 17:333–347

Payne LN (1992) Marek's disease. In: Manual of standards for diagnostic tests and vaccines. Office International Des Epizootie, Paris, pp 629–638

Phillips PA, Biggs PM (1972) Course of infection in tissue of susceptible chickens after exposure to strains of Marek's disease and turkey herpesvirus. J Natl Cancer Inst 49:1367–1373

Pol JMA, Kok GL, Oei HL, de Boer GF (1986) Pathogenicity studies with plaque-purified preparations of Marek's disease virus strain CVI-988. Avian Dis 30:271–275

Prasad LBM (1978) Effect of maternal antibody on viraemic and antibody responses to cell associated and cell free turkey herpesvirus in chickens. Brit Vet J 134:315–321

Pruthi AK, Gupta RKP, Sadana JR (1987) Efficacy of bivalent vaccine against Marek's disease. Res Vet Sci 42:145–149

Purchase HG, Witter RL (1986) Public health concerns from human exposure to oncogenic avian herpesviruses. J Amer Vet Med Assn 189:1430–1436

Rangga-Tabbu C, Cho BR (1982) Marek's disease virus (MDV) antigens in the feather follicle epithelium: difference between oncogenic and nononcogenic MDV. Avian Dis 26:907–917

Rautenschlein S, Sharma JM, Winslow BJ, McMillen JK, Junker DE, Cochran M (1999) Embryo vaccination of turkeys against Newcastle disease infection with recombinant fowlpox virus constructs containing interferons as adjuvants. Vaccine 18:426–433

Reddy SK, Sharma JM, Ahmad J, Reddy DN, McMillen JK, Cook SM, Wild MA, Schwartz RD (1996) Protective efficacy of a recombinant herpesvirus of turkeys as an in ovo vaccine against Newcastle and Marek's diseases in specific pathogen-free chickens. Vaccine 14:469–477

Ricks CA, Avakian A, Bryan T, Gildersleeve R, Haddad E, Ilich R, King S, Murray L, Phelps P, Poston R, Whitfill C, Williams C (1999) In ovo vaccination technology. Adv Vet Med 41:495–515

Riddell C, Milne BS, Biggs PM (1978) Herpes virus of turkey vaccine: Viraemias in field flocks and in experimental chickens. Vet Rec 102:123–125

Rispens BH, Van Vloten J, Mastenbroek N, Maas HJL, Hendrick JL (1972b) Control of Marek's disease in the Netherlands. II. Field trials on vaccination with an avirulent strain (CVI 988) of Marek's disease virus. Avian Dis 16:126–138

Rispens BH, Van Vloten J, Mastenbroek N, Maas HJL, Schat KA (1972a) Control of Marek's disease in the Netherlands. I. Isolation of an avirulent Marek's disease virus (strain CVI 988) and its use in laboratory vaccination trials. Avian Dis 16:108–125

Rosenberger JK (1983) Reovirus interference with Marek's disease vaccination. Proc West Poult Dis Conf pp 50–51

Rosenberger JK, Cloud SS, Olmeda-Miro N (1997) Epizootiology and adult transmission of Marek's disease. In: Fadly AM, Schat KA, Spencer JL (eds) Diagnosis and control of neoplastic diseases of poultry. American Association of Avian Pathologists, Kennett Square, pp 30–32

Ross LJN (1998) Recombinant vaccines against Marek's disease. Avian Pathol 27:S65–S73

Rudd HK (1996) Use of Marek's vaccines – protecting our future. In: Silva RF, Cheng HH, Coussens PM, Lee LF, Velicer LF (eds) Current Research on Marek's Disease. American Association of Avian Pathologists, Inc., Kennett Square, Pennsylvania, pp 347–352

Saijo K, Konno N, Obata F, Sagawa T, Fujikawa Y (1972) The effect of maternal antibody against lyophilized Marek's disease live vaccine (Cell-free HVT). Jap World Vet Poult Ass 33–34

Sarma G, Greer W, Estep C, Winkler DC (1992) Field trial and immunogenicity studies on the polyvalent Marek's disease vaccines in chickens. In: 4th International Symposium on Marek's Disease, 19th World's Poultry Congress, Vol. 1. World's Poultry Science Assn., Amsterdam, pp 310–314

Schat KA (1997) What to do if there is a Marek's disease problem? World Poultry August:18–19

Schat KA, Calnek BW (1978) Characterization of an apparently nononcogenic Marek's disease virus. J Natl Cancer Inst 60:1075–1082

Schat KA, Calnek BW, Fabricant J (1982) Characterization of two highly oncogenic strains of Marek's disease virus. Avian Pathol 11:593–605

Schat KA, Calnek BW, Fabricant J, Graham DL (1985) Pathogenesis of infection with attenuated Marek's disease virus strains. Avian Pathol 14:127–146

Schat KA, Omar AR, Lee LF, Hunt HD (1996) Induction of glycoprotein B (gB)-specific cytotoxic T cells after vaccination with recombinant fowl poxvirus expressing gB. In: Silva RF, Cheng HH, Coussens PM, Lee LF, Velicer LF (eds) Current Research on Marek's Disease. American Association of Avian Pathologists, Inc., Kennett Square, Pennsylvania, pp 432–435

Schmidt U, Werner P, Reinholdt P, Grunert G, Walter I (1984) Effectiveness of Riems MD (Marek's disease) vaccine on laying hens with different virus levels per immunization dose. Monatsh Vetinarmed 39:100–102

Sharma JM (1984) Effect of infectious bursal disease virus on protection against Marek's disease by turkey herpesvirus vaccine. Avian Dis 28:629–640

Sharma JM (1987) Delayed replication of Marek's disease virus following in ovo inoculation during late stages of embryonal development. Avian Dis 31:570–576

Sharma JM, Burmester BR (1982) Resistance to Marek's disease at hatching in chickens vaccinated as embryos with the turkey herpesvirus. Avian Dis 26:134–149

Sharma JM, Coulson BD, Young E (1976) Effect of in vitro adaptation of Marek's disease virus on pock induction on the chorioallantoic membrane of embryonated chicken eggs. Infect Immun 13: 292–295

Sharma JM, Graham CK (1982) Influence of maternal antibody on efficacy of embryo vaccination with cell-associated and cell-free Marek's disease vaccine. Avian Dis 26:860–870

Sharma JM, Lee LF, Wakenell PS (1984) Comparative viral, immunologic and pathologic responses of chickens inoculated with herpesvirus of turkeys as embryos or at hatch. Amer J Vet Res 45:1619–1623

Sharma JM, Witter RL (1983) Embryo vaccination against Marek's disease with serotypes 1, 2 and 3 vaccines administered singly or in combination. Avian Dis 27:453–463

Sondermeijer PJA, Claessens JAJ, Jenniskens PE, Mockett APA, Thijssen RAJ, Willemse MJ, Morgan RW (1993) Avian herpesvirus as a live viral vector for the expression of heterologous antigens. Vaccine 11:349–358

Sonoda K, Sakaguchi M, Okamura H, Yokogawa K, Hamada F, Hirai K (1996) Expression of the NDV-F gene under the control of MDV1-GB promoter in recombinant MDV1. In: Silva RF, Cheng HH, Coussens PM, Lee LF, Velicer LF (eds) Current Research on Marek's Disease. American Association of Avian Pathologists, Inc., Kennett Square, Pennsylvania, pp 408–413

Spencer JL, Grunder AA, Robertson A, Speckmann GW (1972) Attenuated Marek's disease herpesvirus: protection conferred on strains of chickens varying in genetic resistance. Avian Dis 16:94–107

Spencer JL, Robertson A (1972) Influence of maternal antibody on infection with virulent or attenuated Marek's disease herpesvirus. Amer J Vet Res 33:393–400

Thornton DH (1985) Quality control and standardization of vaccines. In: Payne LN (ed) Marek's disease. Martinus Nijhoff, Boston, pp 267–291

Tsukamoto K, Kojima C, Komori Y, Tanimura N, Mase M, Yamaguchi S (1999) Protection of chickens against very virulent infectious bursal disease virus (IBDV) and Marek's disease virus (MDV) with a recombinant MDV expressing IBDV VP2. Virology 257:352–362

Vallejo RL, Bacon LD, Liu HC, Witter RL, Groenen MM, Hillel J, Cheng HH (1998) Genetic mapping of quantitative trait loci affecting susceptibility to Marek's disease virus induced tumors in F_2 intercross chickens. Genetics 148:349–360

van Eck JHH (1997) Should we vaccinate intramuscularly or subcutaneously? World Poultry August: 22–23

Venugopal K, Bland AP, Ross LJN, Payne LN (1996) pathogenicity of an unusual highly virulent Marek's disease virus isolated in the United Kingdom. In: Silva RF, Cheng HH, Coussens PM, Lee LF, Velicer LF (eds) Current Research on Marek's Disease. American Association of Avian Pathologists, Inc., Kennett Square, Pennsylvania, pp 119–124

Vielitz E, Landgraf H (1971) Vaccination tests against Marek's disease by application of an attenuated Marek herpes virus and a herpes virus of turkeys. Deut Tierarztl Woch 78:617–623

Vielitz E, Landgraf H (1972) Vaccination experiments against Marek's disease by application of an attenuated Marek herpesvirus and a turkey herpesvirus. Prog Immunobiol Standard 5:141–148

Vielitz E, Landgraf H (1985) Experiences with monovalent and bivalent Marek's disease vaccines. In: Calnek BW, Spencer JL (eds) Proceedings of the international symposium on Marek's disease. American Association of Avian Pathologists, Kennett Square, PA, pp 570–576

von dem Hagen D, Löliger H-C, Hartmann W (1980) The influence of maternally derived HVT-neutralizing antibody on Marek's disease. In: Biggs PM (ed) Resistance and immunity to Marek's disease. Commission of the European Communities, Luxembourg, pp 271–276

Werner O, Grunert M, Paesch R, Hahnewald R, Magunski S (1992) Comparative evaluation of vaccines with turkey herpesvirus and Marek's disease virus strain CVI-988 in field trials in East Germany. In: 4th International Symposium on Marek's Disease, 19th World's Poultry Congress, Supplement. World's Poultry Science Association, Amsterdam, pp 21–26

Willemart JP (1972) Marek's disease. Study on the efficacy of three vaccines. Rec Med Vet 148:203–216

Witter RL (1982) Protection by attenuated and polyvalent vaccines against highly virulent strains of Marek's disease virus. Avian Pathol 11:49–62

Witter RL (1983) Characteristics of Marek's disease viruses isolated from vaccinated commercial chicken flocks: association of viral pathotype with lymphoma frequency. Avian Dis 27:113–132

Witter RL (1985b) Association in broiler chickens between natural serotype 2 Marek's disease virus infection and leukosis condemnations. In: Calnek BW, Spencer JL (eds) Proceedings of the international symposium on Marek's disease. American Association of Avian Pathologists, Kennett Square, PA, pp 545–554

Witter RL (1985a) Principles of vaccination. In: Payne LN (ed) Marek's disease. Martinus Nijhoff, Boston, pp 203–250

Witter RL (1985c) Review: vaccines and vaccination against Marek's disease. In: Calnek BW, Spencer JL (eds) Proceedings of the international symposium on Marek's disease. American Association of Avian Pathologists, Kennett Square, PA, pp 482–500

Witter RL (1987) New serotype 2 and attenuated serotype 1 Marek's disease vaccine viruses: comparative efficacy. Avian Dis 31:752–765

Witter RL (1988) Protective synergism among Marek's disease vaccine viruses. In: Kato S, Horiuchi T, Mikami T, Hirai K (eds) Advances in Marek's disease research. Japanese Association on Marek's Disease, Osaka, Japan, pp 398–404

Witter RL (1991) Attenuated revertant serotype 1 Marek's disease viruses: safety and protective efficacy. Avian Dis 35:877–891

Witter RL (1992b) Influence of serotype and virus strain on synergism between Marek's disease vaccine viruses. Avian Pathol 21:601–614

Witter RL (1992a) Safety and comparative efficacy of the CVI988/Rispens vaccine strain. In: 4th International Symposium on Marek's Disease, 19th World's Poultry Congress, Vol. 1. World's Poultry Science Assoc., Amsterdam, pp 315–319

Witter RL (1995) Attenuation of lymphoid leukosis enhancement by serotype 2 Marek's disease virus. Avian Pathol 24:665–678

Witter RL (1997) Increased virulence of Marek's disease virus field isolates. Avian Dis 41:149–163

Witter RL (1998) Control strategies for Marek's disease: a perspective for the future. Poult Sci 77:1197–1203

Witter RL, Bacon LD, Nordgren RM (1996) Reproducibility of Efficacy tests for Marek's Disease Vaccines. Proc 45th West Poult Dis Conf 72–74

Witter RL, Burmester BR (1979) Differential effect of maternal antibodies on efficacy of cellular and cell-free Marek's disease vaccines. Avian Pathol 8:145–156

Witter RL, Gimeno I, Reed WM, Bacon LD (1999) An acute form of transient paralysis induced by highly virulent strains of Marek's disease virus. Avian Dis 43:704–720

Witter RL, Hunt HD (1994) Poultry vaccines of the future. Poult Sci 73:1087–1093

Witter RL, Lee LF (1984) Polyvalent Marek's disease vaccines: safety, efficacy and protective synergism in chickens with maternal antibodies. Avian Pathol 13:75–92

Witter RL, Lee LF, Bacon LD, Smith EJ (1979) Depression of vaccinal immunity to Marek's disease by infection with reticuloendotheliosis virus. Infect Immun 26:90–98

Witter RL, Lee LF, Fadly AM (1995) Characteristics of CVI988/Rispens and R2/23, two prototype vaccine strains of serotype 1 Marek's disease virus. Avian Dis 39:269–284

Witter RL, Lee LF, Sharma JM (1990) Biological diversity among serotype 2 Marek's disease viruses. Avian Dis 34:944–957

Witter RL, Li D, Jones D, Lee LF, Kung HJ (1997) Retroviral insertional mutagenesis of a herpesvirus: a Marek's disease virus mutant attenuated for oncogenicity but not for immunosuppression or in vivo replication. Avian Dis 41:407–421

Witter RL, Moulthrop Jr JI, Burgoyne GH, Connell HE (1970b) Studies on the epidemiology of Marek's disease herpesvirus in broiler flocks. Avian Dis 14:255–267

Witter RL, Nazerian K, Purchase HG, Burgoyne GH (1970a) Isolation from turkeys of a cell-associated herpesvirus antigenically related to Marek's disease virus. Amer J Vet Res 31:525–538

Witter RL, Offenbecker L (1978) Duration of vaccinal immunity against Marek's disease. Avian Dis 22:396–408

Witter RL, Offenbecker L (1979) Nonprotective and temperature-sensitive variants of Marek's disease vaccine viruses. J Natl Cancer Inst 62:143–151

Witter RL, Sharma JM, Fadly AM (1980) Pathogenicity of variant Marek's disease virus isolants in vaccinated and unvaccinated chickens. Avian Dis 24:210–232

Witter RL, Sharma JM, Lee LF, Opitz HM, Henry CW (1984) Field trials to test the efficacy of polyvalent Marek's disease vaccines in broilers. Avian Dis 28:44–60

Witter RL, Sharma JM, Offenbecker L (1976) Turkey herpesvirus infection in chickens: induction of lymphoproliferative lesions and characterization of vaccinal immunity against Marek's disease. Avian Dis 20:676–692

Witter RL, Sharma JM, Solomon JJ, Champion LR (1973) An age-related resistance of chickens to Marek's disease: some preliminary observations. Avian Pathol 2:43–54

Witter RL, Silva RF, Lee LF (1987) New serotype 2 and attenuated serotype 1 Marek's disease vaccine viruses: selected biological and molecular characteristics. Avian Dis 31:829–840

Witter RL, Solomon JJ (1972) Experimental infection of turkeys and chickens with herpesvirus of turkeys (HVT). Avian Dis 16:34–44

Yonash N, Bacon LD, Witter RL, Cheng HH (1999) High resolution mapping and identification of new quantitative trait loci (QTL) affecting susceptibility to Marek's disease. Anim Genet 30:126–135

Yoshida I, Yuasa N, Tsubahara H, Horiuchi T, Ishino S (1973) Dose effect of herpesvirus of turkeys on protection of chickens from Marek's disease. Natl Inst An Hlth Quart 13:39–44

Yuasa N, Yoshida I, Taniguchi T, Kawamura H, Wakabayashi T, Yamaguchi S, Takayanagi N, Sato S, Sekiya S, Horiuchi T (1975) Contamination of Marek's disease vaccine with REV group of virus. Jap Vet Poult Assn 8–10

Zander DV (1973) Head start and concurrent infection of a small percentage of birds with a naturally occurring mild Marek's disease virus to reduce "Leukosis" condemnations in broilers. Proc 22nd West Poult Dis Conf 48–58

Zander DV, Hill RW, Raymond RG, Balch RK, Mitchell RW, Dunsing JW (1971) Marek's disease immunization procedures. Proc 20th West Poult Dis Conf 43–63

Zander DV, Hill RW, Raymond RG, Balch RK, Mitchell RW, Dunsing JW (1972) The use of blood from selected chickens as an immunizing agent for Marek's disease. Avian Dis 16:163–178

Zander DV, Raymond RG (1985) Partial flock inoculation with a apathogenic strain of chicken herpes virus of Marek's disease to immunize chicken flocks against pathogenic field strains of MD. In: Calnek BW, Spencer JL (eds) Proceedings of the international symposium on Marek's disease. American Association of Avian Pathologists, Kennett Square, PA, pp 514–530

Zanella A, Bertoldini G, Mambelli N (1970) Attenuation of the pathogenicity of Marek's disease virus. Atti Soc Ital Sci Vet 24:622–624

Zanella A, Granelli G (1974) Marek's disease control: Comparative efficacy of cell-associated and cell-free lyophilized HVT vaccine. Avian Pathol 3:45–50

Zanella A, Marchi R, Tosi P (1985) Polyvalent vaccines for control of Marek's disease. In: Calnek BW, Spencer JL (eds) Proceedings of the international symposium on Marek's disease. American Association of Avian Pathologists, Kennett Square, PA, pp 555–569

Zanella A, Valantines J, Granelli G, Castell G (1975) Influence of strain of chickens on the immune response to vaccination against Marek's disease. Avian Pathol 4:247–253

Zelnik V, Tyers P, Smith GD, Jiang CL, Ross LJN (1995) Structure and properties of a herpesvirus of turkeys recombinant in which US1, US10 and SORF3 genes have been replaced by a lacZ expression cassette. J Gen Virol 76:2903–2907

Zygraich N, Huygelen C (1972) Response of chicks from vaccinated hens to inoculation with turkey herpesvirus. Vet Rec 90:281–282

Immune Responses to Marek's Disease Virus Infection

K.A. Schat and C.J. Markowski-Grimsrud

1 Introduction

Marek's disease (MD), a herpesvirus-induced lymphomatous disease in chickens, has attracted the interest of immunologists since MD virus (MDV) was isolated in 1968 and vaccines became available shortly afterwards (Witter 1985, 2000). The pathogenesis of MD has been reviewed extensively (Calnek 1986, 1998, 2000; Schat 1987b). Infection of chickens with MDV is characterized by several distinct

---

Unit of Avian Health, Department of Microbiology and Immunology, College of Veterinary Medicine, Cornell University, Ithaca, NY 14853, USA

phases in which innate and acquired immune responses play important roles. The first phase is characterized by the replication of MDV in lymphoid cells, which are mostly B lymphocytes. The consequence of the lytic infection is that T lymphocytes become activated. Only activated, but not resting, T cells can be infected with MDV. Schat and Xing (2000) hypothesized recently that a viral homologue of interleukin (IL)-8 (vIL-8) (Liu et al. 1999) may be involved in attracting the activated T cells to the infected B cells (see Sects. 2.1, 6.2) to facilitate the transfer of virus. Intimate contact between B and T cells is important for the transfer of infectious virus from cell to cell (Kaleta 1977), because MDV is highly cell-associated (Schat 1984). Temporal immunosuppression of the humoral immune responses is often one of the consequences of the lytic infection in B lymphocytes. During the second phase, a latent infection is normally established in these activated T cells, although some activated $CD4^+$ and $CD8^+$ T cells may undergo a lytic infection (Baigent et al. 1998). The actual process of establishment and maintenance of latency has not been elucidated, but it is likely that a complex set of interactions, including immune responses and specific cellular and viral genes, is responsible. The lytic and latent phases occur in all MDV-infected chickens. Tumors may develop during the third phase of the infection depending on the genetic resistance and immune status of the host (Bacon 2000; Schat and Davies 2000), and the virulence of the virus strain. A secondary lytic infection with subsequent immunosuppression often precedes the development of tumors. The pathogenesis of MD can be modified by several immunologically based factors such as the presence of maternal antibodies, vaccination status, genetic resistance, stress, and the presence of immunosuppressive agents interfering with innate and acquired immune responses. In this paper, innate and acquired immunity in response to MDV infection and vaccination will be reviewed. In addition, the role of immunosuppression and the relative importance of antiviral versus antitumor immune responses will be discussed.

2 Innate Immune Responses

Infection with MDV results in the activation of innate (nonspecific) and acquired (specific) immune responses. The former will be discussed in this section, while the acquired responses are detailed in Sect. 3. The innate responses include activation of macrophages, natural killer (NK) cells, and soluble factors such as cytokines, nitric oxide (NO), and other soluble mediators. These responses are characterized by rapid activation after infection, lack of specific antigen recognition, and lack of memory. However, the distinction between innate and acquired responses is not always clearly delineated. For example, interferon (IFN)-γ is an important cytokine involved in innate immune activation, but it can also be part of the effector mechanism of cytotoxic T lymphocytes (CTL) (Whitton 1998). Cytokines and soluble mediators, macrophages, and NK cells are important in MDV-infected chickens (Schat 1996) and will be discussed in detail.

2.1 Soluble Mediators and Cytokines

The lytic infection of B and T cells as well as the establishment of latency in, and transformation of, $CD4^+$ and other T cells may cause deregulated expression of interleukins, IFNs, and soluble mediators during all phases of infection. Unfortunately, the interpretation of the older literature is complicated by the fact that until recently, only a few avian cytokine genes had been cloned and sequenced (Schat and Kaiser 1997). For example, it was stated that chickens have only one type of IFN when the IFN type I gene was cloned (Sekellick et al. 1994). It was not until the cloning of the chicken IFN-γ gene by Digby and Lowenthal (1995) that the existence of at least two types of IFNs in chickens was accepted. As a consequence the IFN type I gene has recently been reclassified as IFN-α (Marcus et al. 1999). The importance of cytokines and soluble mediators for MD pathogenesis and immunity will be reviewed for the three distinct phases of pathogenesis.

2.1.1 Cytokines and Soluble Mediators During the Cytolytic Infection

Studies in the 1970s suggested that IFNs could inhibit MDV replication in vitro. The availability of recombinant chicken (rCh)IFN-α and rChIFN-γ confirmed that both types of IFN can interfere with MDV replication (K.A. Schat, unpublished data; Xing and Schat 2000b). However, low levels of NO were also produced in these cultures and it cannot be excluded that the production of NO was responsible for the inhibition by rChIFN-γ. The production of IFNs has been examined in MDV-infected chickens by several researchers. Acid-stable IFN (presumably IFN-α) has been detected after infection of chickens or 18-day-old embryos with MDV or HVT, although there is some disagreement on the induction of IFN by HVT. The discrepancy may be related to the source of tissues used for the detection of IFN-α. Sharma (1989) found that lung tissue was the major source, while Kaletia and Bankowski (1972) used serum and brain extracts to demonstrate IFNs. Xing and Schat (2000a) were unable to detect IFN-α mRNA by RT-PCR in spleens from MDV-infected chickens between 3–15 days postinfection (dpi), but MDV infection upregulated the expression of IFN-γ mRNA as early as 3 dpi. Xing and Schat (2000a) also reported that MDV infection increased the transcription of IL-1β and perhaps IL-8 genes, but infection did not affect IL-2 and IL-6 mRNA production. In addition, infection upregulated the production of iNOS mRNA.

The upregulation of IFN-γ and increased NK cell activity (see Sect. 2.3) are the first immune responses after MDV infection that have been reported. Schat and Xing (2000) hypothesized that the early upregulation of IFN-γ plays an essential role during the early pathogenesis and immune responses of MDV infection. First of all, IFN-γ stimulates the expression of IL-8 receptors on T cells (Tani et al. 1998), which may be essential to attract activated T lymphocytes to virus-infected B lymphocytes through the production of vIL-8 (Liu et al. 1999) (see also Sect. 6.2). IL-8 is a chemoattractant for T cells (Gesser et al. 1996). IFN-γ also stimulates the production of IL-1β and inducible nitric oxide synthase (iNOS) by avian macrophages (Yeh et al. 1999). IL-1 is a versatile molecule: it is a chemoattractant for T cells; it can upregulate expression of IL-2 and therefore stimulates the generation of CTL (Hadden 1988). Although increased production of IL-1β mRNA does not prove that IL-1 production is enhanced in MDV infections, it suggests that it plays a role in the control of MDV, especially since MDV-specific CTL have been described (Sect. 3.2). The finding that iNOS is also upregulated early during the infection is of interest because NO inhibits MDV replication in vitro and in vivo. Interference with NO production by injection of *S*-methylisothiourea increased the number of virus-infected lymphocytes significantly between 3 and 9 dpi (Xing and Schat 2000b).

2.1.2 The Role of Cytokines and Soluble Factors During the Induction and Maintenance of Latency

The induction and maintenance of latency is a hallmark of all herpesvirus infections in their natural hosts. The molecular mechanisms of latency are incompletely understood (Whitley et al. 1998), but immune responses are involved, at least in MD. Infection of 1-day-old, immunologically immature chicks with MDV delayed the onset of latency compared to infection in older chickens, suggesting that immune responses may be important for the establishment of latency (Buscaglia et al. 1988a). Buscaglia and Calnek (1988) found two soluble factors in conditioned medium (CM) from concanavalin (Con)-A-stimulated spleen cells that were able to maintain MDV latency in spleen cell cultures obtained from MDV-infected chickens between 8 and 25 dpi. One factor was named latency maintaining factor (LMF) and was different from IL-2 and IFN-γ. The latter was identified as the second factor. Volpini et al. (1995) confirmed these observations using CM, and both natural and recombinant IFN-α. They also found that the control of latency by these additions was more effective with latently infected spleen cells obtained at 14 dpi than at 7 dpi, suggesting that there may be temporal differences in the host–virus relationship. The mechanism(s) for the maintenance of latency in the presence of CM and IFNs was not determined, but it is tempting to speculate that NO may play a role. The addition of rChIFN-α and -γ to cell cultures can stimulate the production of NO by macrophages (Sekellick et al. 1998; Yeh et al. 1999), which were present in the spleen cell cultures used by Buscaglia and Calnek (1988) and Volpini et al. (1995). Moreover, Xing and Schat (manuscript in preparation) found that virus rescue from NO-producing MD cell lines was increased in the presence of the iNOS inhibitor L-N^G-monomethyl-L-arginine, suggesting that NO may be important in controlling MDV activation in transformed cells and perhaps also in latently infected cells.

2.1.3 The Role of Cytokines and Soluble Mediators in Tumor Cells

MD tumor cells and cell lines are mostly activated T cells expressing major histocompatibility complex class II (MHC-II) antigens and CD4, although $CD8^+$ and $CD4^-CD8^-$ cell lines have been described. Several authors have reported that

MD cell lines produce immunosuppressive substances, which reduce the response of spleen cells to T cell mitogens (Bumstead and Payne 1987; Quere 1992). Unfortunately, these factors have not been characterized, and it is therefore difficult to evaluate the role of these factors in the complex interactions between innate and acquired immune responses.

Expression of immediate-early (IE), early and late MDV genes can be downregulated in some but not all MDV tumor cell lines by CM, rChIFN-α (Volpini et al. 1996), or rChIFN-γ (B.W. Calnek, S.L. Kaplan and K.A. Schat, unpublished data). The mechanisms of downregulation have not been elucidated. The downregulation of IE genes led Kaplan and Schat (S.L. Kaplan and K.A. Schat, unpublished data) to examine the importance of IFN-response elements in the promoter/enhancer region of ICP4. Addition of rChIFN-α and -γ did not influence the expression of luciferase under control of the ICP4 promoter/enhancer.

2.2 Macrophages

Macrophages are of central importance in the interactions between host and pathogen by producing many cytokines and interleukins, phagocytosis of pathogens, and processing and presenting antigens to B and T lymphocytes. The importance of avian macrophages as effector cells in immune responses has recently been reviewed by Qureshi et al. (2000). They noted that research on the importance of these cells has not advanced as far as that for mammals due to the difficulty of harvesting and purifying nonstimulated macrophages. This problem explains, at least in part, why the role of macrophages in the pathogenesis of MDV infection has not been investigated in detail, although in vitro and in vivo studies had shown that macrophages are important in MD (reviewed by Schat and Xing 2000). Several studies clearly showed that MDV was unable to replicate in macrophages in vitro, although MDV could be phagocytosed (Haffer and Sevoian 1979; Haffer et al. 1979; von Bülow and Klasen 1983b). The phagocytosis of MDV by macrophages may be important for transporting MDV to the spleen, where MDV can be found within 36 to 48 h postinfection (Schat and Xing 2000).

One of the interesting paradoxes in MD immunity is the unresponsiveness of spleen cells to T cell mitogens during the lytic infection with all three serotypes of MDV (Schat et al. 1978; Lee et al. 1978a), while MDV antigen-specific CTL are detected during the same period (Omar and Schat 1996). The former is considered evidence for immunosuppression, while the latter is a clear indication of immune responsiveness. Removal of macrophages from spleen cell preparations from MDV-infected chickens restored responsiveness to the mitogens, which led Lee et al. (1978b) to propose that suppressor macrophages were responsible for immunosuppression during the lytic phase of MDV infection. However, Schat and Xing (2000) suggested that these macrophages were actually producing NO that was responsible for the inhibition of mitogen-induced responsiveness. This hypothesis is based on the following observations: (1) Transcription of iNOS mRNA is increased in spleen cells between 6 and 9 dpi (Xing and Schat 2000a).

(2) The levels of circulating NO and NO produced by spleen cells are increased during this period in comparison to control chickens (K.W. Jarosinski and K.A. Schat, unpublished data). (3) Macrophage-derived NO can prevent proliferation of mammalian macrophages through the Jak3/Stat5 pathway (Bingisser et al. 1998). It is also proposed that the inhibition of mitogen responses is beneficial for the host in preventing uncontrolled proliferation of T cells and, therefore, limiting the number of potential T cell target cells for lytic and latent infections. Moreover, the inhibition of MDV replication by NO produced by activated macrophages (Xing and Schat 2000b) may also provide an explanation for a number of other observations that have been made on the role of macrophages in the pathogenesis of MD. For example, Schat and Calnek (1978a) reported that peripheral blood lymphocytes (PBL) from intact, but not from embryonally bursectomized, chickens vaccinated with SB-1 were able to inactivate cell-free virus. Removal of macrophages from the PBL abrogated this effect. It is likely that production of NO by activated macrophages was, at least in part, responsible for the reduction in virus replication. Likewise, the production of NO by activated macrophages is probably responsible for the observations that MDV-activated macrophages inhibit virus replication in chick kidney cells (CKC) (Kodama et al. 1979b) and reduce virus isolation rates from spleen cells (Lee 1979; Powell et al. 1983a). It also likely that this is the explanation for the relative decrease in virus isolation rates when the number of inoculated spleen cells, and thus macrophages, in the inoculum is increased (Calnek et al. 1982). The authors did not define the mechanism for their observation, but they speculated that it was caused by a cell-mediated immune response effect with the higher number of spleen cells containing more effector cells.

In vivo studies also suggest that macrophages are important in the pathogenesis of MD. Treatments increasing the number of activated macrophages such as a single injection with silica (Higgins and Calnek 1976) or repeated injections with brewer's thioglycollate broth (Gupta et al. 1989) reduced virus replication and delayed the onset of MD tumors. In contrast, treatments resulting in the reduction of macrophages with antimacrophage serum (Haffer et al. 1979), levimasole (Kodama et al. 1980), or repeated injections with silica (Gupta et al. 1989) caused increased virus replication and mortality, and a decrease in mean time to death. A likely explanation for these observations is that the treatments either increased or decreased NO production, which would be compatible with increased virus replication in chickens treated with the iNOS inhibitor *S*-methylisothiourea (Xing and Schat 2000b).

The role of macrophages during tumor development is less clear. Sharma (1983/1984) isolated killer cells resembling macrophages from MD tumors, which were able to kill MSB-1 cells in 18-h chromium-release assays (CRA). Similarly, Quere and Dambrine (1988) found effector cells that were able to lyse syngeneic MD tumor cells in 18-h proline-release assays. Removal of macrophages from the spleen cell suspension significantly reduced but did not eliminate the lysis of syngeneic MD cell lines and LSCC-RP9 cells. In vitro-activated macrophages can also restrict MD tumor cell proliferation (Haffer et al. 1979; Lee 1979; von Bülow and KLASEN 1983a) or kill MD cell lines and LSCC-RP9 cells (QURESHI and MILLER 1991) after activation with LPS or cytokines obtained from Con A-stimulated spleen cells.

2.3 Natural Killer Cells

NK cells are rapidly inducible effector cells that form the first line of defense against virus infections and transformed cells. These cells can kill a wide range of target cells when the expression of self-MHC class I antigens is downregulated or altered as a consequence of virus infection or transformation (TAY et al. 1998). Morphologically different populations of cells with NK-like activity have been described in chickens. Granulocytes (MÁNDI et al. 1985, 1987), large granular lymphocytes (SCHAT et al. 1986, 1987), and lymphoblastoid cells (SCHAT et al. 1986; SIEMINSKI-BRODZINA and MASHALY 1991) were able to lyse LSCC-RP9 or LSCC-H32 target cells. Embryonic NK cells express cytoplasmic CD3, but lack expression of T cell receptors (TCR) (GÖBEL et al. 1994). Unfortunately, surface markers specific for avian NK cells are not yet available (GÖBEL et al. 1996). As a consequence, all studies on the importance of NK cells in MDV-infected chickens have been carried out with, at best, partially purified preparations obtained from spleens, peripheral blood cells, and tumors.

SHARMA and COULSON (1979) described the presence of a population of spleen cells with NK-like activity against the MD cell line MSB-1 in specific-pathogen-free (SPF) chickens. The presence of these NK cells increased with age until approximately 7 weeks of age. At the same time, LAM and LINNA (1979) reported that 1-day-old chicks receiving normal spleen cells from 8-week-old donors were protected against challenge with transplantable JMV tumor cells. SHARMA and OKAZAKI (1981) noted that the LSCC-RP9 cell line, developed from an avian leukosis virus subgroup B-induced tumor transplant, was a much better NK-target cell line than MSB-1, and that three other MD cell lines were not lysed by NK cells. HELLER and SCHAT (1985) confirmed that MD cell lines are poor target cells for NK cells and that MD cell lines reduced the lysis of LSCC-RP9 in cold-target inhibition assays. Infection with serotype 1 MDV strains caused increased NK cell activity in some genetic lines of chickens within 1 week after infection (SHARMA 1981; LESSARD et al. 1996) or decreased NK cell activity in a susceptible line, especially when tumors were present (SHARMA 1981). Two types of cytotoxic cells were detected in virus-induced MD tumors: adherent cells and NK-like cells. The former were able to lyse MSB-1 cells in an 18-h CRA, while the latter were present in only a few tumors and lysed only LSCC-RP9 cells (SHARMA 1983/1984). On the other hand, QUERE and DAMBRINE (1988) found NK cells capable of lysing LSCC-RP9 in spleens of MDV-infected chickens after 14 dpi even in the presence of tumors. Vaccination with HVT, SB-1, or HVT + SB-1 increased the NK cell activity significantly as early as 3 dpi (SHARMA 1981; HELLER and SCHAT 1987). The increase occurred if birds were vaccinated before 4 weeks of age, whereas vaccination after 4 weeks actually decreased the level of NK cell cytotoxicity. These observations led Heller and Schat (1987) to speculate that vaccination of young birds enhances the maturation of NK cells.

The cytokine pathway causing activation of chicken NK cells after MDV infection or vaccination has not been yet been elucidated. Keller (1992) found that injection of chickens with supernatant fluids from the MD cell line JMV-1 enhanced NK cell activity, suggesting that cytokines are important for activation. Unfortunately, she was unable to characterize the cytokines. Ding and Lam (1986) reported increased NK cell activity after injecting 10-week-old chickens with an IFN-α inducer, polyinosinic-polycytidylic acid (polyI:C). This finding is compatible with the suggestion that IFN-α/β is important for the activation of mammalian NK cells (Biron 1997). However, the importance of IFN-α for the activation of avian NK cells needs to be questioned. Schat (K.A. Schat, unpublished data) found an actual depression of NK cell activity after injection of young birds with polyI:C. The addition of rChIFN-α to drinking water in quantities sufficient to induce protection against challenge with Newcastle disease virus (Marcus et al. 1999) depressed NK cell activity (K.W. Jarosinski and K.A. Schat, unpublished data). In addition, inoculation of chickens with rMDV expressing IFN-α did not alter the level of NK cell activity compared to controls (K.W. Jarosinski and K.A. Schat, unpublished data).

The importance of increased NK cell activation for protection against MDV infection is not clear. Bumstead (1998) suggested that the MDV1 region on chromosome 1 is important for the control of virus replication, because this region has a clear conservation of synteny with the locus coding for the lectin-NK cell antigen complex. However, data that demonstrate actual lysis of virus-infected target cells by NK cells are lacking. On the other hand, depending on the actual activation pathway, NK cells are also efficient producers of IFN-γ (Biron 1997), which may be an important factor to control MDV replication through the induction of iNOS and production of NO (Xing and Schat 2000a,b). The relevance of NK cells for the control of MD tumor cells is also not clear. As mentioned before, most MD cell lines are poor NK-target cells and actually may inhibit the lysis of LSCC-RP9 (Sharma and Okazaki 1981; Heller and Schat 1985). Moreover, NK cells obtained from tumors or from tumor-bearing animals often have a reduced level of activity (Sharma 1981, 1983/1984). It is likely that the suppression of NK cell activity is caused by chicken fetal antigens (CFA) (Ohashi et al. 1987) which are expressed on many MD tumor cell lines (Murthy et al. 1979).

3 Acquired Immunity to MDV

Acquired immune responses consist primarily of the development of antigen-specific antibodies and CTL. Both types of immune responses have been described in herpesvirus infections in general and are important for protective immunity (Mester and Rouse 1991). Both types of specific immune responses have been described after natural infection with serotype 1 MDV and vaccine strains and will be discussed in the next two sections.

3.1 Humoral Immune Responses

Given the relative complexity of herpesviruses such as MDV, it is not too surprising that antibodies develop in response to the large number of MDV proteins generated during infection. Immunoprecipitation studies using MDV-infected cell lysates and convalescent sera identified 35 virus-specific proteins, of which more than half were glycosylated (VAN ZAANE et al. 1982). However, the actual importance of these antibodies in terms of convalescence and protection from infection remains unclear. Furthermore, the role of active immunity versus protection by passive immunity (i.e., maternal antibodies) is not completely understood. Nevertheless, it has been demonstrated that humoral responses produced during infection contribute, at least in part, to immunity to MDV infection.

Antibodies to several of the MDV glycoproteins have been reported. Glycoprotein B (gB), originally described as "B antigen" (CHURCHILL et al. 1969; CHEN and VELICER 1992), is the MDV homologue to herpes simplex virus (HSV)-encoded gB, and has been identified in all three serotypes of MDV (ROSS et al. 1989; YOSHIDA et al. 1994). MDV serotype 1 gB has been expressed in recombinant baculoviruses (NIIKURA et al. 1992), fowlpox viruses (YOSHIDA et al. 1994), and HVT (ROSS et al. 1993). Antibodies to purified gB can neutralize cell-free MDV (IKUTA et al. 1984), and are believed to play an important role in protection based on studies utilizing a recombinant fowlpox virus expressing gB (rFPV-gB) (NAZERIAN et al. 1992). It should be noted, however, that the role of cell-mediated immunity was not addressed in this study and that gB-specific CTL are also induced by rFPV-gB (OMAR et al. 1998) (see Sect. 3.2). DAVIDSON et al. (1991) reported that MDV was neutralized by monospecific antibodies to two gB monomers with continuous epitopes and a dimer with a conformation-dependent epitope. These antibodies also detected MDV antigens on the membranes of infected cells. These findings are in line with previous studies of HSV, in which gB was shown to induce antibody responses to several continuous and discontinuous epitopes. These antibodies can inhibit virus penetration of host cells, syncytia formation, and cell-to-cell spread (PEREIRA 1994).

The relevance of antibody responses to other MDV glycoproteins is not as well defined. BRUNOVSKIS et al. (1992) used convalescent chicken sera to immunoprecipitate MDV gE and gI, but they failed to detect antibodies against gD. They speculated that gE and gI may be important as immunogens since homologous antibodies against gE and gI can neutralize HSV (GHIASI et al. 1992a,b), pseudorabies virus (PRV) (KIMMAN et al. 1992), and varicella-zoster virus (VZV) (GROSE 1990). In contrast, antibodies to gD probably do not have a major immunologic function since a gD deletion mutant remained infectious and tumorigenic (ANDERSON et al. 1998). This is somewhat surprising, since gD is essential for virus penetration in HSV and PRV infections (LIGAS and JOHNSON 1988; RAUGH and METTENLEITER 1991).

Immunization with baculovirus-expressed MDV glycoprotein C induced antibody production but, in contrast to gB, these antibodies did not protect against MDV challenge (Jang et al. 1996). Information regarding humoral responses to other MDV glycoproteins is lacking at this time, but antibodies to gH and gL may also be relevant to protection based on the HSV and VZV literature.

There is evidence that non-neutralizing antibodies may also play a role in protective immunity. Lee and Witter (1991) found that passive immunization of 1-day-old chicks with sera obtained from chickens vaccinated with inactivated MDV conferred protection from MDV-associated bursal atrophy and early mortality. The sera tested positive for MDV-specific antibodies by ELISA, indirect fluorescent antibody, and agar gel-precipitation assays but not in virus neutralization assays. It was speculated that: (1) virus-neutralizing antibodies may not be essential for protection, and (2) non-neutralizing antibodies may exert other protective mechanisms such as surface coating of lymphocytes to prevent virus spread during MDV lytic infection, probably by blocking viral antigen sites on the cell membrane. It is conceivable that cytophilic antibodies may also function in antibody-dependent, cell-mediated cytotoxicity (ADCC) to augment cytolysis of virus-infected cells. ADCC-like responses have been reported for MDV, but neither the effector cells nor the antigens were characterized (Kodama et al. 1979a; Ross 1980).

Maternal antibodies passed from vaccinated hens to their offspring or transferred by injection have been shown to reduce the severity of MD pathogenesis in terms of morbidity, mortality, and tumor formation (Burgoyne and Witter 1973; Chubb and Churchill 1969; Payne and Rennie 1973). Not surprisingly, the presence of maternal antibodies can also interfere with the efficacy of vaccines administered either in ovo or in neonatal chicks (Calnek and Smith 1972; Eidson et al. 1973; King et al. 1981; Sharma and Graham 1982). Presumably this is due to neutralization of the virus, and accordingly cell-free MDV vaccines are more susceptible to neutralization than cell-associated vaccines, although an appreciable neutralizing effect has been demonstrated against both cell-associated MDV and HVT (Burgoyne and Witter 1973; Calnek and Smith 1972; Witter and Burmester 1979).

A novel approach to study humoral immunity employed anti-idiotypic (anti-Id) antibodies to anti-Marek's disease associated surface antigen (MATSA) (Dandapat et al. 1994). Anti-Id antibodies can mimic the original antigenic determinant, and vaccination with anti-Id to MATSA resulted in reduced MATSA-expressing PBL post-MDV challenge compared to nonimmunized controls. Although MATSA has since been shown to be present on activated T cells (McColl et al. 1987), and is not an MD tumor-specific antigen as previously thought, this approach may prove useful in studying humoral responses to other MDV antigens.

3.2 Cytotoxic T Lymphocyte Responses to MDV

Specific cell-mediated immune responses are known to play a major role in immunity to herpesvirus infections in general. Lysis of infected cells by CTL is a major component of acquired immunity, and is of particular importance in immunity to MDV due to its highly cell-associated nature (SCHAT 1984).

There have been several obstacles in the study of cell-mediated immune responses to MDV, particularly CTL responses. A major problem has been a lack of appropriate target cells for use in chromium release assays (CRA). These assays require cells expressing MHC class I (MHC-I) antigens in conjunction with a high level of MDV antigens. Although MDV is routinely propagated in either CKC or chicken embryo fibroblasts (CEF), neither of these cultures has been found suitable for use as target cells. CEF do not express appreciable levels of MHC-I antigens (DUNON et al. 1990). CKC do express MHC-I, but are not suitable for CTL assays for two reasons. First of all, a high enough multiplicity of infection cannot be obtained using cell-associated MDV as inoculum without causing excessive chromium leakage. Secondly, HVT-infected CKC were lysed by allogeneic as well as, but more rarely, syngeneic effector cells (SCHAT and HELLER 1985). This suggested that effector cells other than CTL, e.g., NK cells, were involved. MDV-transformed cell lines have also been used as target cells, but these cell lines were also lysed in an allogeneic fashion (SCHAT et al. 1982a; POWELL et al. 1983b). One of the first accounts of cell-mediated cytotoxicity was demonstrated against an MD lymphoblastoid cell line, MSB-1 (SHARMA and COULSON 1977), and presumed to be directed primarily against MATSA, which was believed to be an MD-specific tumor-associated antigen (WITTER et al. 1975). However, the effector cells were not characterized, and it was subsequently shown that enzymatic removal of MATSA did not alter cytotoxicity (SCHAT and MURTHY 1980). It was later realized that MATSA was present on nontransformed, activated T cells (MCCOLL et al. 1987). Another early study of T cell-mediated immunity to MDV employed a plaque inhibition test, where MDV (HPRS-16/att)-sensitized PBL were shown to inhibit focus formation by MDV-infected lymphoid cells and to reduce viral spread in CKC cultures pre-infected with MDV (ROSS 1977). Although these studies certainly implied a role for cell-mediated immune responses, the exact nature of the effector cells remains elusive.

In an attempt to circumvent the problems in studying CTL responses, SCHAT and coworkers (1992) developed a molecular-based approach utilizing reticuloendotheliosis virus (REV)-transformed lymphoblastoid cell lines stably transfected with expression vectors containing various MDV-encoded genes. REV-sensitized splenocytes have been previously shown to lyse REV-transformed cell lines in a syngeneic restricted fashion (MACCUBBIN and SCHIERMAN 1986; LILLEHOJ et al. 1988; WEINSTOCK et al. 1989). These cell lines have been used as target cells in CRA to ascertain the importance of specific CTL activity in MDV-infected birds. The cell lines have been developed from both genetically susceptible and resistant inbred, SPF chicken lines (P2a, MHC haplotype $B^{19}B^{19}$, and N2a, MHC haplotype $B^{21}B^{21}$, respectively). Each of these cell lines corresponds to a specific MHC background, and can therefore be used to demonstrate syngeneic versus allogeneic lysis by MDV-sensitized splenocytes. This approach was first used by PRATT et al. (1992b) by transfecting the *Bam*HI A MDV cloned fragment containing the IE ICP4 gene into an REV cell line latently infected with MDV (RECC-CU210). The resulting cell line, CU211, was found to express an additional phosphoprotein, pp38, presumably as a result of transactivation of the gene by the transfected ICP4, and could be lysed by syngeneic MDV-specific effector cells (PRATT et al. 1992a; UNI et al. 1994). The ability of pp38 to specifically elicit CTL responses was confirmed by OMAR and SCHAT (1996), who cloned various IE, early, and late genes into eukaryotic expression vectors and generated cell lines expressing the genes for use as target cells in CRA. Cell lines expressing pp38 as well as the putative oncogene protein Meq and gB were lysed by MDV serotype-1- and 2-sensitized syngeneic splenocytes. In contrast, cell lines expressing ORF A41, ORF-A, and ORF L1 were not recognized (Table 1). All cell lines were lysed by REV-sensitized splenocytes, indicating that transfection of the genes did not alter antigen processing pathways, which has been known to occur during herpesvirus infections and hence interfere with the generation of immune responses (see Sect. 6.2). These effector cells were further characterized as classical, $CD4^{-}CD8^{+}$ $TCR\alpha\beta1^{+}$ CTL (OMAR and SCHAT 1997).

This system can also be used to examine how cell-mediated immune responses differ according to MHC haplotype. The observed genetic resistance to MDV in particular strains of chickens has been linked to specific MHC haplotypes (BRILES et al. 1983). The CTL recognition of specific viral proteins encoded by herpesviruses is dependent upon the MHC antigens expressed, implying a strong role for CTL in immunity to MDV. Since infection of both the genetically susceptible and

Table 1. CTL responses against REV-transformed cell lines expressing immediate-early (*IE*), early (*E*), and late (*L*) genes of serotype 1 Marek's disease virus

Transfected MDV gene	Class of MDV gene: IE, E, or L[a]	Effector cell donor: strain (haplotype)				Reference
		P2a ($B^{19}B^{19}$) Target cell haplotype		N2a ($B^{21}B^{21}$) Target cell haplotype		
		$B^{19}B^{19}$	$B^{21}B^{21}$	$B^{19}B^{19}$	$B^{21}B^{21}$	
ICP4	IE	–[b]	–	–	+[b]	OMAR and SCHAT 1996
ICP22	IE	–	–	–	–	OMAR and SCHAT 1996
ICP27	IE	+	–	–	+	SCHAT and XING 2000
ORF L1	IE or E	–	–	–	–	OMAR and SCHAT 1996
meq	E	+	–	–	+	OMAR and SCHAT 1996
ORF A41	E	–	–	–	–	OMAR and SCHAT 1996
ORF-A	E	–	–	–	–	OMAR and SCHAT 1996
pp38	E	+	–	–	+	OMAR and SCHAT 1996
gB	L	+	–	–	+	OMAR and SCHAT 1996
gC	L	ND	ND	–	+	Markowski et al.[c]
gD	L	–	–	–	–	Markowski et al.[c]
gE	L	+	–	ND	ND	Markowski et al.[c]
gI	L	+	–	ND	ND	Markowski et al.[c]
gH	L	ND	ND	–	+	Markowski et al.[c]
gK	L	–	–	–	–	Markowski et al.[c]
gL	L	–	–	–	–	Markowski et al.[c]

ND, not done;
[a] IE, Immediate-early; E, early; L, late genes;
[b] –, Negative in chromium release assays; +, positive in chromium release assays;
[c] C.J. Markowski, P.H. O'Connell, K.A. Schat, unpublished data.

resistant strains results in a similar primary cytolytic infection, but subsequent reactivation and tumor formation only occurs in the former, it is likely that CTL are especially important in preventing reactivation of virus from the latent state. A particularly exciting finding was that effector cells derived from MDV-infected chickens only exhibited syngeneic CTL activity against ICP4 when the CTL were obtained from genetically resistant N2a and not from susceptible P2a chickens resistance (OMAR and SCHAT 1996). This finding supports the hypothesis that specific cell-mediated responses are an important component in genetic resistance. Two other IE genes were also examined. ICP27 was recognized by CTL from the two genetic strains (SCHAT and XING 2000), while CTL against ICP22 were not detected in either strain (Table 1).

The MDV glycoproteins are of interest, because several HSV glycoproteins have been implicated in protective T cell responses against HSV. The HSV gB is a major determinant of CTL activity during HSV infection (HANKE et al. 1991). Likewise, the MDV-encoded homologue for gB has previously been shown to induce a significant CTL response in addition to neutralizing antibodies (NAZERIAN et al. 1992; OMAR and SCHAT 1996). This CTL response was detected using MDV-sensitized splenocytes or after inoculation with rFPV-gB (OMAR et al. 1998). Vaccination with rFPV-gB resulted in significant protection against challenge against serotype 1 MDV (NAZERIAN et al. 1992), suggesting that the CTL response as well as the antibody response may be important for protection. These studies have been extended to include MDV glycoproteins gC, gD, gE, gI, gH, gK, and gL. Preliminary results indicate that gC, gE, gH, and gI also elicit CTL responses (Table 1) (C.J. Markowski et al., unpublished data).

4 Vaccinal Immunity to MDV

Vaccination with attenuated serotype 1 MDV, nononcogenic serotype 2 (e.g., SB-1), and/or serotype 3 (HVT) strains has been used successfully to prevent MD over the past three decades, and represent one of the few cancer-preventing vaccines in any species (CALNEK 1986). Although vaccination reduces viral replication sharply during the first cytolytic phase of infection (CALNEK et al. 1980; SCHAT et al. 1982b) and prevents tumorigenesis, it does not prevent the establishment of infection and latency (CALNEK et al. 1981). The mechanisms of vaccine-induced immunity are only partially elucidated. It is clear that vaccine-induced immune responses follow a pattern similar to that seen with pathogenic strains regarding development of both virus-neutralizing antibodies (ONUMA et al. 1975; IKUTA et al. 1984) and virus-specific CTL (UNI et al. 1994; OMAR and SCHAT 1996; OMAR et al. 1998). In addition, vaccination with HVT, SB-1, and especially HVT + SB-1, increases NK cell activity (HELLER and SCHAT 1987). Studies using CVI988-vaccinated chickens depleted of either $CD4^+$ or $CD8^+$ T cells prior to challenge with the oncogenic Md-5 strain strongly support the hypothesis that $CD8^+$ CTL are needed for antiviral immunity. Titers of Md-5 were significantly higher in vaccinated CD8-depleted birds than in intact birds at the onset of latency (Morimura et al. 1998). Interestingly, tumor formation or mortality did not occur in any of the vaccinated CD4- or CD8-depleted chickens.

Vaccination may also reduce MDV-induced immunosuppression. For example, vaccination with HVT prevented the impairment of phytohemagglutinin (PHA)-stimulation of T cells usually seen in unvaccinated, MDV-challenged birds (Lee et al. 1978a,b). This impairment had originally been attributed to "suppressor" macrophages, but as discussed earlier, the suppressor cells may actually function through the production of NO (see Sect. 2.2), which will be beneficial in vaccinated birds by reducing viral replication and allowing for more efficient T cell responses.

The question of whether or not vaccination induces antitumor immune responses has not been settled (see Sect. 5). It is likely that the reduction in virus load by antiviral immune responses in combination with the absence of damage to the lymphoid organs (see Sect. 6) translates into a decreased level of MDV-positive target cells for transformation (see the chapter by Calnek, this volume). In addition, it is plausible that the intact, sensitized immune system is able to maintain the latent state of the virus or eliminate cells in which virus becomes reactivated, thus preventing the second lytic phase of infection.

5 Antiviral Versus Antitumor Immune Responses

Immunity to MD has often been cited as an example of an infectious agent eliciting both antiviral and antitumor immune responses. Payne et al. (1976) introduced the concept of the "two-step" immunity to MD, in which the first step consists of the antiviral and the second one of antitumor immune responses. In contrast, Schat (1991) proposed a "one-step" protection concept consisting of only antiviral immune responses. The "two-step" hypothesis was based, in part, on the following two observations. Sharma et al. (1973) suggested that proliferative lesions could regress when older birds are infected, although they did not actually demonstrate lesion regression in individual birds. The second observation was that chickens inoculated with glutaraldehyde-fixed MD tumor cells were protected against MD (Powell 1975). Subsequent studies with glutaraldehyde-fixed MD tumor cells confirmed that inactivated tumor cell lines induced protection against MD, although at lower levels than inactivated virus-infected CKC (Powell and Rowell 1977; Murthy and Calnek 1979). The use of inactivated viral preparations, in contrast to inactivated tumor cells, reduced viral replication during the lytic phase of the pathogenesis, providing some credence to the hypothesis that inactivated tumor cells induced cell-mediated, antitumor immune responses. However, the interpretation of these results is difficult for several reasons. First of all, the generation of CTL depends to a large degree on de novo synthesis of proteins, which are processed in the proteosome to small peptides typically between 8 and 10 amino acids long. These peptides are transferred to the endoplasmic reticulum by the transporter proteins TAP1 and TAP2, where the peptides are bound to the MHC- I antigens. Subsequently the antigen-MHC complex is presented on the cell surface (Koopman et al. 1997). The use of killed, MDV-infected and tumor cells may present enough MHC-antigen complexes to stimulate a CTL response if the treatment with glutaraldehyde does not alter the MHC-antigen complex and the MHC antigens on the killed cells are identical to the MHC of the hosts. The latter is unlikely because chickens were inoculated with non-MHC-defined cells. However, it is possible that the antigens stimulated the production of antibodies and nonspecific macrophage-dependent responses instead of antigen-specific CTL. The second reason is that MD cell lines most likely express MDV-specific antigens in addition to speculative "tumor" antigens, so that immune responses may actually be directed to viral rather than tumor antigens. The MD transplantable tumor JMV and cell lines derived from JMV were often quoted as an example for the induction of antitumor immunity because glutaraldehyde-inactivated, nonproducer JMV cells protected against challenge with JMV (Bülow 1977). However, inoculation of chickens with nononcogenic serotype 2 and 3 viruses as well as inactivated MDV serotype-1-infected CKC protected against challenge with JMV (Bülow 1977; Powell 1978; Schat and Calnek 1978b). Subsequently it was shown that pp38 can be induced in JMV cells (Ikuta et al. 1985) and that VP16 is expressed in these cells (Koptidesova et al. 1995). Based on the current information, it is likely that inoculation with inactivated virus-infected and tumor cells stimulated cytokine production leading to the activation of NK cells, which would be present between 4 and 14 days after vaccination, when the JMV challenge was given. The activated NK cells would not recognize the histoincompatible JMV cells as self and therefore lyse these cells (Kärre 1995). Interestingly, Bülow (1977) had already speculated that the resistance induced against JMV could be related to histocompatibility antigens.

Schierman (1984) and Difronzo and Schierman (1989) inoculated live MHC-defined, nonproducer MD cell lines in syngeneic and allogeneic chickens, resulting in protection against challenge with syngeneic but not allogeneic tumor transplants. These results suggested the possibility of an antitumor immune response, but it is more likely that the inoculated tumor cell line cells expressed viral antigens in the context of the correct MHC-I expression and that the protection was an antiviral response. In support of the latter hypothesis, Tseng et al. (1986) reported that inoculation with syngeneic tumor cells protected against the early phase of MDV replication.

Another problem with the two-step hypothesis is the failure to identify specific tumor antigens on MD cell lines. The identification of MATSA (Witter et al. 1975) was thought to represent a tumor-specific antigen, but this was questioned when McColl et al. (1987) reported the presence of MATSA on activated T cells from SPF chickens (see also Sect. 3.2). Recently, a new antigen, AV37, has been described as a putative tumor antigen associated with MD tumors (Burgess et al. 1996; Ross et al. 1997). However, the exact nature of this antigen is not yet clear and its role, if any, in the immune responses needs to be clarified.

Morimora et al. (1998, 1999) examined immune responses in vaccinated chickens that were depleted of either CD4- or CD8-expressing lymphocytes. CD8 depletion increased the number of $CD4^+$ lymphocytes, but interestingly, no increase in tumors was noted. The authors speculated that perhaps T cells expressing γδ T cell receptors (TCRγδ), natural killer T cells, double negative T cells, or NK cells were able to eliminate transformed cells. Unfortunately, the authors did not determine whether TCRγδ-positive T cells were present after CD8 depletion. The importance of TCRγδ for antiviral or antitumor immune responses in chickens and mammals has not been elucidated and needs further study.

6 Immunosuppression and Immunoevasive Mechanisms

Suppression of immune responses is a common consequence of virus infections and can result from a variety of mechanisms. MDV is certainly not an exception, especially since it can infect both B cells and activated T cells. In addition to reviewing the immunosuppressive effects caused by MDV, a number of strategies employed by the virus to evade immune responses will be discussed. Finally, the interference with MD immunity by two immunosuppressive viruses that have been linked to vaccine breaks will be reviewed.

6.1 MDV-Induced Immunosuppression

MDV-induced immunosuppression involves a number of different mechanisms correlating with the different pathogenic stages of disease. There is a transient immunosuppression of humoral and cell-mediated immune responses, followed by a generalized and often permanent T cell immunosuppression associated with reactivation from latency and lymphomagenesis in susceptible birds (Schat 1987a,b). Susceptibility to secondary infections becomes markedly increased at this stage.

The most direct cause of immunodepression by viruses is the destruction of lymphocytes or their precursors (Adair 1996), and MDV cytolytic infection of B and T cells results in their depletion most likely by induction of apoptosis (O'Brien 1998; Morimura et al. 1996). Analysis of T cell subpopulations from peripheral blood mononuclear cells (PBMC) during MDV infection indicates that $CD4^+$ T cells selectively undergo apoptosis. In contrast, $CD8^+$ T cells are less susceptible to apoptosis, but infection results in downregulation of CD8 expression (Morimura et al. 1995, 1996). Although the mechanistic details have not yet been elucidated, it is certain that both effects have immunosuppressive consequences in terms of cell-mediated immune responses. However, not all immunosuppression in MDV-infected birds can be attributed to lymphoid destruction. This is evidenced by the fact that vaccination with apathogenic strains such as SB-1 or HVT causes immunosuppressive effects, while sparing the host of extensive lymphoid depletion (CALNEK et al. 1979). Vaccination with bivalent HVT and SB-1 or CVI988 in young chicks can result in temporary suppression of B lymphocyte responses such as decreased antibody production, decreased stimulation by B-cell-specific mitogens, and increased susceptibility to infection with pathogenic *E. coli* (FRIEDMAN et al. 1992). Although this has been attributed to a lower level of cytolytic infection, other possibilities exist. B cell dysfunctions, such as impaired IgM production, that were linked to suppressor cells present in spleens of chickens transplanted with MD tumor cell lines have also been reported (KONAGAYA and OKI 1987).

There is considerable evidence that T cell responses such as PHA-induced stimulation can be impaired by a population of suppressor macrophages (LEE et al. 1978a,b), since responses were restored after removal of adherent, macrophage-like cells. This may also explain the early transient suppression of peripheral blood lymphocyte responsiveness to Con-A observed in birds inoculated with the SB-1 strain (SCHAT et al. 1978). As discussed in Sect. 2.1.1, it is likely that the production of NO by these "suppressor" macrophages was responsible for the decreased responsiveness and that this actually is a beneficial response for the host. THEIS (1977) had also found that PHA responses were inhibited in the presence of syngeneic spleen cells from MD lymphoma-bearing chickens. She further characterized three subpopulations of suppressor cells in spleens from chickens inoculated with tumor cells derived from a transplantable MD lymphoma. The macrophage fraction exhibited the most potent suppressor activity (THEIS 1981). BUMSTEAD and PAYNE (1987) provided evidence that these suppressive effects were mediated by soluble inhibitory factors when they identified an immune suppressor factor (SF) in cell culture fluids from MD lymphoblastoid cell lines. In spleen cell cultures, SF apparently stimulates production of prostaglandin, which is a known immunosuppressive factor and may account in part for the diminished mitogen responses. Another possible immunosuppressive factor is CFA, which is found on MD lymphoblastoid cell lines (MURTHY et al. 1979) and has been shown to suppress NK cell activity of splenocytes in vitro and after in vivo treatment (OHASHI et al. 1987). This may explain the earlier observation by HELLER and SCHAT (1985) that MD tumor cell lines are refractory to NK cell-mediated lysis.

There appears, not surprisingly, to be a correlation between virulence of the MDV strain and the degree of immunosuppression. In addition to a decrease in relative weights of lymphoid organs, RIVAS and FABRICANT (1988) reported impaired humoral responses and mitogen stimulation of lymphocytes in vitro over a 3-week period postinfection with various strains of MDV. The overall severity of immunodepression was enhanced by strains of higher pathogenicity (RB-3, Md-5, Md-11). CALNEK et al. (1998) reported a direct correlation between the degree of bursal and thymic atrophy following MDV infection and the virulence of the pathotype of the virus strain. These findings are in accordance with the greater severity of cytolytic infection associated with strains of increased virulence (WITTER et al. 1980; CALNEK et al. 1998). There is also good evidence that mechanisms of immunosuppression are not linked to oncogenicity, based on experiments with mutant MDV clones generated by retroviral insertional mutagenesis (WITTER et al.

1997). While exhibiting a near complete loss of oncogenicity, these clones retained the ability to cause lymphoid atrophy and decreased lymphocyte responsiveness to mitogen stimulation.

6.2 Immunoevasion by MDV

In addition to inducing immunosuppression both directly and indirectly, herpesviruses have evolved a multitude of mechanisms to avoid immune responses (Ploegh 1998). Although there is not much known about MDV-specific mechanisms, well-known strategies for human herpesviruses include downregulation of MHC-I antigens to prevent viral peptide presentation to CTL (Johnson and Hill 1998), and the encoding and expression of viral homologues to several cytokines such as IL-6, IL-10, and IL-17 (Neipel et al. 1997). Human cytomegalovirus encodes an MHC-I homologue, which inhibits NK cell-mediated lysis of infected cells (Reyburn et al. 1997). Other strategies of immunoevasion by mammalian herpesviruses have been described elsewhere (Gooding 1992; Rinaldo 1994; Ploegh 1998). It is plausible that a number of these mechanisms are relevant to MDV as well.

Recently, a viral homologue to the cellular chemokine IL-8 (v-IL8) has been identified in the MDV genome (Liu et al. 1999). MDV v-IL8 also exhibited homology to another recently identified B-cell-homing chemokine (Gunn et al. 1999), which is a chemoattractant specific for B cells. The authors speculated that MDV v-IL8 may serve to attract B cells for the initial cytolytic phase of infection. However, IL-8 is also a known chemoattractant for T lymphocytes (Gesser et al. 1996). Schat and Xing (2000) proposed that a more important role of v-IL8 might be to recruit T cells to the site of primary infection, which then become activated to allow virus transfer from B cells into activated T cells. This event is perhaps most critical for the virus to establish latency and eventual transformation, and spread to the feather follicular epithelium, where infectious cell-free virus is produced and spread to new hosts (Calnek 1986, 2000; Schat 1987b).

Modulation of MHC-I antigens is also a common strategy of herpesviruses to evade immune surveillance. Interestingly, analysis of 40 chicken lines with defined MHC-haplotypes and with varying susceptibilities to MD showed a near perfect correlation between the degree of MHC-I expression on lymphocytes and monocytes and susceptibility to disease. The expression was the lowest in the most resistant ($B^{21}B^{21}$) haplotype and the highest in the most susceptible ($B^{19}B^{19}$) haplotype (Kaufman and Salomonsen 1997). The authors speculated that the resistant line may be inherently capable of increased NK cell activity, which might account in part for the genetic resistance. However, no significant difference in NK cell activity has been seen between these two haplotypes (Sharma 1981; Heller and Schat 1987). In addition, the levels of viral replication between these same strains are very similar during the primary cytolytic phase (Calnek et al. 1979; Abplanalp et al. 1985), when the effects of increased NK cell activity would be most relevant. Moreover, CTL responses against REV antigens are similar in both lines (Omar and SCHAT 1996), suggesting that the lower expression of MHC-I does not interfere with recognition of antigens by CTL. In fact, expression of MHC-I antigens was increased following MDV infection in susceptible B^{13} and B^{12} lines as well as the resistant B^{21} line, although this increase was not maintained for as long in the B^{21} line as in the susceptible lines (CHAUSSE et al. 1995). The authors speculated that the MHC-I molecules may serve as cellular receptors for MDV, but this is highly unlikely since these molecules are expressed on virtually all cells and MDV exhibits a very specific cell tropism. It is more likely a consequence of IFN-γ upregulation in MDV-infected cells (XING and SCHAT 2000a), which may result in the upregulation of MHC-I antigen expression as part of the cellular antiviral response.

The ability to prevent apoptosis is a common characteristic of virus-infected and tumor cells that contributes to immunoevasion. Although MDV infection in the initial cytolytic phase appears to induce apoptosis in the primary lymphoid organs, there is evidence that during the later stage of transformation tumor cells are resistant to apoptosis. The mechanism for this resistance most likely involves the Meq protein, which is related to the *Jun/Fos* oncoprotein gene family (JONES et al. 1992) and is highly expressed in most MD tumor cell lines. Overexpression of *meq* in a rat cell line resulted in transformation of cells that became resistant to apoptosis (LIU et al. 1998). MD tumor cell lines have been found resistant to drug-induced apoptosis (MUSCARELLA and BLOOM 1997). However, the mechanism by which Meq prevents apoptosis is not clear and probably is dependent on an interplay with pp38, a unique MDV-encoded phosphoprotein variably expressed in tumor cell lines and recently linked to the induction of apoptosis in tumor cells (Burgess and Davison, unpublished data quoted by ROSS 1999).

6.3 Interference with MDV Immunity by Other Immunosuppressive Viruses

A number of infectious agents have immunosuppressive effects in chickens and could interfere with the development of MDV immunity. This section will be limited to chicken infectious anemia virus (CIAV) and infectious bursal disease virus (IBDV) – two agents specifically linked to vaccine breaks.

6.3.1 CIAV-Induced Immunosuppression

CIAV, a small, circular, single-stranded DNA virus of the *Circoviridae* family, and the disease caused by CIAV, chicken infectious anemia, have been reviewed by BÜLOW and SCHAT (1997). Vaccinal immunity to MDV is depressed in birds dually inoculated with HVT and CIAV resulting in increased MD mortality (OTAKI et al. 1988a). In addition, a temporary depression of HVT-specific antibody production and decreased PHA responsiveness of splenocytes were noted, implying an impairment of both humoral and T cell immunity (OTAKI et al. 1988a,b).

CIAV infection has been shown to alter cytokine levels such as IL-2, IFN-γ, and IL-1 based on bioassays (MCCONNELL et al. 1993a), which may have a major impact on the generation of antigen-specific CTL. In addition, CIAV has also been shown to suppress macrophage functions (McConnell et al. 1993b). Recent research has provided evidence that subclinical infection with CIAV impairs the development of REV- and MDV-specific CTL (C.J. Markowski, P.H. O'Connell, K.A. Schat, unpublished data), which could explain the increase in MD in vaccinated chickens.

6.3.2 IBDV-Induced Immunosuppression

Infectious bursal disease virus (IBDV) is another widespread immunosuppressive agent belonging to the *Birnaviridae* family (Lukert and Saif 1997). There have been contradictory reports on the influence of IBDV on MD immunity. Giambrone et al. (1976) reported an increase in MD incidence in birds exposed to IBDV, suggesting that MD immune responses could be affected. However, experimental studies, in which chickens were infected with MDV and IBDV showed decreased MD rates compared to chickens infected with MDV alone. In addition, Buscaglia et al. (1988b) also found that infection of latently infected chickens with IBDV actually decreased the number of MDV-infected cells. The finding that treatment with cyclophosphamide (Lu et al. 1976), an immunosuppressive drug causing destruction of bursal cells similar to IBDV infection, or embryonal bursectomy (Schat et al. 1981) also ameliorates MD lesions is strong evidence for the latter. The destruction of B cells by IBDV most likely eliminates targets for the MDV primary cytolytic infection. In addition, infection with IBDV stimulates macrophages to produce cytokines and NO (Kim et al. 1998), which may be important for the maintenance of MDV latency and does not alter the percentages of T cell subsets (Rodenberg et al. 1994). In conclusion, it is unlikely that IBDV infection negatively influences MDV immune responses, especially because IBDV does not affect cell-mediated immune responses, which are essential for immunity to MDV infection.

7 Conclusions and Future Challenges

A considerable amount of knowledge and understanding regarding immunity to MD has emerged since MDV was isolated by Churchill and Biggs (1967) and vaccines were introduced in the early 1970s. The picture that has emerged over the last 30 years indicates that, based on current knowledge, both innate and acquired immune responses are essential components in protective immunity to MD as summarized in Fig. 1. MDV can be detected within 48h in the spleens of chicken intratracheally inoculated with cell-free MDV (K.A. Schat unpublished data, quoted by Schat and Xing 2000). At 3 dpi the first innate immune responses are detected: IFN-γ mRNA production is increased and NK cell-mediated lysis is enhanced. The increased production of IFN-γ is most likely the key event in the

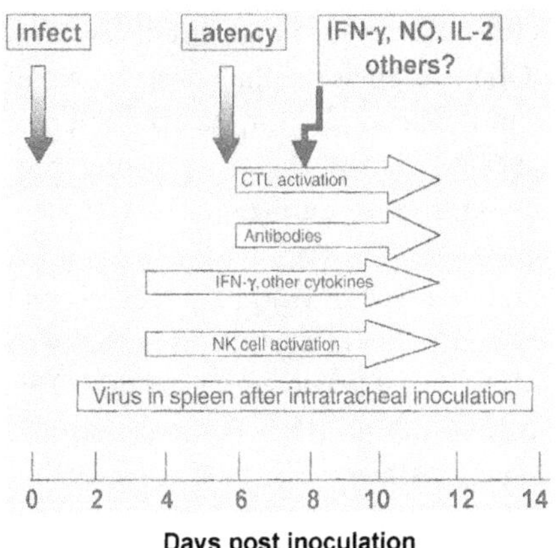

Fig. 1. Development of nonspecific and specific immune responses in chickens after infection with Marek's disease virus. (from SCHAT and XING 2000 with permission)

initiation of the immune responses by upregulating the expression of iNOS and IL-1β mRNA in macrophages. The latter may augment CTL development through the upregulation of IL-2 expression, while the former will lead to the production of NO. The stimulation of NO production by macrophages may help control viral replication and contribute to suppression of T cell proliferation, which curtails the availability of potential target cells for infection. In addition, IFN-γ may be important for the expression of IL-8 receptors on activated T cells, which actually would have a negative effect for the chicken because it allows the establishment of latency. The augmentation in NK cell activity may serve as a first line of defense against virus-infected cells that have altered or downregulated expression of MHC-I. Activated NK cells also may provide an additional source of early IFN-γ (BIRON 1997). Additional, unidentified soluble factors probably contribute to the induction of latency, which is coincident with the appearance of antigen-specific CTL and antibodies. It is likely that the acquired immune responses, particularly CTL, may prevent reactivation from latency, cell-to-cell spread and, therefore, the development of the secondary lytic infection and subsequent tumor formation.

However, a full integration of these components and the relative importance of each is lacking. In order to obtain a more complete picture in the decade ahead, three major areas of immunological research are anticipated. First of all, it is imperative that the role of other cytokines and soluble factors such as NO in the innate as well as acquired immune responses to MD are elucidated. As more chicken cytokines are cloned, this approach will become more feasible. The knowledge of these cytokine pathways will likely allow for ways to augment immune responses relevant to MDV infection, e.g., the engineering of vaccines containing cytokine genes.

Secondly, the identification of the antigenic peptides that induce CTL responses will be instrumental in designing new vaccines. A recombinant fowlpox virus expressing MDV gB (rFPV-gB) induced protection against MD (NAZERIAN et al. 1992; see also Sects. 3.1, 3.2), and it is likely that new knowledge of immunogenic peptides will enable construction of other protective recombinant vaccines. The localization of the gB epitope region for CTL recognition to 100 amino acids

(Schat and Xing 2000) and its further characterization at the nonapeptide level will be an important contribution to this goal. One can envision the construction of recombinant vectors containing several relevant nonapeptides for the generation of CTL, similar to the "string-of-beads" vaccines previously shown to confer protection by CTL in the lymphocytic choriomeningitis virus murine model (Whitton et al. 1993).

Thirdly, the discovery of other MDV genes involved in viral pathogenesis and subversion of immune responses in addition to v-IL8 (Liu et al. 1999) is expected to occur as more sequence information of the MDV genome becomes available. This will be important not only in dissecting the complexity of MDV pathogenesis and immunoevasion, but also for development of counterstrategies such as construction of recombinant viruses crippled by the removal of genes involved in immunoevasive mechanisms.

These are a few of several challenges that lie ahead in order to maintain pace with the rapid evolution of MDV. It is almost certain that novel vaccination strategies will be necessary to provide protection against newly emerging, very-virulent-plus (vv+) strains of MDV that have caused recent vaccine breaks in the field (Witter 2000) and are resistant to the commercial vaccines in current use. The new decade will surely be a challenging, yet exciting time for immunologists whose contributions are relevant not only to MDV but also to other herpesvirus-induced diseases in general.

References

Abplanalp H, Schat KA, Calnek BW (1985) Genetic resistance to Marek's disease in congenic strains of chickens. In: Calnek BW, Spencer JL (eds) Proc Intl Symp on Marek's disease, pp 347–358. American Association of Avian Pathologists, Kennett Square

Adair BM (1996) Virus induced immunosuppression. In: Davison F, Payne LN, Morris TR (eds) Poultry Immunology, pp 301–315. Carfax, Abingdon

Anderson AS, Parcells MS, Morgan RW (1998) The glycoprotein D (US6) homolog is not essential for oncogenicity or horizontal transmission of Marek's disease virus. J Virol 72:2548–2553

Bacon LD (2001) Genetic resistance to Marek's disease. In: Hirai K (ed) Marek's disease. Curr Top Microbiol Immunol 255:121–141

Baigent SJ, Ross LJN, Davison TF (1998) Differential susceptibility to Marek's disease is associated with differences in number, but not in phenotype or location, of pp38+ lymphocytes. J Gen Virol 79: 2795–2802

Bingisser RM, Tilbrook PA, Holt PG, Kees UR (1998) Macrophage-derived nitric oxide regulates T cell activation via reversible disruption of the Jak3/Stat5 signaling pathway. J Immunol 160:5729–5734

Biron C (1997) Activation and function of natural killer cell responses during viral infections. Curr Opin Immunol 9:24–34

Briles WE, Briles RW, Taffs RE, Stone HA (1983) Resistance to malignant lymphoma in chickens is mapped to subregion of major histocompatibility (B) complex. Science 219:977–979

Brunovskis P, Chen X, Velicer LF (1992) Analysis of Marek's disease virus glycoproteins D, I, and E. ProcXIX World's Poultry Congress, Vol. 1, pp 118–121. Ponsen & Looijen, Wageningen

Bülow von V (1977) Cross-protection between JMV Marek's disease derived tumour transplant, Marek's disease and turkey herpesvirus. Avian Pathol 6:353–366

Bülow von V, Klasen A (1983a) Growth inhibition of Marek's disease T-lymphoblastoid cell lines by chicken bone-marrow-derived macrophages activated in vitro. Avian Pathol 12:161–178

Bülow von V, Klasen A (1983b) Effects of avian viruses on cultured chicken bone-marrow-derived macrophages. Avian Pathol 12:179–198
Bülow von V, Schat KA (1999) Chicken infectious anemia. In: Calnek BW, Barnes HJ, Beard CW, McDougald LR, Saif YM (eds) Diseases of Poultry, 10th ed, pp 739–756. Iowa State Univ Press, Ames, IA
Bumstead N (1998) Genomic mapping of resistance to Marek's disease. Avian Pathol 27:S78–S81
Bumstead JM, Payne LN (1987) Production of an immune suppressor factor by Marek's disease lymphoblastoid cell lines. Vet Immunol Immunopathol 16:47–66
Burgess SC, Kaiser P, Davison TF (1996) A novel lymphoblastoid surface antigen and its role in Marek's disease. In: Silva RF, Cheng HH, Coussens PM, Lee LF, Velicer LF (eds) Current research on Marek's disease. Proc 5th Int Symp Marek's Disease, pp 29–39. American Association Avian Pathologists, Kennett Square
Burgoyne GH, Witter RL (1973) Effect of passively transferred immunoglobulins on Marek's disease. Avian Dis 17:824–837
Buscaglia C, Calnek BW (1988) Maintenance of Marek's disease herpesvirus latency in vitro by a factor found in conditioned medium. J Gen Virol 69:2809–2818
Buscaglia C, Calnek BW, Schat KA (1988a) Effect of immunocompetence on the establishment and maintenance of latency with Marek's disease herpesvirus. J Gen Virol 69:1067–1077
Buscaglia C, Calnek BW, Schat KA (1988b) Effect of reticuloendotheliosis virus and infectious bursal disease virus on Marek's disease herpesvirus latency. Avian Pathol 18:265–281
Calnek BW (1986) Marek's disease – a model for herpesvirus oncology. CRC Crit Rev Microbiol 12:293–320
Calnek BW (1998) Lymphomagenesis in Marek's disease. Avian Pathol 27:S54–64
Calnek BW (2001) Pathogenesis of Marek's disease virus infection. In: Hirai K (ed) Marek's disease. Curr Top Microbiol Immunol 255:25–55
Calnek BW, Smith MW (1972) Vaccination against Marek's disease with cell-free turkey herpesvirus: interference by maternal antibody. Avian Dis 16:954–957
Calnek BW, Carlisle JC, Fabricant J, Murthy KK, Schat KA (1979) Comparative pathogenesis studies with oncogenic and non-oncogenic Marek's disease viruses and turkey herpesvirus. Am J Vet Res 40:541–546
Calnek BW, Schat KA, Fabricant J (1980) Influence of vaccinations, in ovo infection, and embryonal bursectomy on Marek's disease pathogenesis. In: Essex M, Todaro G, zur Hauscn H (cds) Viruses in naturally occurring cancer, Vol 7, pp 185–197. Cold Spring Harbor, New York
Calnek BW, Shek WR, Schat KA (1981) Latent infections with Marek's disease virus and turkey herpesvirus. J Natl Cancer Inst 66:585–590
Calnek BW, Shek WR, Schat KA, Fabricant J (1982) Dose-dependent inhibition of virus rescue from lymphocytes latently infected with turkey herpesvirus or Marek's disease virus. Avian Dis 26:321–331
Calnek BW, Harris RW, Buscaglia C, Schat KA, Lucio B (1998) Relationship between the immunosuppressive potential and the pathotype of Marek's disease virus isolates. Avian Dis 42:124–132
Chausse AM, Bernardet N, Musset E, Thoraval P, Vaino O, Dambrine G, Coudert F (1995) Expression of MHC class I antigens following Marek's disease infection. In: Davison TF, Bumstead N, Kaiser P (eds) Adv Avian Immunol Res, pp 135–140. Carfax, Abingdon
Chen X-B, Velicer LF (1992) Expression of the Marek's disease virus homolog of herpes simplex virus glycoprotein B in *Escherichia coli* and its identification as B antigen. J Virol 66:4390–4398
Chubb RC, Churchill AE (1969) Effect of maternal antibody on Marek's disease. Vet Rec 85:303–305
Churchill AE, Biggs PM (1997) Agent of Marek's disease in tissue culture. Nature 215:528–530
Churchill AE, Chubb RC, Baxendale W (1969) The attenuation, with loss of oncogenicity, of the herpes-type virus of Marek's disease (strain HPRS-16) on passage in cell culture. J Gen Virol 4:557–564
Dandapat S, Pradhan HK, Mohanty GC (1994) Anti-idiotype antibodies to Marek's disease-associated tumour surface antigen in protection against Marek's disease. Vet Immunol Immunopathol 40:353–366
Davidson I, Becker Y, Malkinson M (1991) Monospecific antibodies to Marek's disease virus antigen B dimer (200kDa) and monomer (130 and 60kDa) glycoproteins neutralize virus infectivity and detect the antigen B proteins in infected cell membranes. Arch Virol 121:25–139
Difronzo NL, Schierman LW (1989) Transplantable Marek's disease lymphomas. III. Induction of MHC-restricted tumor immunity by lymphoblastoid cells in F_1 hosts. Int J Cancer 44:474–476
Digby MR, Lowenthal JW (1995) Cloning and expression of the chicken interferon-gamma gene. J Interferon Cytokine Res 15:939–945

Ding AHJ, Lam KM (1986) Enhancement by interferon of chicken splenocyte natural killer cell activity against Marek's disease. Vet Immunol Immunopath 11:65–72

Dunon D, Salomonsen J, Skjodt K, Kaufman J, Imhof BA (1990) Ontogenic appearance of MHC class I (B-F) antigens during chicken embryogenesis. Developm Immunol 1:127–135

Eidson CS, Kleven SH, Anderson DP (1973) Efficacy of cell-free and cell-associated herpesvirus of turkeys vaccines in progeny from vaccinated parental flocks. Am J Vet Res 34:869–872

Friedman A, Shalem-Meilin E, Heller ED (1992) Marek's disease vaccines cause temporary B-lymphocyte dysfunction and reduced resistance to infection in chicks. Avian Pathol 21:621–631

Gesser B, Lund M, Lohse N, Vestergaad C, Matsushima K, Sindet-Pedersen S, Jensen SL, Thestrup-Pedersen K, Larsen CG (1996) IL-8 induces T cell chemotaxis, suppresses IL-4, and upregulates IL-8 production by $CD4^+$ T cells. J Leukoc Biol 59:407–411

Ghiasi H, Kaiwar R, Nesburn B, Slanina S, Wechsler SL (1992a) Baculovirus-expressed glycoprotein E (gE) of herpes simplex virus type-1 (HSV-1) protects mice against lethal intraperitoneal and lethal ocular HSV-1 challenge. Virology 188:469–476

Ghiasi H, Kaiwar R, Nesburn AB, Wechsler SL (1992b) Expression of herpes simplex virus type 1 glycoprotein I in baculovirus: preliminary biochemical characterization and protection studies. J Virol 66:2505–2509

Giambrone JJ, Eidson CS, Page RK, Fletcher OJ, Barger BO, Kleven SH (1976) Effect of infectious bursal agent on the response of chickens to Newcastle disease and Marek's disease vaccination. Avian Dis 20:534–544

Göbel TWF, Chen C-LH, Shrimpf J, Grossi CE, Bernot A, Bucy RP, Auffray C, Cooper MD (1994) Characterization of avian natural killer cells and their intracellular CD3 protein complex. Eur J Immunol 24:1685–1691

Göbel TWF, Chen C-LH, Cooper MD (1996) Avian natural killer cells. In: Vainio O, Imhof BA (eds) Immunology and developmental biology of the chicken. Curr Top Microbiol Immunol 212: 107–117

Gooding LR (1992) Virus proteins that counteract host immune defenses. Cell 71:5–7

Grose C (1990) Glycoproteins encoded by varicella-zoster virus: biosynthesis, phosphorylation, and intracellular trafficking. Annu Rev Microbiol 44:59–80

Gunn MD, Ngo VN, Ansel KM, Ekland EH, Cyster JG, Williams LT (1998) A B-cell-homing chemokine made in lymphoid follicles activates Burkitt's lymphoma receptor-1. Nature 391:799–803

Gupta MK, Chauhan HVS, Jha GJ, Singh KK (1989) The role of the reticuloendothelial system in the immunopathology of Marek's disease. Vet Microbiol 20:223–234

Hadden JW (1988) Recent advances in the preclinical and clinical immunopharmacology of intereleukin-2: emphasis on Il-2 as an immunorestorative agent. Cancer Detect Prev 12:537–552

Haffer K, Sevoian M (1979) In vitro studies on the role of the macrophages of resistant and susceptible chickens with Marek's disease. Poultry Sci 58:295–297

Haffer K, Sevoian M, Wilder M (1979) The role of the macrophage in Marek's disease: In vitro and in vivo studies. Int J Cancer 23:648–656

Hanke T, Graham FL, Rosenthal KL, Johnson DC (1991) Identification of an immunodominant cytotoxic T-lymphocyte recognition site in glycoprotein B of herpes simplex virus by using recombinant adenovirus vectors and synthetic peptides. J Virol 65:1177–1186

Heller ED, Schat KA (1985) Inhibition of natural killer activity in chickens by Marek's disease virus-transformed cell lines. In: Calnek BW, Spencer LJ (eds) Proc Int Symp Marek's disease, pp 286–294. American Association of Avian Pathologists, Kennett Square

Heller ED, Schat KA (1987) Enhancement of natural killer cell activity by Marek's disease vaccines. Avian Pathol 16:51–60

Higgins DA, Calnek BW (1976) Some effects of silica treatment on Marek's disease. Infect Immun 13:1054–1060

Ikuta K, Ueda S, Kato S, Hirai K (1984) Identification with monoclonal antibodies of Marek's disease virus and herpesvirus of turkeys related to virus neutralization. J Virol 49:1014–1017

Ikuta K, Ueda S, Kato S, Hirai K (1985) Identification of Marek's disease virus-specific antigens in Marek's disease lymphoblastoid cell lines using monoclonal antibody against virus-specific phosphorylated polypeptides. Int J Cancer 35:257–264

Jang H-K, Kitazawa T, Ono M, Kawaguchi Y, Maeda N, Yokoyama Y, Tohya Y, Niikura M, Mikami T (1996) Protection studies against Marek's disease using baculovirus-expressed glycoproteins B and C of Marek's disease virus type 1. Avian Path 25:5–24

Johnson DC, Hill AB (1998) Herpesvirus evasion of the immune system. In: Whitton JL (ed) Antigen presentation. Curr Top Microbiol Immunol 232:149–177

Jones D, Lee L, Liu J-L, Kung H-J, Tillotson JK (1992) Marek's disease virus encodes a basic-leucine zipper gene resembling the fos/jun oncogenes that is highly expressed in lymphoblastoid tumors. Proc Natl Acad Sci USA 89:4042–4046

Kaleta EF (1977) Vermehrung, Interferenz und Interferoninduktion aviären herpesvirusarten: Beitrag zur Schutzimpfung gegen die Mareksche Krankheit. 2. Mitteilung: Material und Methoden, Ergebnisse, Diskussion und Zusamenfassung. Zbl Vet Med B 24:429–475

Kaleta EF, Bankowski RA (1972) Production of circulating and cell-bound interferon in chickens by type 1 and type 2 plaque-producing agents of the Cal-1 strain of Marek's disease virus and herpesvirus of turkeys. Am J Vet Res 33:573–577

Kärre K (1995) Express yourself or die: peptides, MHC molecules, and NK cells. Science 267:987–979

Kaufman J, Salomonsen J (1997) The minimal essential MHC revisited: both peptide-binding and cell surface expression level of MHC molecules are polymorphisms selected by pathogens in chickens. Hereditas 127:67–73

Keller LH (1992) Protection against reticuloendotheliosis virus strain T tumours is associated with JMV-1 culture supernatant-enhanced natural killer cell activity. Avian Pathol 21:389–399

Kim IJ, Karaca K, Pertile TL, Erickson SA, Sharma JM (1998) Enhanced expression of cytokine genes in spleen macrophages during acute infection with infectious bursal disease virus in chickens. Vet Immunol Immunopathol 61:331–341

Kimman TG, de Wind N, Oei-Lie N, Pol JMA, Berns AJM, Gielkens ALJ (1992) Contribution of single genes within the unique short region of Aujeszky's disease virus (suid herpesvirus type 1) to virulence, pathogenesis and immunogenicity. J Gen Virol 73:243–251

King D, Page D, Schat KA, Calnek BW (1981) Difference between influences of homologous and heterologous maternal antibodies on response to serotype-2 and serotype-3 Marek's disease vaccines. Avian Dis 25:74–81

Kodama H, Sugimoto C, Image F, Mikami T (1979a) Anti-viral immunity against Marek's disease virus infected chicken kidney cells. Avian Pathol 9:33–44

Kodama H, Mikami T, Inoue M, Izawa H (1979b) Inhibitory effects of macrophages against Marek's disease virus plaque formation in chicken kidney cell cultures. J Natl Cancer Inst 63:1267–1271

Kodama H, Mikami T, Izawa H (1980) Effects of levamisole on pathogenesis of Marek's disease. JNCI 65:155–159

Konagaya K, Oki Y (1987) Suppression of IgM production and its mechanism in chicks transplanted with Marek's disease lymphoma cell lines. Jpn J Vet Sci 49:191–193

Koopman J-O, Hämmerling GJ, Momburg F (1997) Generation, intracellular transport and loading of peptides associated with MHC class I molecules. Curr Opin Immunol 9:80–89

Koptidesova D, Kopacek J, Zelnik V, Ross NJL, Pastorekova S, Pastorek JU (1995) Identification and characterisation of a cDNA clone derived from the Marek's disease tumour cell line RPL-1 encoding a homologue of α-transinducing factor (VP-16) of HSV-1. Arch Virol 140:355–362

Lam KM, Linna TJ (1979) Transfer of natural resistance to Marek's disease (JMV) with non-immune spleen cells. I. Studies of cell populations transferring resistance. Int J Cancer 24:662–667

Lee LF (1979) Macrophage restriction of Marek's disease virus replication and lymphoma cell proliferation. J Immunol 123:1088–1091

Lee LF, Witter RL (1991) Humoral immune responses to inactivated oil-emulsified Marek's disease vaccine. Avian Dis 35:452–459

Lee LF, Sharma JM, Nazerian K, Witter RL (1978a) Suppression and enhancement of mitogen response in chickens infected with Marek's disease virus and the herpesvirus of turkeys. Infect Immun 21:474–479

Lee LF, Sharma JM, Nazerian K, Witter RL (1978b) Suppression of mitogen-induced proliferation of normal spleen cells by macrophages from chickens inoculated with Marek's disease virus. J Immunol 120:1554–1559

Lessard M, Hutchings DL, Spencer JL, Lillehoj HS, Gavora JS (1996) Influence of Marek's disease virus strain AC-1 on cellular immunity in birds carrying endogenous viral genes. Avian Dis 40: 645–653

Ligas MW, Johnson DC (1988) A herpes simplex virus mutant in which glycoprotein D sequences are replaced by β-galactosidase sequences binds to but is unable to penetrate cells. J Virol 62:1486–1494

Lillehoj HS, Lillehoj EP, Weinstock D, Schat KA (1988) Functional and biochemical characterization of avian T lymphocyte antigens identified by monoclonal antibodies. Eur J Immunol 18:2059–2065

Liu J-L, Ye Y, Lee LF, Kung HJ (1998) Transforming potential of the herpesvirus oncoprotein MEQ: morphological transformation, serum-independent growth, and inhibition of apoptosis. J Virol 72:388–395

Liu J-L, Lin S-F, Xia L, Brunovskis P, Li D, Davidson I, Lee LF, Kung H-J (1999) MEQ and v-IL8: cellular genes in disguise? Acta Virol 43:94–101

Lu YS, Kermani-Arab V, Moll T (1976) Cyclophosphamide-induced amelioration of Marek's disease in Marek's disease-susceptible chickens. Am J Vet Res 37:687–692

Lukert PD, Saif YM (1997) Infectious Bursal Disease. In: Calnek BW, Barnes HJ, Beard CW, McDougald LR, Saif YM (eds) Diseases of Poultry, 10th ed. pp 721–738. Iowa State Univ Press, Ames

Maccubbin DL, Schierman L (1986) MHC restricted cytotoxic response of chicken T cells: expression, augmentation and clonal expansion. J Immunol 136:12–16

Mándi Y, Seprényi G, Pusztai R, Béládi I (1985) Are granulocytes the main effector cells of natural cytotoxicity in chickens? Immunobiol 170:287–292

Mándi Y, Veromaa T, Baranji K, Miczák A, Béládi I, Toivanen P (1987) Granulocyte-specific monoclonal antibody inhibiting cytotoxicity reactions in the chicken. Immunobiol 174:292–299

Marcus PI, van der Heide L, Sekellick MJ (1999) Interferon action on avian viruses. I. Oral administration of chicken interferon-α ameliorates Newcastle disease. J Interferon Cytokine Res 19: 881–885

McColl K, Calnek BW, Harris WV, Schat KA, Lee LF (1987) Expression of a putative tumor-associated antigen on normal versus Marek's disease virus-transformed lymphocytes. J Natl Cancer Inst 79: 991–1000

McConnell CDG, Adair BM, McNulty MS (1993a) Effects of chicken anemia virus on cell-mediated immune function in chickens exposed to the virus by a natural route. Avian Dis 37:366–374

McConnell CDG, Adair BM, McNulty MS (1993b) Effects of chicken anemia virus on macrophage function in chickens. Avian Dis 37:358–365

Mester JC, Rouse BT (1991) The mouse model and understanding immunity to herpes simplex virus. Rev Inf Dis 13:S935–S945

Morimura T, Hattori M, Ohashi K, Sugimoto C, Onuma M (1995) Immunomodulation of peripheral T cells in chickens infected with Marek's disease virus: involvement in immunosuppression. J Gen Virol 76:2979–2985

Morimura T, Ohashi K, Kon Y, Hattori M, Sugimoto C, Onuma M (1996) Apoptosis and CD8-down-regulation in the thymus of chickens infected with Marek's disease virus. Arch Virol 141: 2243–2249

Morimura T, Ohashi K, Sugimoto C, Onuma M (1998) Pathogenesis of Marek's disease (MD) and possible mechanisms of immunity induced by MD vaccine. J Vet Med Sci 60:1–8

Morimura T, Cho K-O, Kudo Y, Hiramoto Y, Ohashi K, Hattori M, Sugimoto C, Onuma M (1999) Anti-viral and antitumor effects induced by an attenuated Marek's disease virus in CD4- or CD8-deficient chickens. Arch Virol 144:1809–1818

Murthy KK, Calnek BW (1979) Pathogenesis of Marek's disease: effect of immunization with inactivated viral and tumor-associated antigens. Infect Immun 26:547–553

Murthy KK, Dietert RR, Calnek BW (1979) Demonstration of chicken fetal antigen (CFA) on normal splenic lymphocytes, Marek's disease lymphoblastoid cell lines and other neoplasms. Int J Cancer 24:349–354

Muscarella DE, Bloom SE (1997) Involvement of gene-specific DNA damage and apoptosis in the differential toxicity of mitomycin C analogs towards B-lineage versus T-lineage lymphoma cells. Biochem Pharm 53:811–822

Nazerian K, Lee LF, Yanagida N, Ogawa R (1992) Protection against Marek's disease by a fowlpox virus recombinant expressing the glycoprotein B of Marek's disease virus. J Virol 66:1409–1413

Neipel F, Albrecht J-C, Fleckenstein B (1997) Cell-homologous genes in the Kaposi's sarcoma-associated rhadinovirus human herpesvirus 8: determinants of its pathogenicity? J Virol 71:4187–4192

Niikura M, Matsuura Y, Endoh D, Onuma M, Mikami T (1992) Expression of the Marek's disease virus (MDV) homolog of glycoprotein B of herpes simplex virus by a recombinant baculovirus and its identification as the B antigen (gp100, gp60, gp49) of MDV. J Virol 66:2631–2638

O'Brien V (1998) Viruses and apoptosis. J Gen Virol 79:1833–1845

Ohashi K, Mikami T, Kodama H, Izawa H (1987) Suppression of NK activity of spleen cells by chicken fetal antigen present on Marek's disease lymphoblastoid cell lines cells. Int J Cancer 40:378–382

Omar AR, Schat KA (1996) Syngeneic Marek's disease virus (MDV)-specific cell mediated immune responses against immediate early, late, and unique MDV proteins. Virology 222:87–99

Omar AR, Schat KA (1997) Characterization of Marek's disease herpesvirus (MDV)-specific cytotoxic T lymphocytes in chickens inoculated with a nononcogenic vaccine strain of MDV. Immunology 90:579–585

Omar AR, Schat KA, Lee LF, Hunt HD (1998) Cytotoxic T lymphocyte response in chickens immunized with a recombinant fowlpox virus expressing Marek's disease herpesvirus glycoprotein B. Vet Immunol Immunopathol 62:73–82

Onuma M, Mikami T, Hayashi TTA, Okada K, Fujimoto Y (1975) Studies of Marek's disease herpesvirus and turkey herpesvirus specific common antigen which stimulates the production of neutralizing antibodies. Arch Virol 48:85–97

Otaki Y, Nunoya T, Tajma M, Kato A, Nomura Y (1988a) Depression of vaccinal immunity to Marek's disease by infection with the chicken anemia agent. Avian Pathol 17:333–348

Otaki Y, Tajima M, Saito K, Nomura Y (1988b) Immune response of chicks inoculated with chicken anemia agent alone or in combination with Marek's disease virus or turkey herpesvirus. Jpn J Vet Sci 50:1040–1047

Payne LN, Rennie M (1973) Pathogenesis of Marek's disease in chicks with and without maternal antibody. J Natl Cancer Inst 51:1559–1573

Payne LN, Frazier JA, Powell PC (1976) Pathogenesis of Marek's disease. Int Rev Expt Pathol 16: 59–154

Pereira L (1994) Function of glycoprotein B homologues of the family Herpesviridae. Infectious Agents and Dis 3:9–28

Ploegh HL (1998) Viral strategies of immune evasion. Science 280:248–253

Powell PC (1975) Immunity to Marek's disease induced by glutaraldehyde-treated cells of Marek's disease lymphoblastoid cell lines. Nature 257:684–685

Powell PC (1978) Protection against the JMV Marek's disease transplantable tumour by Marek's disease virus-specific antigens. Avian Pathol 7:305–309

Powell PC, Rowell JG (1977) Dissociation of anti-viral and anti-tumor immunity in resistance to Marek's disease virus. J Natl Cancer Inst 59:919–924

Powell PC, Hartley KJ, Mustill BM, Rennie M (1983a) Studies on the role of macrophages in Marek's disease of the chicken. J Ret Soc 34:289–297

Powell PC, Mustill BM, Rennie M (1983b) The role of histocompatibility antigens in cell-mediated cytotoxicity against Marek's disease tumour-derived lymphoblast cell lines. Avian Pathol 12:461–468

Pratt WD, Morgan RW, Schat KA (1992a) Cell-mediated cytolysis of lymphoblastoid cells expressing Marek's disease virus-specific phosphorylated polypeptides. Vet Microbiol 33:93–99

Pratt WD, Morgan RW, Schat KA (1992b) Characterization of reticuloendotheliosis virus-transformed avian T-lymphoblastoid cell lines infected with Marek's disease virus. J Virol 66:7239–7244

Quere P (1992) Suppression mediated in vitro by Marek's disease virus-transformed T-lymphoblastoid cell lines: effect on lymphoproliferation. Vet Immunol Immunopathol 32:149–164

Quere P, Dambrine G (1988) Development of anti-tumoral cell-mediated cytotoxicity during the course of Marek's disease in chickens. Ann Rech Vét 19:193–201

Qureshi MA, Miller L (1991) Signal requirements for the acquisition of tumoricidal competence by chicken peritoneal macrophages. Poultry Sci 70:530–538

Qureshi MA, Heggen CL, Hussain I (2000) Avian macrophage: effector functions in health and disease. Develop Comp Immunol 24:103–120

Raugh I, Mettenleiter TC (1991) Pseudorabies virus glycoproteins gII and gp50 are essential for virus penetration. J Virol 65:5348–5356

Reyburn HT, Mandelboim O, Vales-Gomez M, Davis DM, Pazmany L, Strominger JL (1997) The class I MHC homologue of human cytomegalovirus inhibits attack by natural killer cells. Nature 386: 514–517

Rinaldo CR Jr (1994) Modulation of major histocompatibility complex antigen expression by viral infection. Am J Path 144:637–650

Rivas AL, Fabricant J (1988) Indications of immunodepression in chickens infected with various strains of Marek's disease virus. Avian Dis 32:1–8

Rodenberg J, Sharma JM, Belzer SW, Nordgren RM, Naqi S (1994) Flow cytometric analysis of B cell and T cell subpopulations in specific-pathogen-free chickens infected with infectious bursal disease virus. Avian Dis 38:16–21

Ross LJN (1977) Antiviral T cell-mediated immunity in Marek's disease. Nature 268:644–646

Ross LJN (1999) T-cell transformation by Marek's disease virus. Trends Microbiol 7:22–29

Ross LJN (1980) Mechanism of protection conferred by HVT. In: Biggs PM (ed) Resistance and immunity to Marek's disease, pp 289–300. EEC publications, Luxembourg

Ross LJN, Binns MM, Tyers P, Pastorek J, Zelnik V, Scott S (1993) Construction and properties of a turkey herpesvirus recombinant expressing the Marek's disease virus homologue of glycoprotein B of herpes simplex virus. J Gen Virol 74:371–377

Ross LJN, Sanderson M, Scott SD, Binns MM, Doel T, Milne B (1989) Nucleotide sequence and characterization of the Marek's disease virus homologue of glycoprotein B of herpes simplex virus. J Gen Virol 70:1789–1804

Ross N, O'Sullivan G, Rothwell C, Smith G, Burgess SC, Rennie M, Lee LF, Davison TF (1997) Marek's disease virus *Eco*R1-Q gene (*meq*) and a small RNA antisense to ICP4 are abundantly expressed in CD4+ cells carrying a novel lymphoid marker, AV37, in Marek's disease lymphomas. J Gen Virol 78:2191–2198

Schat KA (1984) Characteristics of the virus. In: Payne LN (ed) Marek's disease. Scientific basis and methods of control, pp 77–112. Martinus Nijhoff, Boston/The Hague

Schat KA (1987a) Immunity in Marek's disease and other tumors. In: Toivanen A, Toivanen P (eds) Avian Immunology: Basis and Practice, Vol 2, pp 101–128. CRC Press, Boca Raton

Schat KA (1987b) Marek's disease – A model for protection against herpesvirus-induced tumors. Cancer Surveys 6:1–37

Schat KA (1991) Importance of cell-mediated immunity in Marek's disease and other viral tumor diseases. Poultry Sci 70:1165–1175

Schat KA (1996) Immunity to Marek's disease, lymphoid leukosis and reticuloendotheliosis. In: Davison F, Payne LN, Morris TR (eds) Poultry Immunology, pp 209–234. Carfax, Abingdon

Schat KA, Calnek BW (1978a) In vitro inactivation of cell-free Marek's disease herpesvirus by immune peripheral blood lymphocytes. Avian Dis 22:693–697

Schat KA, Calnek BW (1978b) Protection against Marek's disease-derived tumor transplants by the nononcogenic SB-1 strain of Marek's disease virus. Infect Immun 22:225–232

Schat KA, Davies C (2000) Resistance to Viral Diseases. In: Axford RFE, Bishop SC, Nicholas F, Owen JB (eds) Breeding for disease resistance in farm animals, 2nd ed, pp 271–300. CAB, Wallingford

Schat KA, Heller ED (1985) A chromium-release assay for the study of cell-mediated immune responses to Marek's disease antigens. In: Calnek BW, Spencer LJ (eds) Proc Int Symp Marek's disease, pp 306–316. American Association of Avian Pathologists, Kennett Square

Schat KA, Kaiser P (1997) Avian cytokines. In: Schijns V, Horzinek MC (eds) Cytokines in Veterinary Medicine, pp 289–300. CAB, Wallingford

Schat KA, Murthy KK (1980) In vitro cytotoxicity against Marek's disease lymphoblastoid cell lines after enzymatic removal of Marek's disease tumor-associated surface antigen. J Virol 34:130–135

Schat KA, Xing Z (2000) Specific and nonspecific immune responses to Marek's disease virus. Develop Comp Immunol 24:201–221

Schat KA, Schultz RD, Calnek BW (1978) Marek's disease: Effect of virus pathogenicity and genetic susceptibility on response of peripheral blood lymphocytes to concanavalin-A. In: Bentvelzen P, Hilgers J, Yohn DS (eds) Adv Comp Leukosis Res, pp 183–185. Elsevier, Amsterdam

Schat KA, Calnek BW, Fabricant J (1981) Influence of the bursa of Fabricius on the pathogenesis of Marek's disease. Infect Immun 31:199–207

Schat KA, Shek WR, Calnek BW, Abplanalp H (1982a) Syngeneic and allogeneic cell-mediated cytotoxicity against Marek's disease lymphoblastoid tumor cell lines. Int J Cancer 35:187–194

Schat KA, Calnek BW, Fabricant J (1982b) Characterization of two highly oncogenic strains of Marek's disease virus. Avian Pathol 11:593–605

Schat KA, Calnek BW, Weinstock D (1986) Cultivation and characterization of avian lymphocytes with natural killer cell activity. Avian Pathol 15:539–556

Schat KA, Calnek BW, Weinstock D (1987) Cultured avian lymphocytes with natural killer cell activity. In: Weber WT, Ewert DL (eds) Avian Immunology, pp 157–169. Alan R. Liss, New York

Schat KA, Pratt WD, Morgan RW, Weinstock D, Calnek BW (1992) Stable transfection of reticuloendotheliosis virus-transformed lymphoblastoid cell lines. Avian Dis 36:432–439

Schierman LW (1984) Transplantable Marek's disease lymphomas. II. Variable tumor immunity induced by different lymphoblastoid cells. JNCI 73:423–428

Sekellick MJ, Ferrandino AF, Hopkins DA, Marcus PI (1994) Chicken interferon gene: cloning, expression, and analysis. J Interferon Res 14:71–79

Sekellick MJ, Lowenthal JW, O'Neill TE, Marcus PI (1998) Chicken interferon type I and II enhance synergistically the antiviral state and nitric oxide secretion. J Interferon Cytokine Res 18: 407–414

Sharma JM (1981) Natural killer cell activity in chickens exposed to Marek's disease virus: inhibition of activity in susceptible chickens and enhancement of activity in resistant and vaccinated chickens. Avian Dis 25:882–893

Sharma JM (1983/1984) Presence of adherent cytotoxic cells and non-adherent natural killer cells in progressive and regressive Marek's disease tumors. Vet Immunol Immunopath 5:125–140

Sharma JM (1989) In situ production of interferon in tissues of chickens exposed as embryos to turkey herpesvirus and Marek's disease virus. Am J Vet Res 50:882–886

Sharma JM, Coulson BD (1977) Cell-mediated cytotoxic response to cells bearing Marek's disease tumor-associated surface antigen in chickens infected with Marek's disease virus. J Natl Cancer Inst 58: 1647–1651

Sharma JM, Coulson BD (1979) Presence of natural killer cells in specific-pathogen-free chickens. JNCI 63:527–531

Sharma JM, Graham CK (1982) Influence of maternal antibody on efficacy of embryo vaccination with cell-associated and cell-free Marek's disease vaccine. Avian Dis 26:860–870

Sharma JM, Okazaki W (1981) Natural killer cell activity in chickens: target cell analysis and effect of antithymocyte serum on effector cells. Infect Immun 31:1078–1085

Sharma JM, Witter RL, Burmester BR (1973) Pathogenesis of Marek's disease in old chickens: lesion regression as the basis for age-related resistance. Infect Immun 8:715–724

Sieminski-Brodzina LM, Mashaly MM (1991) Characterization by scanning and transmission electron microscopy of avian peripheral blood mononuclear cells exhibiting natural killer-like (NK) activity. Develop Comp Immunol 15:181–188

Tani K, Su SB, Utsunomiya I, Oppenheim JJ, Wang JM (1998) Interferon-gamma maintains the binding and functional capacity of receptors for IL-8 on cultured human T cells. Eur J Immunol 28:502–507

Tay CH, Szomolanyi-Tsuda E, Welsh RM (1998) Control of infections by NK cells. In: Kärre K, Colonna M (eds) Specificity, function, and development of NK cells. Curr Top Microbiol Immunol 230:193–220

Theis GA (1977) Effects of lymphocytes from Marek's disease-infected chickens on mitogen responses of syngeneic normal chicken spleen cells. J Immunol 118:887–894

Theis GA (1981) Subpopulations of suppressor cells in chickens infected with cells of a transplantable lymphoblastic leukemia. Infect Immun 34:526–534

Tsjeng CK, Fletcher OJ, Schierman LW (1986) Preferential protection against Marek's disease pathogenesis by immunisation with syngeneic virus-nonproducer lymphoblastoid cells. Avian Pathol 15:557–567

Uni Z, Pratt WD, Miller MM, O'Connell PH, Schat KA (1994) Syngeneic lysis of reticuloendotheliosis virus-transformed cell lines transfected with Marek's disease virus genes by virus-specific cytotoxic T cells. Vet Immunol Immunopath 44:57–69

Volpini LM, Calnek BW, Sekellick MJ, Marcus PI (1995) Stages of Marek's disease virus latency defined by variable sensitivity to interferon modulation of viral antigen expression. Vet Microbiol 47:99–109

Volpini LM, Calnek BW, Sneath B, Sekellick MJ, Marcus PI (1996) Interferon modulation of Marek's disease virus genome expression in chicken cell lines. Avian Dis 40:78–87

Weinstock D, Schat KA, Calnek BW (1989) Cytotoxic T lymphocytes in reticuloendotheliosis virus-infected chickens. Eur J Immunol 19:267–272

Whitley RJ, Kimberlin DW, Roizman B (1998) Herpes simplex viruses. Clin Infect Dis 26:541–553

Whitton JL (1998) An overview of antigen presentation and its central role in the immune response. In: Whitton JL (ed) Antigen presentation. Curr Top Microbiol Immunol 232:1–13

Whitton JL, Sheng N, Oldstone MBA, McKee TA (1993) A "string-of-beads" vaccine, comprising linked minigenes, confers protection from lethal-dose virus challenge. J Virol 67:348–352

Witter RL (1985) Principles of vaccination. In: Payne LN (ed) Marek's disease. Scientific basis and methods of control, pp 203–250. Martinus Nijhoff, Boston/The Hague

Witter RL (2001) Protective efficacy of Marek's disease vaccines. In: Hirai K (ed) Marek's disease. Curr Top Microbiol Immunol 255:57–90

Witter RL, Burmester BR (1979) Differential effect of maternal antibodies on efficacy of cellular and cell-free Marek's disease vaccines. Avian Pathol 8:145–156

Witter RL, Stephens EA, Sharma JM, Nazerian K (1975) Demonstration of a tumor-associated surface antigen in Marek's disease. J Immunol 115:177–183

Witter RL, Sharma JM, Fadly AM (1980) Pathogenicity of variant Marek's disease virus isolants in vaccinated and unvaccinated chickens. Avian Dis 24:210–232

Witter RL, Deshan L, Jones D, Lee LF, Kung H-J (1997) Retroviral insertional mutagenesis of a herpesvirus: a Marek's disease virus mutant attenuated for oncogenicity but not for immunosuppression or in vivo replication. Avian Dis 41:407–421

Xing Z, Schat KA (2000a) Expression of cytokine genes in Marek's disease virus-infected chickens and chicken embryo fibroblasts. Immunology 100:70–76

Xing Z, Schat KA (2000b) Inhibitory effects of nitric oxide and gamma interferon on in vitro and in vivo replication of Marek's disease virus. J Virol 74:3605–3612

Yeh H-Y, Winslow BJ, Junker DE, Sharma JM (1999) In vitro effects of recombinant chicken interferon-γ on immune cells. J Interferon Cytokine Res 19:687–691
Yoshida S, Lee LF, Yanagida N, Nazerian K (1994) The glycoprotein B genes of Marek's disease virus serotypes 2 and 3: identification and expression by recombinant fowlpox viruses. Virology 200: 484–493
Zaane van D, Brinkhof JMA, Westenbrink F, Gielkens ALJ (1982) Molecular-biological characterisation of Marek's disease virus. I. Identification of virus-specific polypeptides in infected cells. Virology 121:116–132

Genetic Resistance to Marek's Disease

L.D. Bacon, H.D. Hunt, and H.H. Cheng

1 Introduction

Marek's disease (MD) is economically one of the most significant diseases in chickens (Purchase 1985). It is of interest that numerous estimates of heritability of resistance to MD are relatively high compared to the resistance to other diseases in chickens or other diseases in domestic livestock. Therefore, it is valid to define and understand the factors controlling genetic resistance, particularly when one can select and use this resistance in conjunction with other methods to control MD, e.g., vaccination and management. Six areas influencing the genetics of MD

USDA Agricultural Research Service, Avian Disease and Oncology Laboratory, 3606 East Mount Hope Road, East Lansing, MI 48823, USA resistance will be reviewed. Exemplary references will be cited as space does not permit exhaustive citations. This review will generally be limited to the period since a DNA alpha-herpesvirus (Marek's disease virus or MDV) was shown to incite MD (Churchill et al. 1967; Solomon et al. 1968). The isolation of MDV permitted the differentiation of MD (Calnek and Witter 1997) from other avian tumors, e.g., those caused by RNA retroviruses inducing lymphoid leukosis (LL), myelocytomatosis (ML) (Payne and Fadly 1997), or reticuloendotheliosis (RE) (Witter 1997). Gene nomenclature in this paper will adhere to recently established guidelines (Crittenden et al. 1996).

2 Heritability

Numerous studies have demonstrated that, following MDV challenge, MD occurs at different rates in different chicken strains (Cole 1985). While experiments comparing various lines have generally not provided statistical estimates of heritability, several experiments involving sire families within lines or line crosses have established valuable criteria. Von Krosigk et al. (1972) utilized 37,998 commercial crossline test White Leghorn chickens from 2504 paternal half-sib families that were exposed to MDV in three countries. Heritability of MD by contact exposure was 0.10, compared with 0.21 for intraperitoneal injections, and the average heritability at 50% mortality was 0.14. Interestingly, the injection of infected blood was preferable to contact exposure for identifying genetic differences in MD resistance. Friars et al. (1972) utilized broilers from a female breeder line that were inoculated with MDV-infected blood and estimated realized heritability of MD resistance at 0.67 ± 0.30, based on an index of full- and half-sib family means. This very high value was associated with a large standard error, indicating that the real value may have been substantially lower. However, Gavora et al. (1974) also estimated a high average heritability of resistance to MD (0.61) among sire families of two Leghorn strains selected for egg production that were inoculated with the BC-1 strain of MDV. A study of MD mortality in cockerels of a brown-egg line is of interest because it tabulates the heritability of cumulative MD mortality from pooled data of two single crosses biweekly from 12 to 28 weeks of age (Flock et al. 1992). The cockerels received an oculonasal inoculation of a virulent MD virus at hatch. The highest heritability value (0.18) was at 18 weeks when the mortality value was close to 50%.

Another large study used 23 genetic groups of experimental and commercial meat and egg chickens that were challenged with moderately virulent BC-1 or highly virulent RB-1B MDV (Ameli et al. 1992). Heritability estimates were obtained from the sire component of variance in three unselected White Leghorn control strains. The heritabilities of MD incidence and MD mortality were 0.06 and 0.13 after BC-1 challenge, whereas after the more virulent RB-1B challenge the estimates were significantly higher (0.62 and 1.00). This study, as well as several of the earlier citations, measured correlations between MD incidence and egg production and growth traits. The studies frequently conclude there is a desirable negative correlation between MD incidence and egg production or egg weight, whereas positive correlations may exist between MD incidence and age at first egg, total blood spots, and egg specific gravity.

3 Selection

Most selection schemes purposely avoid inbreeding, whereas one unique exception included inbreeding. Therefore, selection without inbreeding will be reviewed separately.

3.1 Selection Avoiding Inbreeding

Since at least the early 1900s, most commercial poultry breeders have tried to avoid inbreeding due to its deleterious influence on reproductive fitness traits. Therefore, most attempts to select breeder strains for MD resistance have avoided inbreeding. Two options have been utilized to select for MD resistance avoiding inbreeding: (1) mass selection, i.e., breeding from survivors; or (2) family selection utilizing pedigreed offspring as breeders.

3.1.1 Mass Selection

Several researchers, and perhaps commercial breeders as well, have shown that mass selection improves resistance to MD by exposing chickens to high levels of infectious material, e.g., by growing chicks in an infected environment, or by inoculation of tumor material or MDV. Exposure to MDV must be high for selection to be effective and affect MD resistance in subsequent generations (Cole 1985).

About 32 years *before* the identification of the MDV, innovative experiments were initiated at Cornell University to select chickens with good economic traits that were resistant to the "avian leukosis complex" (Hutt et al. 1944; Cole 1985). From 1935 to 1969, two strains (C and K) were selected for resistance to tumors, at the same time another strain (S) was selected for susceptibility to tumors. The C and S strains were initiated in 1935 using males and females from the same population. The C and K strains were derived from the same female families in 1936. However, the K strain was initiated by mating females to males (three in 1936 and several more in 1938) from a commercial strain (Kimber) that had been selected for increased viability over a period of 8 years (Hutt et al. 1941; R.K. Cole, personal communication). The selection scheme involved growing the chicks in an infected environment, and in close proximity to adults where a few would develop tumors.

When the selected chickens were reared in the infected environment, the yearly tumor mortality in the susceptible strain S ultimately ranged from 50 to over 60%, in contrast to approximately 1–2% in the resistant K and C strains. Based on necropsy records, it was retrospectively concluded that most of the tumor mortality was attributable to MD. Following identification of MDV, the S strain was routinely highly susceptible to tumors compared to the K or C strains after infection with virulent JM (COLE 1968), mildly virulent BC-1, or highly virulent RB-1B (AMELI et al. 1992) strains of MDV. The difference in MD tumor development between the S and K strains decreased using the more virulent RB-1B MDV. Earlier studies had also demonstrated JM would differentiate resistance in some of 14 chicken strains that was not detected using GA or RB-1B MDV (SCHAT et al. 1981).

An example of mass selection following the identification of MDV was provided by MAAS et al. (1981), who developed two Plymouth Rock lines that were resistant to MD. For nine generations breeders were selected from survivors of a heavy exposure to the virulent Dutch K MDV (two generations by inoculation of infected duck embryo fibroblast cells, and seven by contact exposure). By the sixth generation the susceptibility to MD was reduced from 76% to 8% (averaged for sex and line), but there was no change in resistance thereafter. In a similar experiment, mass selection was conducted for three generations in the Cornell Randombred stock following inoculation of a tumor inoculum containing the JM virus. This resulted in a reduction in susceptibility to MD from 40% to less than 10% (COLE 1985). Thus, mass selection can clearly result in an increase in MD resistance in various strains. However, mass selection is not recommended to commercial breeders because of the initial loss of a high proportion of the population, and this drastically reduces the opportunity to select for important economic traits.

3.1.2 Family Selection

Family selection requires the production of pedigreed (wing-banded) chicks from pedigreed parents. The classic example of the use of family selection for the development of MD resistant and susceptible lines was the development of the Cornell lines N and P from the Regional Cornell Randombred (RCR) control stock (COLE 1968, 1985). Initially, the semen from each of 15 males was artificially inseminated into five unrelated hens to produce two hatches of 2338 pedigreed chicks that were challenged with the JM strain of MDV at 2 days of age. The chicks were necropsied upon death or after the 8-week test period, and the incidence of MD was 51.3%. To reproduce the first selected generation, the sires (5) and dams (21) which had produced families the most resistant to MD were selected as breeders for line N. The P line was established by using those sires (5) and dams (20) whose progenies had been the most susceptible. In subsequent generations, following JM challenge of test chicks, five sires were selected to produce 15 sons (as well as daughters) to reproduce the next generation in each line. To reduce inbreeding, each of the five selected sires was the parent of at least two of the 15 sons tested. The females (25) selected to produce the next generation included many whose tested progeny had not involved mating to one of the five males selected for re-use. Their potential value was suggested by the fact that their family had been much better than the average for the five from the given sire. The five dams for each of the five sires used to produce the next generation included several whose family-test results played a major role in selection of the sire. Thus, the progeny produced were full sibs of those used in the JM challenge test (R.K. Cole, personal communication). After JM challenge, MD was observed in 51–53% of the RCR controls in each generation. However, after 4 years of selection the incidence of MD in the resistant N line was 3.6% in contrast to 96% in the susceptible line P. Chicks from a fifth generation of selection showed no further improvement in response.

A number of other studies using variations of family selection and MDV exposure in a number of different breeds of chickens have also been successful in increasing resistance to MD (Cole 1985). However, the differential in resistance to MD was not as great in other selected strains as seen between lines N and P. It is instructive to review the criteria utilized in selecting lines N and P. Important parameters may include: (a) the number of breeders and offspring analyzed, and (b) the utilization of an MDV infection level that will produce an intermediate incidence of MD in the stock to be selected, a very high incidence of MD in a susceptible strain, and a low incidence in a resistant strain.

3.2 Family Selection with Inbreeding

In 1939, the United States Department of Agriculture opened a laboratory at East Lansing, Michigan [originally the Regional Poultry Research Laboratory, presently the Avian Disease and Oncology Laboratory (ADOL)], to define the etiological cause of tumors in chickens, and to establish methods, including genetics, for controlling tumors. In order to identify individual genes controlling resistance, geneticist Nelson Waters envisioned a long-range breeding program where lines of chickens would be inbred during selection for resistance or susceptibility to tumors; then, by making interline crosses, one would differentiate any major genes influencing defined types of tumors. A technical bulletin summarizes the unique selection, inbreeding, and crossing of the lines for the identification of genes influencing resistance to avian tumors (Stone 1975), and a current review describes additional lines established from the original inbred lines (Bacon et al. 2000). Fifteen lines were initiated and gradually inbred during selection for resistance or susceptibility to tumors. During development, one group of pedigreed chicks in each line was kept in quarantine isolation and used for line reproduction, whereas another group was injected with suspensions of tumor material and reared in a non-quarantined environment. The non-quarantine inoculated birds provided an estimate of the genetic resistance or susceptibility of their unexposed sibs. Three of the original lines, or their sub-lines, are presently maintained. These basic inbred lines include line 6 selected for tumor resistance, and lines 7 and 15 selected for tumor susceptibility. By 1952 the lines were over 95% inbred (Waters and Fontes 1960). Shortly after the identification of MDV, less than 1% of line 6 chickens were shown to develop MD after infection with the GA or JM strains of MDV, and this compared to 76–80% of line 7, and 55–30% of respectively infected line 15I chickens (Crittenden et al. 1972). The hybrid 15_5X7_1 F_1 chicken is genetically uniform, possessing hybrid vigor that is lacking in the inbred lines, and it is highly susceptible to MD (in the designation $15I_5$ the *I* indicates inbred and the subscript *5* a sub-line). This hybrid is generally used in MDV challenge or vaccine studies at the ADOL. Chickens produced from F_2 and backcross matings involving line 6 with lines 7 or 15 provided evidence that there are several major genes influencing resistance to MD (Crittenden et al. 1972; Stone 1975).

4 Chicken Genes Influencing MD

The rapid development of strains with improved MD resistance following selection, and results from MDV infection of F_2 chickens of selected inbred lines, both suggested a few major genes influence MD. It was envisioned that if these genes are identified then selection for certain alloalleles in other strains might be beneficial. One example of a complex of genes with a major influence on MD is the major histocompatibility complex (*MHC*) (*B* system of haplotypes). *Non-MHC* genes are also clearly influencing MD resistance, but only a few have been defined. In this review, the various mechanisms postulated to control genetic resistance are not described in detail, in part because the precise genes (even in the *MHC*) determining resistance have not been identified. However, even though the exact genes have not been identified, certain differences between genetically selected resistant and susceptible lines following MDV infection have allowed investigators to predict certain mechanisms that are determining resistance. The various mechanisms proposed to control genetic resistance to MD have been evaluated (Calnek 1985), and most remain plausible. The *MHC*-linked resistance is generally thought to involve regulatory components of cell-mediated and/or humoral immune responsiveness, or to involve differences in the specificity of immune response to MDV or tumor cells. The *non-MHC* genes are thought to affect parameters such as cellular interactions, numbers or differences in infected target cells, and regulation of cytokines and innate immunity. For example, new evidence suggests chickens of MD resistant and susceptible strains both have predominantly $CD4^+$ thymus cells in tissue lesions during the early cytolytic and latent phases of MDV infection. While $CD4^+$ cells remain predominant in lesions of susceptible strain chickens during the time of tumor development, $CD8^+$ cells become predominant in lesions of resistant chickens (Burgess et al. 2001). This supports other data which indicate that the $CD8^+$ cells are cytolytic and may specifically kill MD tumor cells (Schat and Davies 2000). The challenge for geneticists is to identify the precise genes that control mechanisms of MD resistance, often defined by making comparisons between MDV-infected chickens of selected resistant and susceptible lines, e.g., the genes determining response of CD8 cells to tumors.

4.1 *Non-MHC* Genes

Only a few genes outside the *MHC* have been correlated with resistance to MD, and their influence has generally not been demonstrated in more than one or two strains. Therefore, we will only briefly identify these genes. Evidence indicated that the ADOL MD resistant line 6 and susceptible line 7 had the same *B*2* haplotype; therefore reciprocal immunizations of lymphoid cells were made between the lines to establish whether they differed for lymphocyte antigens. Two loci were identified for thymic alloantigens, i.e., *TH1* (primarily expressed on thymocytes), and *LY4* (primarily expressed on peripheral thymus cells). Another locus (*BU1*) determined antigens on bursal lymphocytes. An initial study indicated that the *LY4* allele from line 6 was associated with MD resistance in 6_1X7_2 F_2 and F_3 progeny (Fredericksen et al. 1977). Later, the influence of *LY4* and *TH1* antigens on MD resistance was studied in 6_3X15_1 F_7 chickens. In that cross, one *LY4/TH1* genotype was more resistant to MD than the other three genotype groups (Fredericksen et al. 1982). Thus, antigenic products of the *LY4* locus appear to specifically interact with products of *TH1* alleles to influence MD. The *BU1* allele of line 6 was associated with resistance to MD in progeny of a $(6_1 \times 7_2) \times 7_2$ backcross population (Tregaskes et al. 1996). MD resistance may also be associated with cholinesterase (Bachev and Lalev 1990) or alkaline phosphatase genotypes (Iotova et al. 1990).

There is indirect evidence that chicken endogenous viral genes (avian leukosis virus subgroup E viral genes abbreviated *ALVE*) may influence MD resistance. The Cornell lines selected for susceptibility (S) and resistance (K) to MD have different *ALVE* genes, or frequencies of *ALVE* genes. Strain S has two *ALVE* genes coding for complete infectious ALVE that were lacking in strain K (Aggrey et al. 1998; Kuhnlein et al. 1989a). Also, *ALVE6* (coding for ALVE envelope only) existed at a lower frequency in two sets of lines selected for MD resistance than in non-selected control strains (Kuhnlein et al. 1989b). However, the presence of *ALVE3* or *ALVE2* in semicongenic line 0 X 15B F_1 chickens did not influence MD (Crittenden 1991). Genome mapping research (Sect. 6) has identified several *non-MHC* quantitative trait loci that are linked to MD resistance, but the specific genes affecting resistance will require additional analysis.

4.2 *MHC* Genes

The genes of the polymorphic *EaB* blood group locus (Briles et al. 1950) were shown to act as the *MHC* (Schierman and Nordskog 1961). The *EaB* locus was subsequently subdivided into several loci containing Class I (*BF*) and Class II (*BL*) genes that are similar to those of mammalian species, and unique Class IV (*BG*) genes (Pink et al. 1977; Miller et al. 1988; Guillemot and Auffray 1989). The *BG*, two *BF* (*FI* and *FIV*), and two *BL* (*LI* and *LII*) loci have been linked to one haplotype that segregates independently of additional Class I (*FV* and *FVI*) and Class II (*LIII*, *LIV* and *LV*) genes that are contained in a haplotype termed *RfpY*

(Briles et al. 1993; Miller et al. 1996). Although the *B* and *RfpY* haplotypes segregate independently, they are physically on the same microchromosome. In a chicken line segregating for haplotypes of both systems, only the *B* haplotype acted as the *MHC* based on skin-graft rejection and mixed lymphocyte response (Pharr et al. 1996).

In 1967, an abstract indicated that alleles at the *B* blood group locus influenced resistance to a natural exposure to MD. Single cross-chicks inheriting *B*19* from a *B*19/*21* paternal parent were about twice as susceptible to MD as those inheriting *B*21* (Hansen et al. 1967). Shortly thereafter, another abstract indicated hybrid chickens with *B*2* and *B*6* were more resistant to MD than those with *B*13* or *B*19* (Briles and Oleson 1971). Compelling evidence for a *B* influence on MD resistance was derived from analyses of the Cornell lines that were selected for MD susceptibility (P) or resistance (N) as described in Sect. 4.1 (Briles et al. 1977). Line P had a gene frequency of 0.97 for *B*19* and 0.03 for *B*13* and lacked *B*21*, whereas line N contained only *B*21*. In progeny of *B*19/*21* F_1 dams backcrossed to line P *B*19/*19* males, the *B*19/*19* offspring had 58–80% MD in contrast to 12–17% MD in the *B*19/*21* siblings. Also, when F_1 *B*19/*21* sires were backcrossed to line P *B*19/*19* hens, the *B*19/*19* offspring had 60–70% MD in contrast to 0–9% MD in the *B*19/*21* siblings. Thus, the *B*21* allele was strongly resisting MD compared to *B*19* when chickens were challenged with the mild JM virus. This suggested that during the 4 years of selection for resistance to MD in the Regional Randombred stock there may have been an indirect selection for *B* haplotype. Analyses within the Regional Cornell Randombred strain substantiated this possibility. Chickens homozygous or heterozygous for genotypes involving alleles *B*3*, *B*5*, *B*13*, *B*15*, *B*19*, or *B*27*, developed 40–93% MD. However, if these genotypes were heterozygous for *B*21*, only 4–6% developed MD, and chickens homozygous for *B*21/*21* had 0% MD (Briles et al. 1980).

A number of references have previously compared the influence of various *B* haplotypes on MD in different strains of chickens. The *B*5*, *B*13* and *B*19* haplotypes are frequently associated with susceptibility to MD in contrast to the *B*2* and *B*6* alleles which are frequently more resistant, or the *B*21* allele which is often the most resistant (Longenecker et al. 1976; Bacon 1987). Recently, the *B*11* haplotype from Ancona chickens, and the *B*23* haplotype from a New Hampshire strain, were highly resistant to the very virulent MD strain RB-1B (Miller et al. 1996; Schat et al. 1994). It is also remarkable the lines selected for MD resistance in Ottawa, Canada, also differ from control strains for *MHC* genes (Lakshmanan et al. 1997). However, while one can make a general conclusion that certain genes have major influences on MD resistance in several populations, there are exceptions to these general trends. For example, the Cornell strains C and K were both selected for MD resistance for many years but strain C possessed the *B*6*, *B*13*, and *B*15* haplotypes (Bacon et al. 1981) and strain K had *B*15* (Briles and Briles 1982). Moreover, the selected inbred ADOL MD resistant line 6 and susceptible line 7 possess the same *B*2* haplotype, even at the DNA level of expressed Class I *BF1V* and Class II *BLBII* genes (Hunt and Fulton 1998; Pharr et al. 1998). These data support the postulation that genes outside the *MHC* influence MD resistance, and suggest that some genes may interact to complement the *B* haplotype influence. Interesting data from line crosses provide evidence for an interaction. Hartmann studied MD resistance in White Leghorns produced by crossing line M (segregating for *B*2*, *B*13*, *B*14*, and *B*21*) with two unrelated lines homozygous for different *B* haplotypes; i.e., line G (*B*2*) and R (developed from MD resistant Cornell line K containing *B*15*) (HARTMANN 1989). In the G line cross, F_1 *B*2/*21* progeny had the greatest MD resistance (9%), while the *B*2/*2* chickens were moderately resistant (27%MD) compared to *B*2/*13* and *B*2/*14* chickens (42–43% MD mortality). These results are expected, based on literature cited in this section. In contrast, when alleles from line M were inherited in chickens from the R line cross, the *B* haplotype did not influence MD. Indeed, *B*15/*21* chickens had the highest incidence of MD (30%), similar to the MD incidence in chickens that were *B*2/*15* (21%), *B*13/*15* (29%), or *B*14/*15* (17%). Thus, while one can frequently expect to improve resistance to MD in commercial poultry populations by selecting for *B* haplotypes, this needs to be evaluated on the basis of the effects achieved in crossbreds of the particular breeding lines (e.g., see BLANKERT et al. 1990).

Shortly after the *B* haplotype was shown to influence MD resistance, chickens were identified and bred that were recombinants between the *BG* and the *BF/BL* regions of the *B* haplotype. An important study clearly demonstrated that MD resistance is attributable to genes in the *BF/BL* region of the *B* haplotype (BRILES et al. 1983), and others (PLACHY et al. 1984; HEPKEMA et al. 1993) verified this. A later study disclosing differences in MD susceptibility of two *BF-BG* recombinant haplotypes, both serologically identified as *BF2-BG23*, suggests the possible influence of sub-regional segments of the *MHC* (SCHAT et al. 1994). A significant influence of *RfpY* on MD was present among chicks segregating for haplotypes derived from stock in which the *RfpY* was originally identified (WAKENELL et al. 1996). However, *RfpY* genes were not linked to MD resistance using chickens from crosses between ADOL lines 6 and 7, or between Cornell lines N and P (BACON et al. 1996; VALLEJO et al. 1998), or from four lines selected for nine generations for multitrait immunocompetence (LAKSHMANAN and LAMONT 1998). Thus, compelling evidence links the *BF/BL* region genes to MD resistance, but additional recombinants or transgenic technology will be required to establish whether *BF* and/or *BL*, or other genes in the *MHC*, are responsible.

5 Congenic Chickens

Geneticists use congenic lines to establish the role of different alleles of a polymorphic gene on traits. If a major gene is known to affect a number of traits of importance, then congenic lines may be established as briefly described below for the chicken *MHC*. On the other hand, if one wants to isolate individual alleles at unidentified polymorphic loci that affect an important trait that differs between inbred lines, then recombinant congenic strains will be useful. Both types of congenic chickens have been developed to define genes (and their mechanisms) influencing resistance to MD.

5.1 MD Resistance in *B* Congenic Chickens

Several sets of *B* congenic lines have been developed with the principal interest of understanding the role of the chicken *MHC* on immune responses and resistance to different diseases, particularly resistance to MD. *MHC* congenic lines are developed by introducing variant *MHC* genes into an inbred line in an F_1 cross, and then the heterozygote (identified by blood-typing) is backcrossed to the inbred line for five to ten generations or more. Heterozygotes are then mated to produce chickens homozygous for the introduced gene to develop the congenic line. After five generations of backcrossing, each congenic line is approximately 95% identical to the parental inbred line, and after ten generations identity is over 99.9% (excluding the introduced gene). At the University of California at Davis (UCD) seven *B* haplotypes were introduced into inbred line 003 using five generations of backcrossing. Two-day-old cockerels of the *B* congenic lines were challenged with three strains of MDV over 3 years, i.e., virulent strains JM-10 or GA-5, or the very virulent strain RB-1B. Although there was some variation in line resistance using different viruses, the 003.*B* congenic line with *B*2* averaged the most resistant, followed closely by the lines with *B*21* and *B*Q* (a haplotype that is highly similar or identical to *B*21* in the *BF/BL* region). Chickens from the other 003.*B* congenic lines with *B* haplotypes, *15*, *18*, *19*, or *24*, and line 003 with *B*17*, were all substantially more susceptible to MD (Schat et al. 1981; Abplanalp et al. 1985).

At the ADOL, seven *B* haplotypes were introduced into inbred line $15I_5$ using 10–11 backcross generations. After five backcross generations the initial studies on MD resistance were conducted using the JM virus. Only 0–3% of the 15.*B* congenic chickens with *B*21* developed MD tumors in contrast to 60%–88% MD in the lines with *B* haplotypes *2* (from line 6_1 or 7_2), *5*, *12*, *13*, or *19*, and *B*15* from line $15I_5$. However, after challenge with the very virulent Md5 virus, 88% or more of the chickens in all the 15.*B* congenic lines developed MD (Bacon and Witter 1992). Thus, while there were *B* haplotype effects on MD in line 003.*B* congenics that were fairly consistent with virulent and very virulent viruses, in line $15I_5$ (that was previously selected for MD susceptibility) the *B* haplotype effects were only detectable using a mildly virulent virus.

5.2 Vaccinal Immunity Against MD in *B* Congenic Chickens

In the United States, essentially all commercial chickens are vaccinated against MD. Moreover, shortly after MD vaccines were developed it was demonstrated that strains that were genetically more resistant to MD were also generally better immunized by the vaccines than susceptible strains (Spencer et al. 1974; Zanella et al. 1975). Therefore, a series of experiments were conducted to determine whether the *B* haplotype influences vaccinal immunity against MD. Chickens from the 15.*B* congenic lines received MDV vaccines at hatch and were challenged with the very virulent Md5 virus on day 6. When seven 15.*B* congenic lines were vaccinated with serotype 3 HVT, the chickens with *B*5* and *B*12* were more resistant than ones with *B*2*, *B*13*, or *B*19*, while those with *B*15* and *B*21* had intermediate resistance. Thus, following HVT vaccination the *B*5* and *B*12* haplotypes determined MD resistance that was not identified by challenge of unvaccinated chicks (BACON and WITTER 1992). A second experiment utilized five of the *B* congenic lines and serotype 1 (R2/23), 2 (301B/1), and 3 (HVT) vaccines in a similar protocol. Chickens with *B*2* and *B*13* developed less protection against MD compared to ones with *B*15* following vaccination with any of the vaccines. However, the chickens with *B*5* and *B*21* developed variable protection. Those with *B*5* developed good protection using serotype 2 and 3 vaccines, but poor protection using serotype 1 vaccine; in contrast, those with *B*21* developed poorer protection after serotype 2 vaccination than after serotype 1 or 3 vaccination (BACON and WITTER 1993). A third experiment utilized the same *B* congenic lines and protocol but with two vaccines of serotypes 1 (R2/23 and Rispens) and 2 (301B and SB-1), as well as the serotype 3 vaccine (HVT). The serotype 1 vaccines were again preferable for chickens with *B* haplotypes *2*, *13*, *15* or *21*, but serotype 2 vaccines were again more protective for *B*5* chickens. Thus, the *B* haplotype influence on MD vaccine efficacy was dependent on the serotype of the vaccine (BACON and WITTER 1994a). The *B* influence on vaccinal immunity was also demonstrated in *B* heterozygotes, initially in chickens produced by 15.*B* congenic x line 7_1 F_1 matings (BACON and WITTER 1995). In a second experiment, *B*2/*19* White Leghorns from a commercial strain-cross developed less protection than *B*2/*21* chickens following HVT, or bivalent (HVT + 301B/1) vaccination (BACON and WITTER 1994b).

5.3 MD Resistance in Recombinant Congenic Strain Chickens

Polymorphic non-linked genes controlling a trait can be separated by development of recombinant congenic strains (RCSs). This permits individual gene identification and study. RCSs are developed by limited backcross matings between two inbred strains known to differ for a trait(s), followed by approximately 20 generations of subsequent brother–sister matings (DEMANT and HART 1986; DEMANT et al. 1989). RCSs of chickens are being developed between two ADOL inbred lines that have the same *MHC*, but that are resistant (6_3) or susceptible (7_2) to MD. An F_1 and two backcross generations were made using 6_3 as the recurrent parental female line. Each RCS should contain a different 12.5% of the line 7_2 genome. Nineteen RCSs are presently in the seventh inbreeding generation of development, and theoretically they should have 64% homozygosity of introduced genes. In the first generation after full-sib matings, chickens from 17 of the RCS were challenged with the JM strain of MDV. Three RCSs had evidence for MD tumor susceptibility in contrast to others that were resistant and similar to line 6_3 (Bacon et al. 1996). Subsequent generations of these RCSs are becoming increasingly inbred, and the individual strains and strain crosses are undergoing further challenges with MDV to identify *non-MHC* genes influencing MD susceptibility (Yonash et al. 1998). The genomes of the RCSs have been analyzed and several quantitative trait loci have been linked to MD susceptibility (see Sect. 6).

6 Chicken Genomics for MD Resistance

The generation of serological reagents that detected protein variation made it possible to test whether a particular gene or gene product is associated with MD resistance. Unfortunately, the number of distinct proteins that can be surveyed through this method is very limited. However, the development of numerous DNA-based markers eliminates this problem. And as the majority of these genetic markers are mapped (known location), hypervariable (easy to distinguish between individuals or lines), and easy to score by semiautomated equipment (e.g., DNA sequencers), it is becoming possible to systematically survey the entire chicken genome. Thus, in less than 10 years, the field of genomics or whole genome genetics has greatly enhanced our ability to identify and understand genes and pathways that influence resistance to MD. Further improvements are expected as new technologies continue to "trickle down" from the Human Genome Project.

6.1 QTL Mapping

As described earlier, resistance to MD is complex as many genes of varying effect control it. Genomics identifies these quantitative trait loci (QTL) or regions of the genome that contain one or more of these genes by associating genotypic variation with phenotypic variation. Or in other words, genetic markers that are linked to an MD resistance gene that show a non-random inheritance of particular alleles with MD resistant individuals in a designed pedigree will be revealed.

Thus far, two genome-wide scans for QTL conferring MD resistance have been performed. Both used ADOL developed Lines 6 (MD resistant) and 7 (MD susceptible) as parents because these lines are nearly 100% inbred, differ greatly with respect to MD phenotype, and *non-MHC* genes could be readily identified as the *MHC* is fixed and would not confound the analysis. In the ADOL study, 272 unvaccinated F_2 progeny were challenged with JM strain MDV and measured for MD as well as a variety of MD-associated traits such as viral titer, number of tumors, and length of survival (Vallejo et al. 1998; Yonash et al. 1999). Using 135+ markers that cover 2500+ cM or ~65% of the chicken genome, 14 QTL (7 significant and 7 suggestive) were defined that explain one or more MD-associated trait as determined by several QTL algorithms. The QTL were of small to moderate effect as they explained 2–10% of the variance, with additive gene substitution effects from 0.01 to 1.05 phenotypic standard deviations. Collectively, the QTL explained up to 75% of the genetic variance. In general, resistance was dominant, although the resistant allele in three cases came from the susceptible line indicating that inferior parental stocks may provide valuable genes. Interestingly, with respect to the line 7 allele, 10 of the 14 QTL displayed nonadditive gene action; 3 with overdominance and 7 being recessive. Theoretically, nonadditive QTL should be among the most useful for marker-assisted selection (MAS). Looking at all possible two-way interactions, at least two significant epistatic interactions with MD QTL were found.

In the second QTL scan, a (line 6 × 7) × 7 backcross population was used at Compton and the MDV challenge strain was HPRS16 (Bumstead 1998). Particular attention was directed at viral load kinetics using a quantitative PCR technique (Bumstead et al. 1997). Using markers from the genetic map as well as selected probes, a significant region on chromosome 1 was identified.

We should note that the QTL identified by Bumstead et al. (1998) did not match any of the 14 QTL identified in the ADOL study. While initially disconcerting, lack of congruency has also been observed in maize, even where identical seed stock, number of progeny, and genetic markers have been employed (Beavis 1998). The most likely explanation besides the differences in mating structure and MDV strain is that with QTL of small effect, only a few QTL will be identified in any given experiment, which creates sampling problems. This explanation is supported by a retrospective study (Kearsey and Farquhar 1998) where 47 studies in plants identified an average of 4 QTL with each QTL explaining ~6% of the phenotypic variation. This also suggests that the size of effect attributed to any QTL is probably overestimated. Furthermore, it is difficult to resolve a QTL below a 20-cM interval (Darvasi and Soller 1997), which is equivalent to ~6 MB of DNA and would include ~300 genes. Despite all of the above limitations, genome-wide QTL scans are valuable as they identify regions in the genome for further analysis.

One approach that is growing in popularity for identifying the causative gene for a QTL is comparative mapping. The evidence is growing (e.g., Smith and Cheng 1998; Groenen et al. 2000; Burt et al. 2000) that a surprisingly high amount of conserved synteny exists between the chicken and mammalian genomes, especially human. As the human and mouse genomes have considerably more mapped genes, one can identify potential positional candidate genes by making the relevant alignment. For example, a QTL on chromosome 1 is equivalent to a region on mouse chromosome 6 that contains the *CMV1* gene conferring resistance to murine cytomegalovirus. Similarly, IL-8 [chemotactic for monocytes; MDV contains a related gene, v-IL8 (Liu et al. 1999)], SPP1 (also known as the cytokine Eta-1 for early T lymphocyte activation-1), and the linked cytokine cluster are particularly attractive as candidates for a QTL on chromosome 4. This ability will only improve as more genes get mapped in more species.

A different approach that we are beginning to employ uses the 6C.7 RCS. As described earlier, RCSs genetically dissect a complex trait into a series of single gene traits. Genetic characterizations of all the 19 RCSs by microsatellite markers suggest each of the three susceptible strains carry a different QTL. Mating these RCSs to line 6 and backcrossing again to line 6 should enable us to fine map the QTL as has been done in mice (e.g., MOEN et al. 1996). RCSs also provide the opportunity to examine specific epistatic interactions. Finally, association of functional characterization to MD incidence in the RCS may be another powerful way to identify biological pathways conferring MD resistance.

6.2 DNA Microarrays

Another recent technological advance is DNA microarrays. In this technique, DNA containing known and undefined genes is spotted at very high density to specific locations on a solid support to generate a microarray or DNA chip. In the "classical" or most widely used form, mRNA is reverse transcribed and hybridized to the microarrays followed by quantification of the amount of material bound by spot. Thus, one is essentially performing a Northern blot only multiplied several 1000-fold. As this gives a gene expression "fingerprint," by comparing samples from two or more treatments, it is possible to associate the expression of particular genes with the phenotype. Since the original publication by Pat Brown's group at Stanford (SCHENA et al. 1995), this technology has been highly touted and a number of papers have successfully demonstrated the ability of this procedure to provide insights on complex biological systems.

We have begun to test DNA microarrays containing chicken cDNAs in an attempt to identify genes and pathways involved in MD resistance. Using arrays containing ~1200 cDNAs from a T cell-enriched library (Tirunagaru et al., in press; see the University of Delaware web site at udgenome.ags.udel.edu/chickest/array.htm), the expression of genes in peripheral blood lymphocytes (PBLs) from uninfected and MDV-infected line 6 and 7 chicks were examined. The preliminary results suggest that up to 25% of the genes that are detectable show differential expression between the resistant and susceptible lines. For instance, for *MHC* and related genes, *MHC* class I, genes with high homology to the human HLA-DR alpha promoter binding protein and tapasin are all induced upon MDV infection in line 7. However, the same genes exhibit little or no change in line 6 following MDV infection. Conversely, *MHC* class II and RACK, an *MHC* complex protein, are upregulated in line 6 following infection but not in line 7. Similarly for T Cell receptors, T cell receptor (TCR) ζ and γ chain, and CD3ε are induced by MDV infection in line 7 but not in line 6. The DNA microarray results are consistent with what is already known, i.e., line 7 (MD susceptible) is readily stimulated by mitogens while line 6 (MD resistant) shows minimal stimulation (LEE and BACON 1983).

While the small number of genes examined limits the power of this preliminary experiment, it demonstrates that DNA microarrays yield a wealth of information as every gene provides some information. Mapping of the differentially expressed genes may be an attractive method for identifying positional candidate genes for the MD QTL. Even if a particular gene does not lie in an MD QTL region, each gene yields clues as to what relevant biological pathway is involved and, thus, other genes in the pathway can be tested. With several groups identifying more unique cDNAs, this technology will improve rapidly.

6.3 Proteomics

Besides variation in DNA or RNA, it is becoming more feasible to systematically examine protein variation and interactions associated with MD resistance. For example, the yeast two-hybrid system developed by FIELDS and SONG (1989) screens a cDNA library that has been fused to the activation domain (AD) of a transcriptional activator to identify proteins that interact with bait (protein of interest) that is fused with the DNA-binding domain (BD). As the AD and BD do not need to be physically connected to promote transcription, if two proteins interact, a reporter gene is expressed if AD and BD are brought into close proximity of each other.

Using this system, it is possible to examine chicken (host) and MDV (pathogen) protein interactions. For instance, the MDV SORF2 gene was used as bait since SORF2 overexpression in the RM1 strain may account for the change in phenotype from the parental JM strain (JONES et al. 1996). Two different proteins were found to specifically interact with SORF2, which were later confirmed by other biochemical assays (LIU 2000). One protein could not be identified based on its DNA sequence; however, the other protein was growth hormone. This is significant as previous studies have shown an association between growth hormone and MD resistance (KUHNLEIN et al. 1997). Given that the complete sequence has been determined for MDV (LEE et al. 2000), it is not difficult to imagine a systematic approach to identify all chicken-MDV interactions similar to the one employed in yeast (UETZ et al. 2000).

Likewise, great strides are being made to examine all the proteins that are expressed in a particular cell or tissue by 2-D gel electrophoresis (e.g., SHEVCHENKO et al. 1996; JUNGBLUT et al. 1998). Methods now exist to identify unique proteins by automated digestion with trypsin, mass spectrometry, and protein database searches. Once these techniques become more available, a comprehensive analysis of MD-associated cells would be very wise.

6.4 Future Implications

Since the development of blood typing reagents, geneticists have been making claims that they can develop quick and fast assays to select for genetic resistance to MD. Unfortunately, the reality often speaks otherwise. This may soon change as the ability to understand structure (DNA), expression (RNA and protein), and function for the majority of the chicken genome rapidly improves and translates into economical tools that breeders can incorporate. QTL and candidate genes have already been identified, and many of these are being tested in commercial flocks. So it is no longer the question as to whether genomics will aid in our biological understanding of the chicken but when and how, and what impact this information will have on genetic improvement for MD resistance.

7 Transgenics and MD Resistance

Pathogen-derived resistance (PDR) is the ability to confer resistance against an otherwise devastating virus by introducing a single pathogen-derived or virus-targeted sequence into the DNA of the host (Beachy 1997; Salter and Crittenden 1991; Wilson 1993). Pathogen-derived viral resistance is successfully used in the plant breeding industry to increase disease resistance. Although the plant breeding industry has benefited the most from PDR, this phenomenon was first described using transgenic chickens resistant to the avian leukosis virus (ALV) subgroup A (Crittenden and Salter 1992). PDR was serendipitously discovered during attempts to produce transgenic chickens by inoculating fertile eggs with ALV. In these experiments, a germ line integration occurred that produced a mutant form of the virus. This germ line mutant expressed the subgroup A viral envelope protein, but no other ALV proteins were produced. The transgenic chicken line produced from this germ line chimera was called alv6 and is highly resistant to infection by ALV subgroup A virus (Federspiel et al. 1991). The mechanism of this resistance is similar to viral receptor interference in which the envelope protein produced in the transgenic alv6 chicken binds to the normal cell surface receptor for the virus and effectively blocks the natural route of infection. This resistance is very strong and is capable of withstanding large doses of injected virus (Salter and Crittenden 1991).

Several herpesvirus genomes, including MDV, have been sequenced and the genes involved in replication have been extensively studied. The proteins required for herpesvirus replication are grouped into immediate-early (IE or α genes), early (E or β genes), and late (L or γ genes) categories which describe their temporal expression in the replication cycle. Viral protein 16 (VP16 or α TIF), which is incorporated into the virion, begins the process by activating the transcription of the IE genes including infected cell protein 4 (ICP4). The ICP4 polypeptide of HSV-1 is composed of 1298 amino acids (McGeoch et al. 1986). ICP4 is an essential regulatory protein inducing most β and γ genes (Gu and DeLuca 1994; Papavassiliou et al. 1991; Smith et al. 1993) while repressing itself, ORF-P, and α0 genes during the replication cycle (Lagunoff and Roizman 1995; Leopardi et al. 1995; Lium et al. 1996). ICP4 has several functional domains including a dimerization domain, primary and secondary DNA-binding domains, and at least two transactivation domains (DeLuca and Schaffer 1988; Shepard et al. 1989). Viral replication can be inhibited in tissue culture cells by expressing a mutant form of ICP4 lacking the transactivation domains (Shepard et al. 1990). This 530 amino acid form of ICP4 retains the dimerization, DNA binding, and autoregulatory domains. In vitro, this mutant ICP4 is capable of forming heterodimers with the wild-type ICP4 resulting in a dominant inhibitory phenotype with regard to viral growth and ICP4 activities. Transgenic mice, displaying PDR to HSV, have been produced using this mutant form of ICP4 (SMITH and DELUCA 1992). The viral titers of herpesvirus challenged transgenic mice were reduced fivefold compared to control levels (1.2 vs 2.3 mean log titer). Although the transgenic mice were significantly more resistant to infection, there were adverse effects attributed to the transgene. The transgenic mice were consistently lower in weight at 3 weeks of age compared to nontransgenic littermates. The promoter used to express the mutant ICP4 was thought to be active only upon viral infection. However, small but significant levels of expression were detected during embryonic development. This expression, coupled with the DNA and autoregulatory domains remaining in the mutant, is thought to be responsible for the adverse loss of weight (SHEPARD et al. 1990; SMITH and DELUCA 1992). Thus, with respect to practical applications, the ICP4 inhibitory mutant will require modification such that it does not cause adverse effects while still conferring resistance to viral infection. Recently, the ICP4 dimerization domain has been localized to a 34-amino-acid segment (GALLINARI et al. 1994). A 148-amino-acid region of ICP4 containing this 34-amino-acid segment is capable of dimerization with wild-type ICP4. However, it remains to be determined whether this minimal construct will inhibit viral replication or alter embryonic development and economic traits. Numerous other essential MD viral proteins could represent alternate or co-targets for inhibiting MDV infection in a transgenic application. Basic research into the mechanisms of MD viral protein function and improved methods to produce transgenic chickens will be needed to investigate the usefulness of this application to MDV disease resistance.

References

Abplanalp H, Schat KA, Calnek BW (1985) Resistance to Marek's disease of congenic lines differing in major histocompatibility haplotypes to three virus strains. In: Calnek BW, Spencer JL (eds) Proc. International symposium on Marek's disease. American Association of Avian Pathologists, Kennett Square, PA, pp 347–385

Aggrey SE, Kuhnlein U, Gavora JS, Zadworny D (1998) Association of endogenous viral genes with quantitative traits in chickens selected for high egg production and susceptibility or resistance to Marek's disease. Br Poultry Sci 39:39–41

Ameli H, Gavora JS, Spencer JL, Fairfull RW (1992) Genetic resistance to two Marek's disease viruses and its relationship to production traits in chickens. Can J Anim Sci 72:213–225

Bachev N, Lalev M (1990) Resistance to Marek's disease of fowls of different cholinesterase genotypes in relation to phagocyte activity. Zhivotnov'dni Nauki 27:52–56

Bacon LD (1987) Influence of the major histocompatibility complex on disease resistance and productivity. Poultry Sci 66:802–811

Bacon LD, Hunt HD, Cheng HH (2000) A review of the development of chicken lines to resolve genes determining resistance to diseases. Poultry Sci 79:1082–1093

Bacon L, Motta J, Cheng H, Vallejo R, Witter R (1996) Use of recombinant congenic chicken strains to define non-MHC genes influencing Marek's disease susceptibility. In: Silva RF, Cheng HH, Coussens PM, Lee LF, Velicer LF (eds) Current research on Marek's disease. American Association of Avian Pathologists, Kennett Square, PA, pp 63–68

Bacon LD, Polley CR, Cole RK, Rose NR (1981) Genetic influences on spontaneous autoimmune thyroiditis in (CSXOS)F2 chickens. Immunogenetics 12:339–349

Bacon LD, Vallejo R, Cheng H, Witter R (1996) Failure of Rfp-Y genes to influence resistance to Marek's disease. In: Silva RF, Cheng HH, Coussens PM, Lee LF, Velicer LF (eds) Current research on Marek's disease. American Association of Avian Pathologists, Kennett Square, PA, pp 69–74

Bacon LD, Witter RL (1992) Influence of turkey herpesvirus vaccination on the B-haplotype effect on Marek's disease resistance in 15.B-congenic chickens. Avian Dis 36:378–385

Bacon LD, Witter RL (1993) Influence of B-haplotype on the relative efficacy of Marek's disease vaccines of different serotypes. Avian Dis 37:53–59

Bacon LD, Witter RL (1994a) Serotype specificity of B-haplotype influence on the relative efficacy of Marek's disease vaccines. Avian Dis 3:65–71

Bacon LD, Witter RL (1994b) B haplotype influence on the relative efficacy of Marek's disease vaccines in commercial chickens. Poultry Sci 73:481–487

Bacon LD, Witter RL (1995) Efficacy of Marek's disease vaccines in Mhc heterozygous chickens: Mhc congenic X inbred line F1 matings. J Heredity 86:269–273

Beachy RN (1997) Mechanisms and applications of pathogen-derived resistance in transgenic plants. Curr Opin Biotechnol 8:215–220

Beavis WB (1998) QTL analyses: power, precision, and accuracy. In: Paterson AH (ed) Molecular Dissection of Complex Traits. CRC Press, New York, pp 145–162

Blankert JJ, Albers GAA, Briles WE, Vrielink-van Ginkel M, Groot AJC, te Winkel GP, Tilanus MGJ, van der Zijpp AJ (1990) The effect of serologically defined major histocompatibility complex haplotypes on Marek's disease resistance in commercially bred white leghorn chickens. Avian Dis 54:818–823

Briles WE, Briles RW (1982) Identification of haplotypes of the chicken major histocompatibility complex (B). Immunogenetics 15:449–459

Briles WE, Briles RW, McGibbon WH, Stone HA (1980) Identification of B-alloalleles associated with resistance to Marek's disease. In: Biggs PM (ed) Resistance and immunity to Marek's disease. ECSC-EEE-EAEC, Brussels, Luxembourg, pp 349–416

Briles WE, Briles RW, Taffs RE, Stone HA (1983) Resistance to a malignant lymphoma in chickens is mapped to subregion of major histocompatibility (B) complex. Science 219:977–979

Briles WE, Goto RM, Auffray C, Miller MM (1993) A polymorphic system related to but genetically independent of the chicken major histocompatibility complex. Immunogenetics 37:408–414

Briles WE, McGibbon WH, Irwin MR (1950) On multiple alleles affecting cellular antigens in the chicken. Genetics 35:633–652

Briles WE, Oleson WL (1971) Differential depletion of B blood genotypes under stress of Marek's disease. Poultry Sci 50:1558 (Abstract)

Briles WE, Stone HA, Cole RK (1977) Marek's disease: effects of B histocompatibility alloalleles in resistant and susceptible chicken lines. Science 195:193–195

Bumstead N (1998) Genomic mapping of resistance to Marek's disease. Avian Path 27:S78–S81

Bumstead H, Sillibourne J, Rennie M, Ross N, Davison TF (1997) Quantification of Marek's disease virus in chicken lymphocytes using the polymerase chain reaction with fluorescence detection. J Virol Methods 65:75–81

Burgess SC, Basaran BH, Davison TF (2001) Resistance to Marek's disease herpesvirus-induced lymphoma is multiphasic and dependant on host genotype. Vet Path, vol 38

Burt DW, Bruley C, Dunn IC, et al. (2000) The dynamics of chromosome evolution in birds and mammals. Nature 402:411–413

Calnek BW (1985) Genetic resistance. In: Payne LN (ed) Marek's disease. Martinus Nijhoff Publishing, Boston, pp 293–328

Calnek BW, Witter RL (1997) Neoplastic diseases/Marek's disease. In: Calnek BW (ed) Diseases of poultry, 10th Edition. Iowa State University Press, Ames, IA, pp 369–413

Churchill AE, Biggs PM (1967) Agent of Marek's disease in tissue culture. Nature 215:528–530

Cole RK (1968) Studies on genetic resistance to Marek's disease. Avian Diseases 12:9–28

Cole RK (1985) Natural resistance to Marek's disease – a review. In Calnek BW, Spencer JL (eds) Proceedings international symposium on Marek's disease. American Association of Avian Pathologists, Kennett Square, PA, pp 318–329

Crittenden LB (1991) Retroviral elements in the genome of the chicken: implications for poultry genetics and breeding. Crit Rev Poult Biol 3:73–109

Crittenden LB, Bitgood JJ, Burt DW, Ponce de Leon FA, Tixier-Bouchard M (1996) Nomenclature for naming loci, alleles, linkage groups and chromosomes to be used in poultry genome publications and databases. Genet Sel Evol 28:289–297

Crittenden LB, Muhm RL, Burmester BR (1972) Genetic control of susceptibility to the avian leukosis complex: II. Marek's disease. Poult Sci 51:261–267

Crittenden LB, Salter DW (1992) A transgene, alv6, that expresses the envelope of subgroup A avian leukosis virus reduces the rate of congenital transmission of a field strain of avian leukosis virus. Poult Sci 71:799–806

Darvasi A, Soller M (1997) A simple method to calculate resolving power and confidence interval of QTL map location. Behav Genet 27:125–132

DeLuca NA, Schaffer PA (1988) Physical and functional domains of the herpes simplex virus transcriptional regulatory protein ICP4. J Virol 62:732–743

Demant P, Hart AAM (1986) Recombinant congenic strains – A new tool for analyzing genetic traits determined by more than one gene. Immunogenetics 24:416–422

Demant P, Olmen LCJM, Oudshoorn-Snoek M (1989) Genetics of tumor susceptibility in the mouse: MHC and non-MHC genes. Ann Rev Genet 7:117–179

Federspiel MJ, Crittenden LB, Provencher LP, Hughes SH (1991) Experimentally introduced defective endogenous proviruses are highly expressed in chickens. J Virol 65:313–319

Fields S, Song O (1989) A novel genetic system to detect protein-protein interactions. Nature 340:245–246

Flock DK, Voss M, Velitz E (1992) Genetic analysis of Marek's disease mortality in brown-egg type cockerels. In: Proc 19th Worlds Poultry Cong Amsterdam, pp 181–188

Fredericksen TL, Gilmour DG, Bacon LD, Witter RL, Motta J (1982) Tests of association of lymphocyte alloantigen genotypes with resistance to viral oncogenesis in chickens. 1. Marek's disease in F7 progeny derived from 63 X 151 crosses. Poultry Sci 61:2322–2326

Fredericksen TL, Longenecker BM, Pazderka F, Gilmour DG, Ruth RF (1977) A T-cell antigen system of chickens: Ly-4 and Marek's disease. Immunogenetics 5:535–552

Friars GW, Chambers JR, Kennedy A, Smith AD (1972) Selection for resistance to Marek's disease in conjunction with other economic traits in chickens. Avian Dis 16:2–10

Gallinari P, Wiebauer K, Nardi MC, Jiricny J (1994) Localization of a 34-amino-acid segment implicated in dimerization of the herpes simplex virus type 1 ICP4 polypeptide by a dimerization trap. J Virol 68:3809–3820

Gavora JS, Grunder AA, Spencer JL, Gowe RS, Robertson A, Speckmann GW (1974) An assessment of effects of vaccination on genetic resistance to Marek's disease. Poultry Sci 53:889–897

Groenen MAM, Cheng HH, Bumstead N, et al. (2000) A consensus linkage map of the chicken genome. Genome Res 10:137–147

Gu B, DeLuca N (1994) Requirements for activation of the herpes simplex virus glycoprotein C promoter in vitro by the viral regulatory protein ICP4. J Virol 68:7953–7965

Guillemot F, Auffray C (1989) Molecular biology of the chicken major histocompatibility complex. Poultry Biol 2:255–275

Hansen MP, Van Zandt JM, Law GRJ (1967) Differences in susceptibility to Marek's disease in chickens carrying two different B locus blood group alleles. Poultry Sci 46:1268 (Abstract)

Hartmann W (1989) Evaluation of "major genes" affecting disease resistance in poultry in respect to their potential for commercial breeding in recent advances. In: Avian immunology research, Alan R Liss, Inc. New York, pp 221–231

Hepkema BG, Blankert JJ, Albers GAA, Tilanus MGJ, Egberts E, van der Zijpp AJ, Hensen EJ (1993) Mapping of susceptibility to Marek's disease within the major histocompatibility (B) complex by refined typing of white leghorn chickens. Animal Genetics 24:283–287

Hunt HD, Fulton JE (1998) Analysis of polymorphisms in the major expressed class I locus (B-FIV) of the chicken. Immunogenetics 47:456–467

Hutt FB, Cole RK, Bruckner JH (1941) Four generations of fowls bred for resistance to neoplasms. Poultry Sci 20:514–526

Hutt FB, Cole RK, Ball M, Bruckner JH, Ball RF (1944) A relation between environment to two weeks of age and mortality from lymphomatosis in adult fowls. Poultry Sci 23:396–404

Iotova I, Marinov B, Georgieva V, Stoyanchev T, Semerdzhiev V (1990) The relationship of the resistance of the different alkaline phosphatase genotypes of broiler-type fowls with carcass quality and resistance to Marek's disease. Zhivotnov'dni Nauki 27:60–65

Jones D, Brunovskis P, Witter R, Kung H-J (1996) Retroviral insertional activation in a herpesvirus: transcriptional activation of US genes by an intergrated long terminal repeat in a Marek's disease virus clone. J Virol 70:2460–2467

Jungblut PR, Otto A, Favor J, Lowe M, Muller EC, Kastner M, Sperling K, Klose J (1998) Identification of mouse crystallins in 2D protein patterns by sequencing and mass spectrometry. Application to cataract mutants. FEBS Lett 435:131–137

Kearsey MJ, Farquhar AGL (1998) QTL analysis in plants; where are we now? Heredity 80:137–142

Kuhnlein U, Gavora JS, Spencer JL, Bernon DE, Sabour M (1989a) Incidence of endogenous viral genes in two strains of White Leghorn chickens selected for egg production and susceptibility or resistance to Marek's disease. Theor Appl Genet 77:26–32

Kuhnlein U, Ni L, Weigen S, Gavora JS, Fairfull W, Zadworny D (1997) DNA polymorphisms in the chicken growth hormone gene: response to selection for disease resistance and association with egg production. Anim Genet 28:116–123

Kuhnlein U, Sabour M, Gavora JS, Fairfull RW, Bernon DE (1989b) Influence of selection for egg production and Marek's disease resistance on the incidence of endogenous viral genes in white leghorns. Poultry Sci 68:1161–1167

Lagunoff M, Roizman B (1995) The regulation of synthesis and properties of the protein product of open reading frame P of the herpes simplex virus 1 genome. J Virol 69:3615–3623

Lakshmanan N, Gavora JS, Lamont SJ (1997) Major histocompatibility complex class II DNA polymorphisms in chicken strains selected for Marek's disease resistance and egg production or for egg production alone. Poultry Sci 76:1517–1523

Lakshmanan N, Lamont SJ (1998) Rfp-Y region polymorphism and Marek's disease resistance in multitrait immunocompetence-selected chicken lines. Poultry Sci 77:538–541

Lee LF, Bacon LD (1983) Ontogeny and line differences in the mitogenic response of chicken lymphocytes. Poultry Sci 62:579–584

Lee LF, Wu P, Sui D, Ren D, Kamil J, Kung H-J, Witter RL (2000) The complete UL sequence and the overall genomic organization of the GA strain of Marek's disease virus. Proc Natl Acad Sci USA 97:6091–6096

Leopardi R, Michael N, Roizman B (1995) Repression of the herpes simplex virus 1 alpha 4 gene by its gene product (ICP4) within the context of the viral genome is conditioned by the distance and stereoaxial alignment of the ICP4 DNA binding site relative to the TATA box. J Virol 69:3042–3048

Liu J-L, Lin S-F, Xia L, Brunovskis P, Li D, Davidson I, Lee LF, Kung H-J (1999) Meq and v-IL8; cellular genes in disguise? Acta Virol 43:94–101

Liu H-C (2000) Identification of a host protein interacting with the Marek's disease virus SORF2 protein. Ph.D. dissertation, Michigan State University

Lium EK, Panagiotidis CA, Wen X, Silverstein S (1996) Repression of the alpha0 gene by ICP4 during a productive herpes simplex virus infection. J Virol:70:3488–3496

Longenecker BM, Pazderka F, Gavora JS, Spencer JL, Ruth RF (1976) Lymphoma induced by herpesvirus: resistance associated with a major histocompatibility gene. Immunogenetics 3:401–407

Maas HJL, Antonisse AJ, Man Der Zijpp AJ, Groenendal JE, Kok GL (1981) The development of two White Plymouth Rock lines resistant to Marek's disease by breeding from survivors. Avian Pathol 10:137–150

McGeoch DJ, Dolan A, Donald S, Brauer DHK (1986) Complete DNA sequence of the short repeat region in the genome of herpes simplex virus type 1. Nucleic Acids Res 14:1727–1744

Miller MM, Abplanalp H, Goto R (1988) Genotyping chickens for the B-G subregion of the major histocompatibility complex using restriction fragment length polymorphisms. Immunogenetics 28:374–379

Miller MM, Goto RM, Taylor RL Jr, Zoorob R, Auffray C, Briles RW, Briles EW, Bloom SE (1996) Assignment of Rfp-Y to the chicken MHC/NOR microchromosome and evidence for high frequency recombination associated with the nucleolar organizer region. Proc Natl Acad Sci 93:3958–3962

Moen CJA, Groot PC, Hart AAM, Snoek M, Demant P (1996) Fine mapping of colon tumor susceptibility (*Scc*) genes in the mouse, different from the genes known to be somatically mutated in colon cancer. PNAS 93:1082–1086

Payne LN, Fadly AF (1997) Neoplastic diseases/Leukosis/Sarcoma Group. In: Calnek BW (ed) Diseases of Poultry, 10th. Edition. Iowa State University Press, Ames, IA, pp 414–466

Papavassiliou AG, Wilcox KW, Silverstein SJ (1991) The interaction of ICP4 with cell/infected-cell factors and its state of phosphorylation modulate differential recognition of leader sequences in herpes simplex virus DNA. Embo J 10:397–406

Pharr GT, Dodgson JB, Hunt HH, Bacon LD (1998) Class II MHC cDNAs in 15I5 B-congenic chickens. Immunogenetics 47:350–354

Pharr GT, Gwynn AV, Bacon LD (1996) Histocompatibility antigen(s) linked to Rfp-Y (Mhc-like) genes in the chicken. Immunogenetics 45:52–58

Pink JR, Droege W, Hala K, Miggiano VC, Ziegler AA (1977) A three-locus model for the chicken major histocompatibility complex. Immunogenetics 5:203–216

Plachy J, Jurajda V, Benda V (1984) Resistance to Marek's disease is controlled by a gene within the B-F region of the chicken major histocompatibility complex in Rous sarcoma regressor or progressor inbred lines of chickens. Folia biologica (Praha) 30:251–258

Purchase GR (1985) Clinical disease and its economic impact. In: Payne LN (ed) Marek's disease, scientific basis and methods of control. Martinus Nkjhoff Publishing, Boston, pp 17–42

Salter DW, Crittenden LB (1991) Insertion of a disease resistance gene into the chicken germline. Biotechnology 16:125–131

Schat KA, Calnek BW, Fabricant J (1981) Influence of oncogenicity of Marek's disease virus on evaluation of genetic resistance. Poultry Sci 60:2559–2666

Schat KA, Davies CJ (2000) Viral diseases. In: Axford RFE, Bishop SC, Nicholas FW, Owen JB (eds) Breeding for disease resistance in farm animals. CAB International 2000, pp 271–300

Schat KA, Taylor RL Jr, Briles WE (1994) Resistance to Marek's disease in chickens with recombinant haplotypes of the major histocompatibility (B) complex. Poultry Sci 73:502–508

Schena M, Shalon D, Davis RW, Brown PO (1995) Quantitative monitoring of gene expression patterns with a complementary DNA microarray. Science 270:467–470

Schierman LW, Nordskog AW (1961) Relationship of blood type to histocompatibility in chickens. Science 134:1008–1009

Shepard AA, Imbalzano AN, DeLuca NA (1989) Separation of primary structural components conferring autoregulation, transactivation, and DNA-binding properties to the herpes simplex virus transcriptional regulatory protein ICP4. J Virol 63:3714–3728

Shepard AA, Tolentino P, DeLuca NA (1990) Trans-dominant inhibition of herpes simplex virus transcriptional regulatory protein ICP4 by heterodimer formation. J Virol 64:3916–3926

Shevshenko A, Jensen ON, et al. (1996) Linking genome and proteome by mass spectrometry: large-scale identification of yeast proteins from two dimensional gels. Proc Natl Acad Sci USA 93:14440–14445

Smith CA, Bates P, Rivera-Gonzalez R, Gu B, DeLuca NA (1993) ICP4, the major transcriptional regulatory protein of herpes simplex virus type 1, forms a tripartite complex with TATA-binding protein and TFIIB. J Virol 67:4676–4687

Smith CA, DeLuca NA (1992) Transdominant inhibition of herpes simplex virus growth in transgenic mice. Virology 191:581–588

Smith EJ, Cheng HH (1998) Mapping chicken genes using preferential amplification of specific alleles. Micro Comp Genom 3:13–20

Solomon JJ, Witter RL, Nazerian K, Burmester BR (1968) Studies on the etiology of Marek's disease. I. Propagation of the agent in cell culture. Proc Soc Exp Biol Med 127:173–177

Spencer JL, Gavora JS, Grunder AA, Robertson A, Speckmann GW (1974) Immunization against Marek's disease: influence of strain of chickens, maternal antibody, and type of vaccine. Avian Dis 18:33–44

Stone HA (1975) Use of highly inbred chickens in research. USDA Agricultural Research Service Technical Bulletin No. 14, Washington, DC

Tregaskes CA, Bumstead N, Davison TF, Young JR (1996) Chicken B-cell marker chB6 (Bu-1) is a highly glycosylated protein of unique structure. Immunogenetics 44:212–217

Uetz P, Giot L, et al. (2000) A comprehensive analysis of protein-protein interactions in *Saccharomyces cerevisiae*. Nature 403:623–627

Vallejo RL, Bacon LD, Liu H-C, Witter RL, Groenen MAM, Hillel J, Cheng HH (1998) Genetic mapping of quantitative trait loci affecting susceptibility to Marek's disease virus induced tumors in F2 intercross chickens. Genetics 148:349–360

Von Krosigk CM, McClary CF, Vielitz E, Zander DV (1972) Selection for resistance to Marek's disease and its expected effects on other important traits in white leghorn strain crosses. Avian Dis 16:11–19

Wakenell PS, Miller MM, Goto RM, Gauderman WJ, Briles WE (1996) Association between the Rfp-Y haplotype and the incidence of Marek's disease in chickens. Immunogenetics 44:242–245

Waters NF, Fontes AK (1960) Genetic resistance of inbred lines of chickens to Rous sarcoma virus. J Natl Cancer Inst 25:351–357

Wilson TM (1993) Strategies to protect crop plants against viruses: pathogen-derived resistance blossoms. Proc Natl Acad Sci USA 90:3134–3141

Witter RL (1997) Neoplastic diseases/Reticuloendotheliosis. In: Calnek BW (ed) Diseases of poultry, 10th edition. Iowa State University Press, Ames, IA, pp 414–466

Yonash N, Bacon LD, Witter RL, Cheng HH (1998) Developing recombinant congenic strains (RCS) in chickens as a tool to study genetic resistance to Marek's disease (MD). Proc 6th World Congress on Genetics Applied to Livestock Production 27:331–334

Zanella A, Valantines J, Granelli G, Castelli G (1975) Influence of strains of chickens on the immune response to vaccination against Marek's disease. Avian Pathol 4:247–253

The Genomic Structure of Marek's Disease Virus

R.F. Silva[1], L.F. Lee[1], and G.F. Kutish[2]

1 Introduction

All herpesviruses contain linear, double-stranded DNA genome surrounded by a 100-nm diameter icosahedral protein capsid core consisting of 162 capsomers. In thin-section electron micrographs, an electron transparent region, termed the tegument, surrounds the capsid, which in turn is surrounded by a lipid-containing envelope. Variations in tegument thickness and the state of the envelope result in reports of enveloped herpesviruses varying from 120nm to over 300nm in diameter (Roizman 1996). The herpesviruses were originally classified into families or subtypes (alpha-, beta- and gamma-herpesviruses) based upon their biological characteristics. This delineation was simple and required a few easily measured assays. Eventually, this delineation was too simple and inadequate. Herpesvirus classification now involves not only biological characterization, but tissue tropism, genomic organization, and protein comparisons.

Marek's disease virus (MDV) is a typical herpesvirus with a 100-nm capsid surrounded by a tegument and envelope. In cell culture, 150–160-nm enveloped

[1] USDA, Agricultural Research Service, Avian Disease and Oncology Laboratory, East Lansing, MI 48823, USA
[2] USDA, Agricultural Research Service, Plum Island Animal Disease Center Greenport, New York, USA

particles can be seen, while 273–400-nm particles were reported in infected feather follicle epithelium (CALNEK et al. 1970). Based upon virus neutralization and agar-gel precipitation analysis, three MDV serotypes are recognized. Serotype 1 viruses include the oncogenic MDVs and their cell-culture attenuated variants. The serotype 2 MDVs include the naturally occurring nononcogenic chicken MDVs, while the nononcogenic turkey herpesviruses (also referred to as HVT) are classified as serotype 3 viruses (BÜLOW and BIGGS 1975a,b). Primarily based upon its ability to produce lymphoid tumors in chickens, MDV was originally believed to be related to the Epstein-Barr virus and was classified as a gamma-herpesvirus (ROIZMAN 1990). However, the MDV genomic structure, and its rapid spread in cell culture, resulted in MDV being reclassified as an alpha-herpesvirus (ROIZMAN et al. 1992).

This review will concentrate on examining the structural organization of the three MDV serotypes, comparing and contrasting the three serotypes as well as comparing the MDV genome with other herpesviruses. Except where pertinent to describing genome organization, specific open reading frames (ORFs) or genes will not be discussed. In some cases, the same ORFs have been given different names in Md5 and GA. Whenever possible, both names will be used. At the time of writing, the HVT nomenclature is not firmly established and as a result we will refer to the HVT ORFs using their GA homologue names.

2 Overall Organization of the MDV Genome

Early on, characterization and classification of MDV was based upon restriction endonuclease (RE) maps and partial DNA sequence data. Recently, two complete genomes of serotype 1 MDVs, portions of a serotype 2 MDV, and the complete genome of a serotype 3 MDV were sequenced and made available for this review.

The three MDV serotypes have distinctly different RE patterns (HIRAI et al. 1979, 1981; IGARASHI et al. 1987; ROSS et al. 1985; SILVA and BARNETT 1991). However, their genomes are collinear, and based upon low stringency hybridizations, they share significant DNA homology (FUKUCHI et al. 1985a; GIBBS et al. 1984) as well as several cross-reacting antigens (IKUTA et al. 1983; SILVA and LEE 1984), suggesting a common evolutionary origin.

The 180kb, double-stranded DNA structure of MDV consists of long (UL) and short (US) unique regions, each flanked by inverted repeats (TRL, IRL, IRS, and TRS). The RE map of MDV indicates that the genomic organization resembles

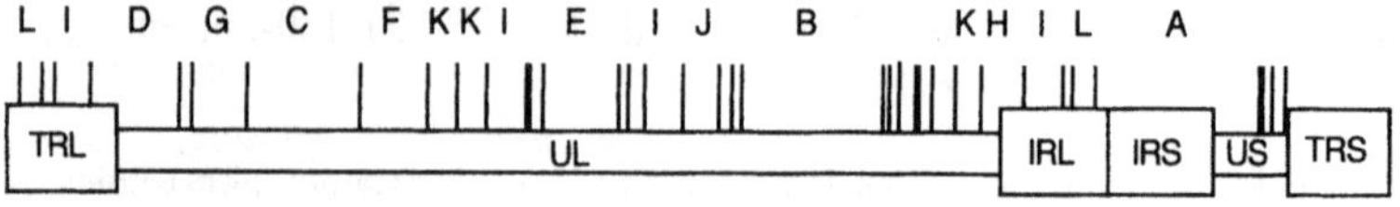

Fig. 1. *Bam*HI map of Serotype 1 MDV

HSV, an alpha-herpesvirus (Fig. 1) (Fukuchi et al. 1984; Igarashi et al. 1987; Ono et al. 1992). In addition, early DNA sequencing indicated that the MDV and HVT genomes may be collinear with and closely related to the alpha-herpesviruses, herpes simplex virus (HSV), and varicella-zoster virus (VZV) (Buckmaster et al. 1988; Ross and Binns 1991). As more MDV DNA sequence data became available, more refined comparisons were made. Utilizing DNA sequence data, dinucleotide relative abundance, and aligning four protein sequences in 13 herpesvirus genomes, Karlin et al. (1994) concluded MDV is most closely related to the alpha-herpesviruses but has some attributes of gamma-herpesviruses as well.

During the latter part of 1999 and the beginning of 2000, the genomes of several MDVs were completely sequenced. Lucy Lee's group at the ARS lab in East Lansing sequenced the UL region from the GA strain (virulent strain) of serotype 1 MDV. By joining the newly created UL sequences with published GA sequences, it was possible to reconstruct the complete sequence of the GA virus (Lee et al. 2000). The Md5 strain (very virulent strain) (Tulman et al. 2000) and FC126 (HVT) (Afonso et al. 2000) were sequenced by Dan Rock's group at the ARS lab at the Plum Island Animal Disease Center. Unfortunately, at the time of writing this review the complete sequence for HPRS24 (serotype 2 MDV) was not available. Only the sequence of the UL and US of HPRS24 were available from GenBank (Hatama et al. 1999; Izumiya et al. 1999a,b; Kato et al. 1999a,b; Tsushima et al. 1999). Consequently, most of the discussions will focus on serotypes 1 and 3.

The DNA sequence data indicate that the three MDV serotypes consist of an HSV-like organization with UL and US regions flanked by inverted repeats (TRL, IRL and IRS, TRS respectively), thus confirming the earlier RE analysis (Fig. 2). Although the three serotypes have an overall similar organization, the lengths of individual viruses differ between the serotypes (Table 1).

3 Comparing GA and Md5

The GA genome is about 174,000bp long, while Md5 is over 3800bp longer. The longer TRL and IRLs in Md5 account for about 2800bp of this difference in length. The increased length of the TRL and IRL appear to be due to more copies of a direct repeat sequence that is located at the terminus of the TRL and in the IRL, close to IRS. The two ULs are of similar length, with Md5 possibly being 87bp longer. The inverted repeats surrounding US are each approximately 626bp longer in Md5 than in GA, while the US of GA is over 300bp longer than the US in Md5 (Table 1).

3.1 Unique Short

When compared to GA, the junction between the IRS and US has moved into the US in Md5. As a result, GA has one copy of SORF2 (MDV087), located totally

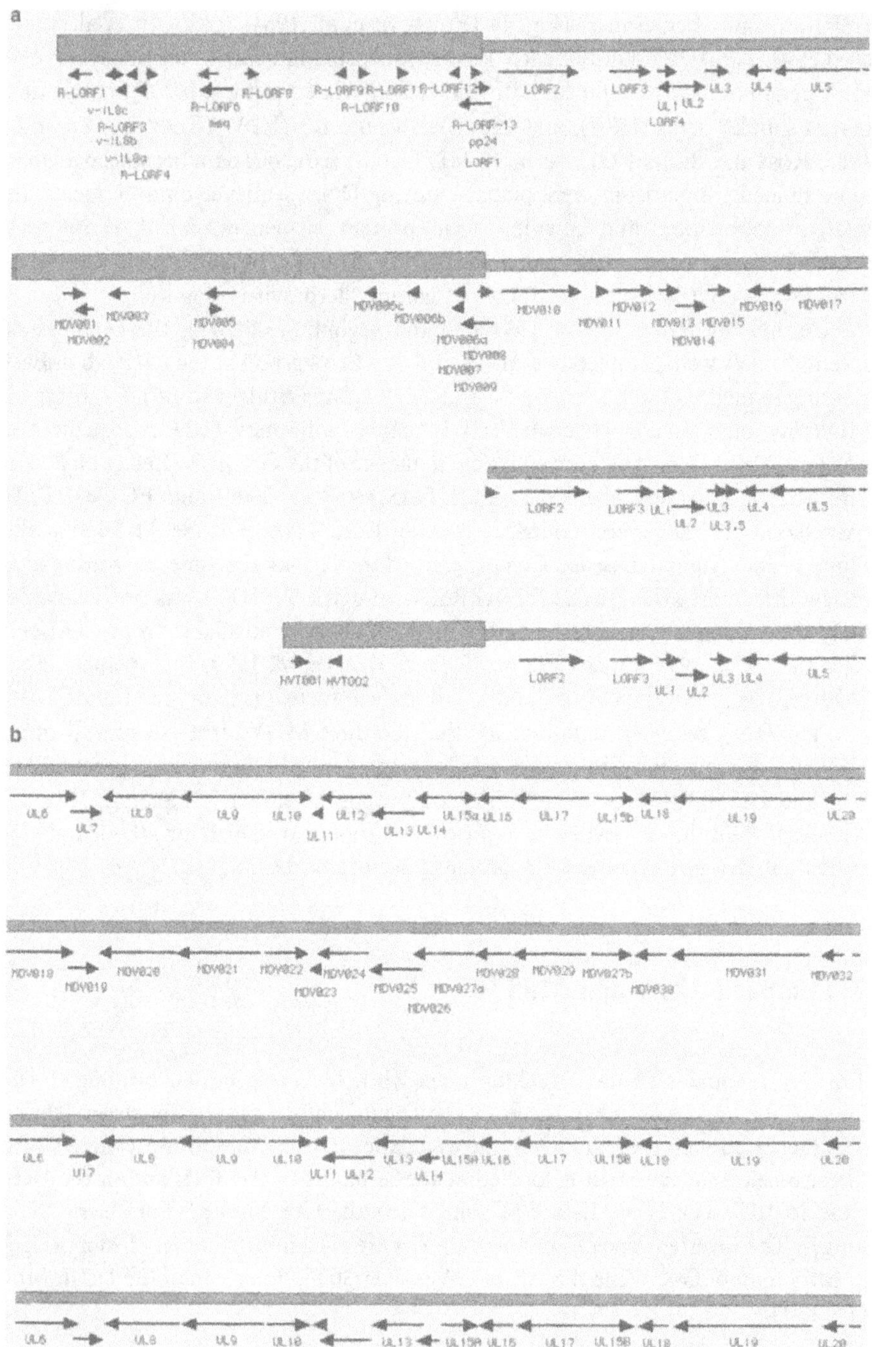

Fig. 2a–h. ORF map of MDVs. Comparable regions of each of the viruses are aligned. From *top to bottom*, the viruses are GA (serotype 1), Md5 (serotype 1), HPRS24 (serotype 2), and FC126 (HVT or serotype 3). The ORF nomenclature for HVT is temporary, with homologues of GA being used when appropriate and unique HVT nomenclature used when no known homologues exist

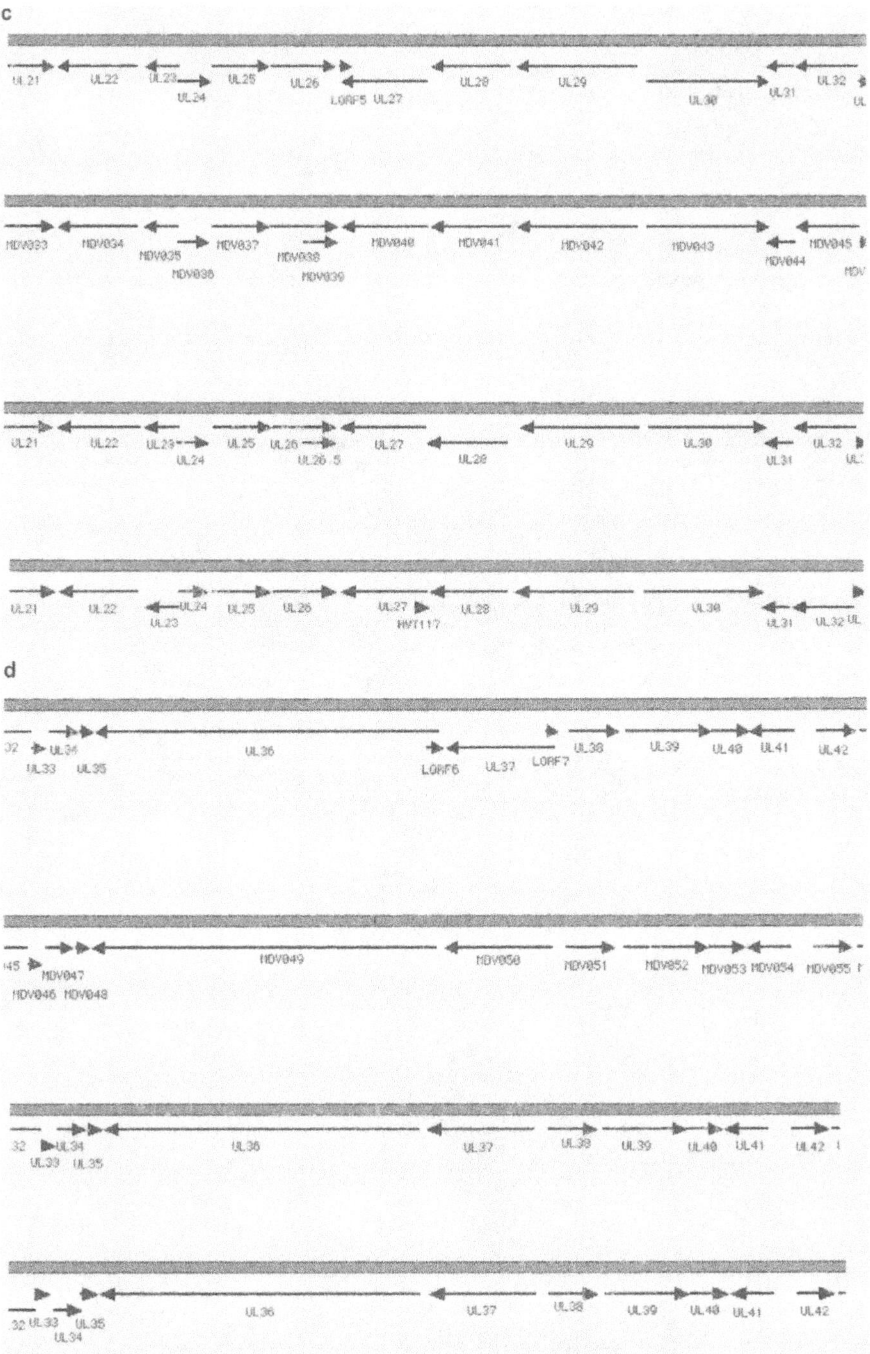
c
UL21 UL22 UL23 UL24 UL25 UL26 LORF5 UL27 UL28 UL29 UL30 UL31 UL32
MDV033 MDV034 MDV035 MDV036 MDV037 MDV038 MDV039 MDV040 MDV041 MDV042 MDV043 MDV044 MDV045
UL21 UL22 UL23 UL24 UL25 UL26 UL26.5 UL27 UL28 UL29 UL30 UL31 UL32
UL21 UL22 UL24 UL23 UL25 UL26 UL27 HVT117 UL28 UL29 UL30 UL31 UL32
d
UL34 UL33 UL35 UL36 LORF6 UL37 LORF7 UL38 UL39 UL40 UL41 UL42
MDV047 MDV046 MDV048 MDV049 MDV050 MDV051 MDV052 MDV053 MDV054 MDV055
UL34 UL33 UL35 UL36 UL37 UL38 UL39 UL40 UL41 UL42
UL33 UL35 UL34 UL36 UL37 UL38 UL39 UL40 UL41 UL42

Fig. 2c,d.

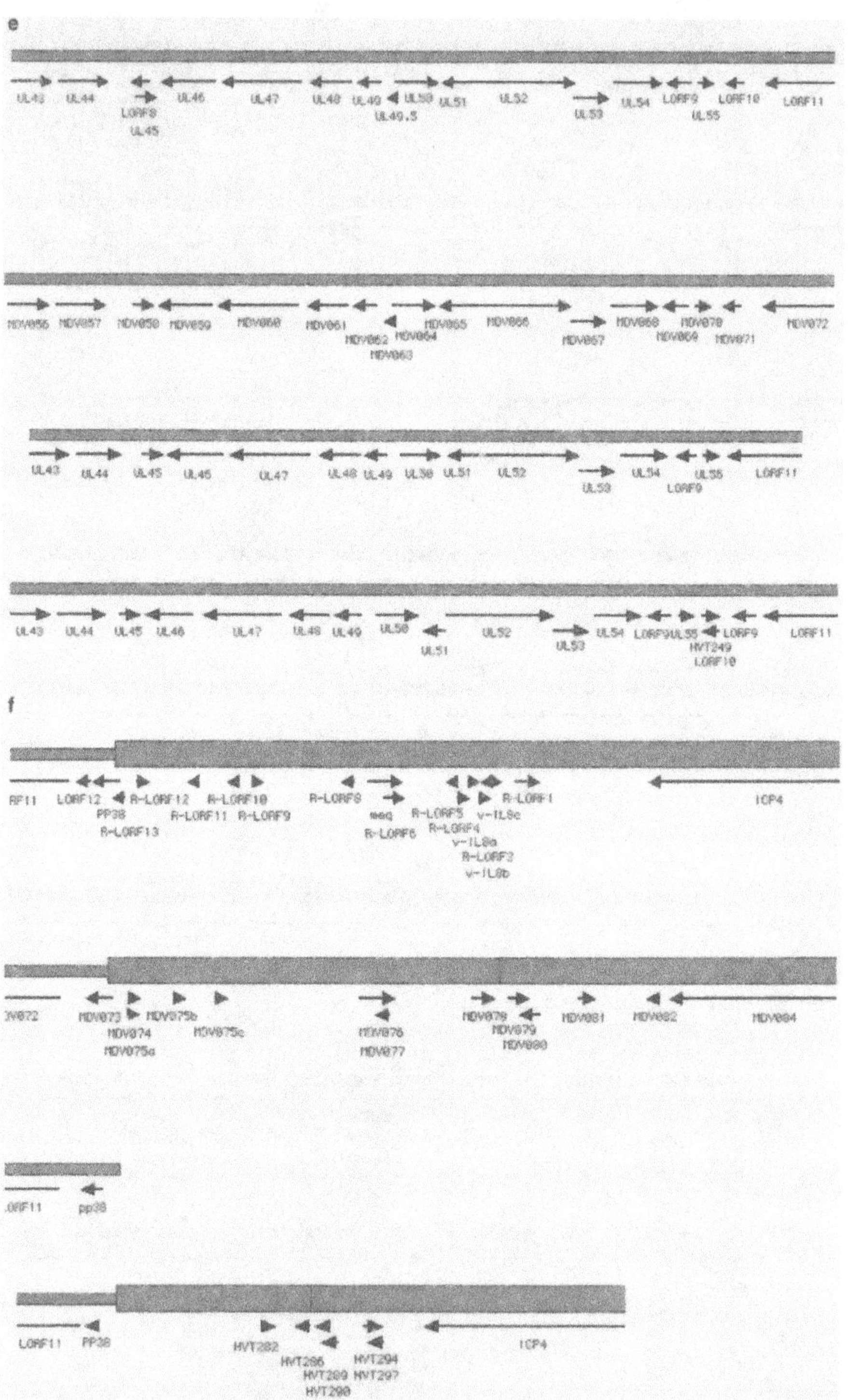

Fig. 2e,f.

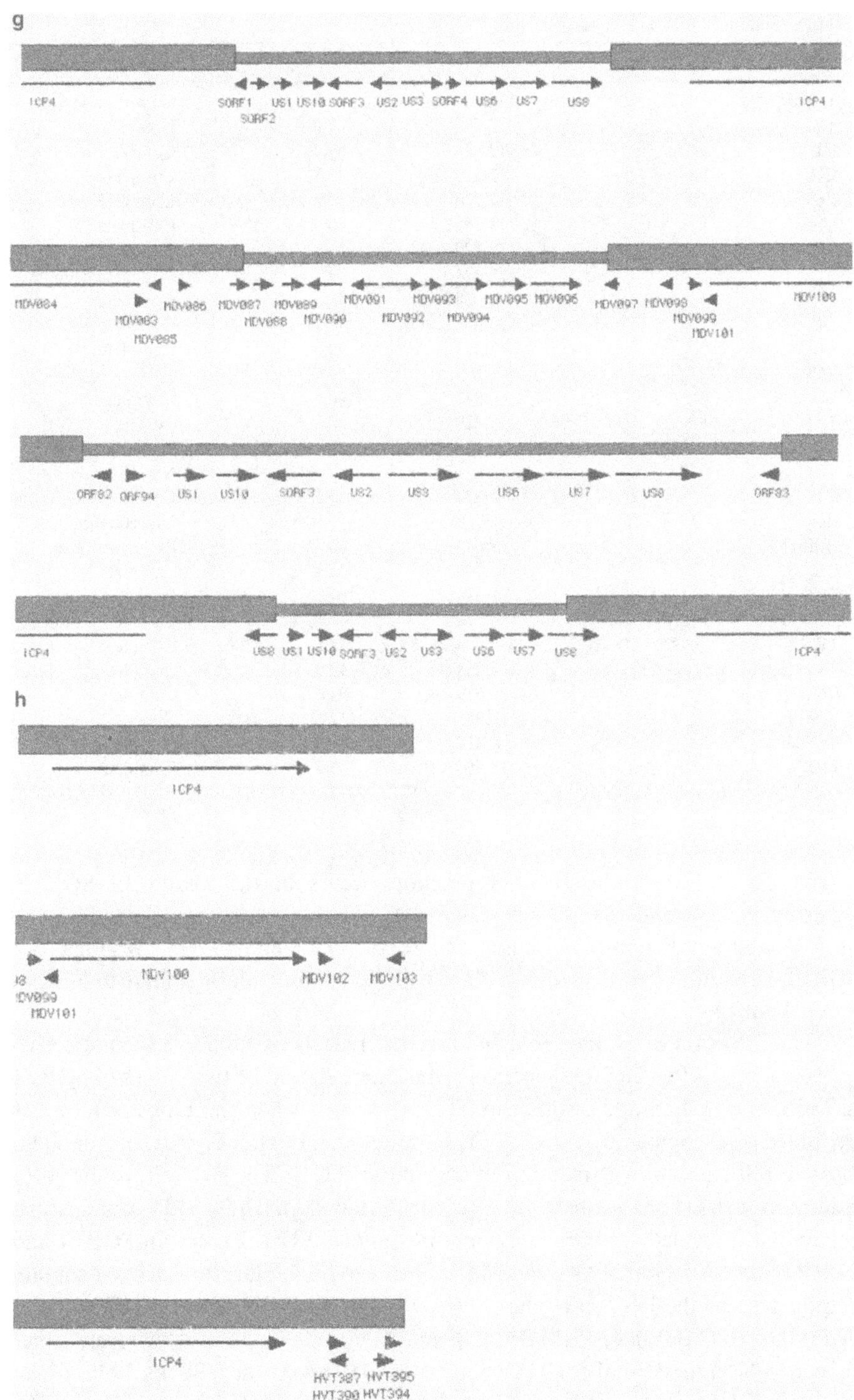
g
ICP4
SORF1 US1 US10 SORF3 US2 US3 SORF4 US6 US7 US8
SORF2
ICP4
MDV084
MDV083
MDV085
MDV086 MDV087 MDV089 MDV091 MDV093 MDV095 MDV096
MDV088 MDV090 MDV092 MDV094
MDV097 MDV098
MDV099
MDV101
MDV100
ORF02 ORF94 US1 US10 SORF3 US2 US3 US6 US7 US8 ORF03
ICP4
US8 US1 US10 SORF3 US2 US3 US6 US7 US8
ICP4
h
ICP4
MDV100
MDV102 MDV103
MDV101
ICP4
HVT387 HVT395
HVT390 HVT394

Fig. 2g,h.

Table 1. Summary of MDV genome lengths

Virus	Serotype	Length (bp)	G+C (%)	Analyzed
GA	1	174,040	43.9	Complete genome
TRL		12,585	48.8	Complete
UL		113,476	41.7	Complete
IRL		12,579	48.8	Complete
IRS		12,120	50.5	Complete
US		11,160	40.9	Complete
TRS		12,120	50.5	Complete
Md5	1	177,874	44.1	Complete genome
TRL		14,028	49.5	Complete
UL		113,563	41.7	Complete
IRL		13,943	49.5	Complete
IRS		12,264	49.9	Complete
US		10,847	40.6	Complete
TRS		13,229	50.3	Complete
HPRS24	2	126,123	50.7	Partial genome
TRL				No sequence
UL		110,637	50.5	Complete?
IRL				No sequence
IRS		1,688	64.1	Partial sequence
US		12,109	49.2	Complete
TRS		1,689	64.2	Partial sequence
FC126	3	159,160	47.6	Complete genome
TRL		5,909	52.6	Complete
UL		111,868	45.3	Complete
IRL		5,659	52.6	Complete
IRS		13,553	52.8	Complete
US		8,617	43.5	Complete
TRS		13,554	55.7	Complete

in the US. In Md5, the IRS-US junction occurs in the middle of SORF2, resulting in a second SORF2-like ORF (MDV097) that has its 5′ SORF2 end in TRS and its 3′ non-SORF2-like end in the right end of US. Whether this SORF2-like ORF is expressed and if so, whether its function is similar to SORF2 is not known.

The inclusion of SORF1 into the inverted repeats of Md5 is a second consequence of the shifted IRS-US junction. Thus, there should be two copies of SORF1 in Md5, one in IRS and another in TRS. However, when the DNA sequence is examined and compared with GA, Md5 contains an extra T, in a run of 7 Ts, located 120 bases downstream from the initial AUG. The extra T results in a reading frame shift and a new TAA termination codon. Thus, Md5 contains two copies of a truncated, and probably nonfunctional SORF1. Except for SORF1 and SORF2, the US region of GA and Md5 appears to be similar. Both viruses contain homologues of the HSV US genes, 1 (MDV088), 10 (MDV089), 2 (MDV091), 3 (MDV092), 6 (MDV094), 7 (MDV095), and 8 (MDV096), all of which were either shown to be nonessential for HSV replication (Roizman and Sears 1996) or, in the case of US6 (gD), shown to be expressed in vivo but nonessential for MDV replication (Anderson et al. 1998; Niikura et al. 1999). Homologues of the HSV and equine herpesvirus (EHV) US 4, 5, 9, 11, and 12 are missing.

3.2 Inverted Repeats

As previously mentioned, the inverted repeats of Md5 are longer than the repeats in GA. It is not known whether this is an actual difference in size or reflects the difficulty the DNA sequencers of the GA virus might have had in sequencing through many copies of direct repeats found at the termini and junction of IRL and IRS. In other respects, the inverted repeats of GA and Md5 appear similar. Located in the inverted repeats are some known MDV-specific genes that may be important in oncogenesis. The possible function and role of some of these unique genes, such as *meq* (MDV005) (Liu et al. 1997; Qian et al. 1996) and v-IL8 (MDV003) (Liu et al. 1999), will be discussed in other chapters.

Except for the two large ICP4 (MDV084) ORFs in IRS and TRS, the inverted repeats of both viruses consist of multiple short ORFs interspersed among what appears to be long noncoding regions. In many cases, it is difficult to predict which of these ORFs might actually represent spliced or unspliced transcripts. Utilizing codon bias, hexon codon usage, intron splice site detection algorithms, and hidden Markov model analysis (Baldi et al. 1994) on the sequences in the repeats failed to detect likely protein coding ORFs. Consequently, ORFs shown mapping to the inverted repeats in Fig. 2 are a conservative estimate of possible functional genes. Whether much of this region is truly not transcribed remains to be determined. Although MDV contains only a few known spliced genes [v-IL8 (Liu et al. 1999), transcripts antisense to ICP4 (Cantello et al. 1997), LORF2 (MDV010) (H.-J. Kung, personal communication), 1.8kb family of messages (Peng et al. 1992), UL15 (Dolan et al. 1991), and pp14 (Hong and Coussens 1994)], all but UL15 and MDV010 (LORF2) are located in the inverted repeats. Thus, it is also possible that the inverted repeats contain many, as yet, unrecognized spliced genes with short exons.

3.3 Unique Long

Earlier studies had found serotype 1 MDV and HVT to be homologous and collinear with HSV (Buckmaster et al. 1988). Thus, it was expected that the complete DNA sequencing of MDV and HVT would reveal large segments containing homologues of HSV genes. What was unexpected was the degree of collinearity. Both GA and Md5 contain the homologues of HSV UL1–UL55. Not only are these homologues present in UL, but the UL1–UL54 homologues are exactly in the same order and orientation as in HSV (Roizman and Sears 1996). Even the unusual splicing pattern of UL15, with UL16 and UL17 encoded in the UL15 intron, is maintained in MDV. The single exception is the insertion of a novel hypothetical protein coding ORF (designated MDV069 in Md5 or LORF9 in GA) between the UL54 and UL55 homologues. In HSV, UL1–UL55 consist of a mixture of early and late genes involved in HSV DNA replication and structural proteins of the capsid and envelope. Presumably, the MDV homologues perform the same function. Based upon blast scores and predicted homology, the

UL1–UL54 genes of MDV appear to be more related to EHV-1 genes (31–61% identity) (Telford et al. 1992) than to HSV genes (22–49% identity).

It is interesting to note that while the central portion of the UL contains the highly collinear homologues of HSV UL1–UL54, the UL regions adjacent to the TRL and IRL contain a number of large ORFs that have no homology with known HSV genes. Furthermore, the regions adjacent to the repeats in UL are also the areas where several reticuloendotheliosis viruses (REV) have preferentially inserted (Isfort et al. 1992). The REV insertions suggest that the MDV ORFs located here may not be required for MDV replication in cell culture. The preferential insertion of REV sequences in these locations may also indicate that the regions at the ends of UL are recombinant hot spots. This may explain why the central portion of the UL contains the highly conserved, mostly essential MDV genes while the non-HSV-like genes accumulate adjacent to the inverted repeats.

Beginning at the left end of UL, several MDV-specific ORFs were detected. A moderately long, potentially 333-amino-acid-encoding ORF of unknown function (LORF1 or MDV009) is located adjacent to the TRL. Next to this ORF is a long ORF (LORF2 or MDV010) that potentially could code for a 756-amino-acid phospholipase protein. LORF2 is presumably spliced and if shown to be expressed and confirmed to be a lipase would be the first example of a lipase being encoded by a herpesvirus. From an earlier report, it appears that REV sequences may have been inserted into LORF2, suggesting LORF2 is nonessential for replication (Isfort et al. 1992). Finally, an ORF (MDV012 or LORF3) potentially coding for a 384-amino-acid protein of an unknown function is located just to the left of UL1. REV sequences were also found in LORF3 (Isfort et al. 1992), indicating that LORF3 may also be nonessential for replication. At the other end of the UL, another MDV-specific ORF (LORF9 or MDV069) that potentially could code for a 269-amino-acid protein is located between UL54 and UL55. Between UL55 and the IRL are two large MDV-specific ORFs. LORF10 (MDV071) potentially codes for a 194-amino-acid protein that has some homology with EHV-4 gene 3. LORF11 (MDV072) is an ORF that could code for a 903-amino-acid protein that has no known homology with other herpesviruses. Isfort, and his group (1992), also reported that REV sequences were present in LORF11, again suggesting that this ORF may be nonessential for MDV replication.

3.4 Origin of Replication

HSV has three origins of replication (Ori), one each in IRS and TRS (OriS) and another in the UL, between ORFs UL29 (MDV042) and UL30 (MDV043) (OriL) (Roizman and Sears 1996). Both GA and Md5 have two putative origins of replication in TRL and IRL. Both of the MDV putative origins of replication are similar to OriS in HSV in that they contain the expected AT-rich hairpin loop and putative UL9 (MDV021) binding sites. A functioning Ori was mapped to the TRL and IRL regions in a serotype 2 MDV (Camp et al. 1991). Homologous sequences, suggestive of an origin of replication, are also present in the TRL and IRL of HVT.

Unexpectedly, no Ori-like sequences could be detected between what would be MDV homologues of the HSV UL29 and UL30 genes.

4 Comparing the Three Serotypes

The genomes of all three serotypes are similarly arranged. However, there are several significant differences. The TRL and IRL of HVT is less than 6000bp long versus 12,579–14,028bp long in GA or Md5. Based upon RE analysis, the TRL and IRL of the serotype 2 MDV (HPRS24) would be even shorter than the TRL and IRL of HVT. In contrast, the HVT IRS and TRS are longer than GA (approximately 13,600 vs 12,120bp in GA) and are similar in length to the TRS and IRS of Md5 (13,229 and 13,943bp respectively). Using RE data, HPRS24 is expected to have even longer IRS and TRS repeats. Even more interesting is the observation that while the ORFs in the inverted repeats of GA and Md5 are homologous to each other, they are unrelated to any ORFs in the inverted repeats of HVT. This will be discussed further in a later section.

In HVT, the US is so short (8617bp) that US-repeat junction occurs in the middle of US8 (MDV096). Consequently, the carboxy half of the US8 homologue is present as two copies, one in IRS and the other as the normal carboxy portion of US8 located in TRS. Apparently, the US-repeat junction can vary to some extent. It will be interesting to determine how variable this junction is in other MDV strains and whether the duplication or partial duplication of some US genes will affect biological characteristics.

HVT is slightly different in another respect. At the extreme left end of the UL, there is a stretch of 266 bases that are inversely repeated in a region near the IRL. In addition, a LORF9-like gene is located between LORF10 and LORF11. It is unknown what role this LORF9-like duplication has on HVT replication.

5 ORFs Unique to Serotype 1 MDV

In this section, we will concentrate on ORFs that appear to be present in Md5 and GA, but absent in either serotype 2 MDV or HVT. As was mentioned earlier, the inverted repeats are unique for each serotype. Except for ICP4, there does not appear to be any homology between serotype 1 MDV repeats and the repeats in HVT. HVT does not contain a homologue of *meq* or v-IL8. Furthermore, HVT does not have a 132-bp-like repeat region (Fukuchi et al. 1985b; Maotani et al. 1986; Silva and Witter 1985), nor an equivalent of the 1.8kb family of messages that are associated with the 132-bp repeats (Chen and Velicer 1991; Kopacek et al. 1993; Peng et al. 1992). We do not have the DNA sequences from the inverted repeats of a serotype 2 MDV, and do not know if the serotype 2 MDV repeats significantly differ from the repeats in serotype 1 and 3 viruses. However, a blastx search of GenBank, using segments from a 4-kb DNA fragment located in the IRL of a serotype 2 MDV (Camp et al. 1993), failed to find any significant matches. Thus, the serotype 2 MDV repeats may also be unique.

We can only speculate why the inverted repeats in the different serotypes are so dissimilar. It is possible that because there are two copies of each inverted repeat, any mutation introduced into one repeat will have one of two outcomes (Harland and Brown 1989; Silva and Witter 1992). If there is selective pressure or the mutation is beneficial, then the mutation will be copied into the other half of the inverted repeat pair. If the mutation is detrimental, then the mutation will be lost and the parental genotype will be restored. Thus, the redundancy built into the repeats permits the "testing" of multiple mutations without the immediate result of a single spontaneous deleterious mutation adversely affecting virus replication. Beneficial mutations are efficiently incorporated permanently into the genome. Such a system might allow the inverted repeats of serotype 1, 2, and 3 to rapidly diverge to the point where they now share little homology.

In general, genes that appear to be serotype specific are located in the repeat regions. Exceptions to this rule are SORF2 and SORF4. Further research needs to be done on these unique ORFs.

6 Perspectives

The recent availability of the complete DNA sequence from two oncogenic serotype 1 MDVs, HVT and the partial DNA sequence from a serotype 2 virus, has allowed us for the first time to clearly see how the MDV viruses are organized and how they relate to one another. As expected, the sequence data provided some surprises as well. The extreme high collinearity of some MDV genes with HSV genes, and the suggestion that MDV may be more closely related to EHV than to HSV or VZV, was a surprise. We knew from earlier experiments that some genes located in the repeats in serotype 1 viruses were not present in HVT. However, the sequence data not only confirmed this finding but extended it by showing that repeat regions of serotype 1 MDV are completely different from the repeats in HVT.

Probably the most important result of having the complete DNA sequence of several MDVs is that we now can identify potentially interesting ORFs to study. The identified ORFs can be divided into three classes. The first class consists of the MDV genes that are homologous to other known alpha-herpesviruses. The MDV homologues of UL1–UL54 fall into this class. While it is not always certain, it can be assumed that these genes play a similar role in MDV replication as they do in the other herpesviruses. Consequently, genes in this class would not be particularly interesting for MDV researchers to study.

A second class would consist of the MDV genes that are shared by all three serotypes, but are not found in other herpesvirus genomes. Presumably, these are the genes most likely to explain why MDV is a lymphotropic virus that infects and replicates in chickens, thus differing from other alpha-herpesviruses. Examples are pp38 (MDV073) (Cui et al. 1991), or a homologue of pp38 that exists in all three serotypes, but not in other herpesviruses. SORF 1 and 3 and the MDV-specific ORFs in UL that flank the highly conserved UL1–UL54 are also located in all three serotypes. In particular, SORF2 is known to be homologous to genes in avian adenovirus and fowlpox virus (Brunovskis and Velicer 1995). Similarly, the potential phospholipase gene (LORF2 or MDV010) is also homologous to a gene in an avian adenovirus (GenBank accession #AF007578). It is interesting to speculate that these two genes may play a critical role in allowing these viruses to replicate in birds.

The third class consists of the ORFs that are uniquely present in Md5 and GA, but not present in HVT or serotype 2 MDVs. Presumably, genes in this class would be responsible for the pathogenic and oncogenic potential of serotype 1 MDVs. As we have seen, the repeats are serotype-1-specific and represent fertile ground to search for these genes. It is not clear whether SORF1 is real or an artifact. However, SORF2 and SORF4 appear to be only present in serotype 1 MDV. Both of these ORFs deserve further study.

In conclusion, the recent availability of DNA sequences has ushered in a golden age for MDV virologists. Now we can pick and choose potentially relevant genes to study, instead of relying upon a blind shotgun approach that has typically been done in the past. New genes, combined with new techniques such as yeast two-hybrid systems, microarrays, and more efficient means to mutagenize the MDV genome should result in an explosion of new knowledge regarding MDV replication and pathogenesis.

Acknowledgements. We thank Lonnie Milam for his help in identifying and characterizing some of the MDV and HVT ORFs. We are also grateful to Blanca Lupiani for providing a table comparing the Md5 and GA ORFs.

References

Afonso CL, Tulman ER, Lu Z, Zsak L, Rock DL, Kutish GF (2000) The genome of herpesvirus of turkeys. (submitted for publication)

Anderson AS, Parcells MS, Morgan RW (1998) The glycoprotein D (US6) homolog is not essential for oncogenicity or horizontal transmission of Marek's disease virus. J Virol 72(3):2548–2553

Baldi P, Chauvin Y, Hunkapiller T, McClure MA (1994) Hidden Markov models of biological primary sequence information. Proc Natl Acad Sci USA 91(3):1059–1063

Brunovskis P, Velicer LF (1995) The Marek's disease virus (MDV) unique short region: alphaherpesvirus-homologous, fowlpox virus-homologous, and MDV-specific genes. Virology 206:324–338

Buckmaster AE, Scott SD, Sanderson MJ, Boursnell MEG, Ross NLJ, Binns MM (1988) Gene sequence and mapping data from Marek's disease virus and herpesvirus of turkeys: Implications for herpesvirus classification. J Gen Virol 69:2033–2042

Bülow VV, Biggs PM (1975a) Differentiation between strains of Marek's disease virus and turkey herpesvirus by immunofluorescence assays. Avian Pathol 4:133–146

Bülow VV, Biggs PM (1975b) Precipitating antigens associated with Marek's disease virus and a herpesvirus of turkeys. Avian Pathol 4:147–162

Calnek BW, Adldinger HK, Kahn DE (1970) Feather follicle epithelium: a source of enveloped and infectious cell-free herpesvirus from Marek's disease. Avian Dis 14(2):219–233

Camp HS, Coussens PM, Silva RF (1991) Cloning, sequencing, and functional analysis of a Marek's disease virus origin of replication. J Virol 65:6320–6324

Camp HS, Silva RF, Coussens PM (1993) Defective Marek's disease virus DNA contains a gene encoding a potential nuclear DNA binding protein and a HSV a-like sequence. Virology 196:484–495

Cantello JL, Parcells MS, Anderson AS, Morgan RW (1997 Feb) Marek's disease virus latency-associated transcripts belong to a family of spliced RNAs that are antisense to the ICP4 homolog gene. J Virol 71(2):1353–1361

Chen X, Velicer LF (1991) Multiple bidirectional initiations and terminations of transcription in the Marek's disease virus long repeat regions. J Virol 65:2445–2451

Cui Z-Z, Lee LF, Liu JL, Kung HJ (1991) Structural analysis and transcriptional mapping of the Marek's disease virus gene encoding pp 38, an antigen associated with transformed cells. J Virol 65:6509–6515

Dolan A, Arbuckle M, McGeoch DJ (1991) Sequence analysis of the splice junction in the transcript of herpes simplex virus type 1 gene UL15. Virus Res 20(1):97–104

Fukuchi K, Sudo M, Lee Y-S, Tanaka A, Nonoyama M (1984) Structure of Marek's disease virus DNA: Detailed restriction enzyme map. J Virol 51:102–109

Fukuchi K, Sudo M, Tanaka A, Nonoyama M (1985a) Map location of homologous regions between Marek's disease virus and herpesvirus of turkey and the absence of detectable homology in the putative tumor-inducing gene. J Virol 53:994–997

Fukuchi K, Tanaka A, Schierman LW, Witter RL, Nonoyama M (1985b) The structure of Marek's disease virus DNA: The presence of unique expansion in nonpathogenic viral DNA. Proc Natl Acad Sci USA 82:751–754

Gibbs CP, Nazerian K, Velicer LF, Kung H-J (1984) Extensive homology exists between Marek's disease herpesvirus and its vaccine virus, herpesvirus of turkeys. Proc Natl Acad Sci USA 81: 3365–3369

Harland J, Brown SM (1989) A herpes simplex virus type 2 variant in which a deletion across the L-S junction is replaced by single or multiple reiterations of extraneous DNA. J Gen Virol 70: 2121–2138

Hatama S, Jang HK, Izumiya Y, Cai JS, Tsushima Y, Kato K, Miyazawa T, Kai C, Takahashi E, Mikami T (1999) Identification and DNA sequence analysis of the Marek's disease virus serotype 2 genes homologous to the herpes simplex virus type 1 UL20 and UL21. J Vet Med Sci 61(6):587–593

Hirai K, Ikuta K, Kato S (1979) Comparative studies on Marek's disease virus and herpesvirus of turkey DNAs. J Gen Virol 45:119–131

Hirai K, Ikuta K, Kato S (1981) Restriction endonuclease analysis of the genomes of virulent and avirulent Marek's disease viruses. Microbiol Immunol 25:671–681

Hong Y, Coussens PM (1994) Identification of an immediate-early gene in the Marek's disease virus long internal repeat region which encodes a unique 14-kilodalton polypeptide. J Virol 68:3593–3603

Igarashi T, Takahashi M, Donovan J, Jessip J, Smith M, Hirai K, Tanaka A, Nonoyama M (1987) Restriction enzyme map of herpesvirus of turkey DNA and its collinear relationship with Marek's disease virus DNA. Virology 157:351–358

Ikuta K, Ueda S, Kato S, Hirai K (1983) Most virus-specific polypeptides in cells productively infected with Marek's disease virus or herpesvirus of turkeys possess cross-reactive determinants. J Gen Virol 64:961–965

Isfort R, Jones D, Kost R, Witter R, Kung H-J (1992) Retrovirus insertion into herpesvirus in vitro and in vivo. Proc Natl Acad Sci USA 89:991–995

Izumiya Y, Jang HK, Cai JS, Nishimura Y, Nakamura K, Tsushima Y, Kato K, Miyazawa T, Kai C, Mikami T (1999a) Identification and sequence analysis of the Marek's disease virus serotype 2 gene homologous to the herpes simplex virus type 1 UL52 protein. J Vet Med Sci 61(6):683–687

Izumiya Y, Jang HK, Sugawara M, Ikeda Y, Miura R, Nishimura Y, Nakamura K, Miyazawa T, Kai C, Mikami T (1999b) Identification and transcriptional analysis of the homologues of the herpes simplex virus type 1 UL30 to UL40 genes in the genome of nononcogenic Marek's disease virus serotype 2. J Gen Virol 80(Pt 9):2417–2422

Karlin S, Mocarski ES, Schachtel GA (1994) Molecular evolution of herpesviruses: genomic and protein sequence comparisons. J Virol 68:1886–1902

Kato K, Jang HK, Izumiya Y, Cai JS, Tsushima Y, Miyazawa T, Kai C, Mikami T (1999a) Identification and sequence analysis of the Marek's disease virus serotype 2 homologous genes of the herpes simplex virus type 1 UL25, UL26 and UL26.5 genes. J Vet Med Sci 61(7):787–793

Kato K, Jang HK, Izumiya Y, Cai JS, Tsushima Y, Miyazawa T, Kai C, Mikami T (1999b) Identification of the Marek's disease virus serotype 2 genes homologous to the glycoprotein B (UL27), ICP18.5 (UL28) and major DNA-binding protein (UL29) genes of herpes simplex virus type 1. J Vet Med Sci 61(10):1161–1165

Kopacek J, Ross LJN, Zelnik V, Pastorek J (1993) The 132bp repeats are present in RNA transcripts from 1.8kb gene family of Marek disease virus-transformed cells. Acta Virol 37:191–195

Lee LF, Wu P, Sui D, Ren D, Kamil J, Kung H-J, Witter RL (2000) The complete unique long sequence and the overall genomic organization of the GA strain of Marek's disease virus. Proc Natl Acad Sci USA 97:6091–6096

Liu J-L, Lin S-F, Xia L, Brunovskis P, Li D, Davidson I, Lee LF, Kung H-J (1999) MEQ and v-IL8: cellular genes in disguise? Acta Virologica 43:94–101

Liu JL, Lee LF, Ye Y, Qian Z, Kung HJ (1997) Nucleolar and nuclear localization properties of a herpesvirus bZIP oncoprotein, MEQ. J Virol 71(4):3188–3196

Maotani K, Kanamori A, Ikuta K, Ueda S, Kato S, Hirai K (1986) Amplification of a tandem direct repeat within inverted repeats of Marek's disease virus DNA during serial in vitro passage. J Virol 58:657–660

Niikura M, Witter RL, Jang H-K, Ono M, Mikami T, Silva RF (1999) MDV glycoprotein D is expressed in the feather follicle epithelium of infected chickens. Acta Virologica 43:159–163

Ono M, Katsuragi-Iwanaga R, Kitazawa T, Kamiya N, Horimoto T, Niikura M, Kai C, Hirai K, Mikami T (1992) The restriction endonuclease map of Marek's disease virus (MDV) serotype 2 and collinear relationship among three serotypes of MDV. Virology 191:459–463

Peng F, Bradley G, Tanaka A, Lancz G, Nonoyama M (1992) Isolation and characterization of cDNAs from BamHI-H gene family RNAs associated with the tumorigenicity of Marek's disease virus. J Virol 66:7389–7396

Qian Z, Kahn J, Brunovskis P, Lee LF, Kung H-J (1996) Transactivation and DNA-binding activities of MEQ. In: Silva RF, Cheng HH, Coussens PM, Lee LF, Velicer LF (eds) Current Research on Marek's Disease. American Association of Avian Pathologists, Kennett Square, PA, pp 257–264

Roizman B (1990) Herpesviridae: a brief introduction. 2 ed. In: Fields BN, Knipe DM (eds) Virology. Raven Press, Ltd., New York, pp 1787–1793

Roizman B (1996) Herpesviridae. 3rd ed. In: Fields BN, Knipe DM, Howley P, Chanock RM, Hirsh MS, Melnick JL, Monath TP, Roizman B (eds) Fields Virology. Raven Publishers, Philadelphia, Vol. 2, pp 2221–2230

Roizman B, Desrosiers RC, Fleckenstein B, Lopez C, Minson AC, Studdert MJ (1992) The family Herpesviridae: An update. Arch Virol 123:425–449

Roizman B, Sears AE (1996) Herpes Simplex Viruses and Their Replication. 3rd ed. In: Fields BN, Knipe DM, Howley P, Chanock RM, Hirsh MS, Melnick JL, Monath TP, Roizman B (eds) Fields Virology Raven Publishers, Philadelphia, Vol. 2, pp 2231–2295

Ross LJN, Binns MM (1991) Properties and evolutionary relationships of the Marek's disease virus homologues of protein kinase, glycoprotein D and glycoprotein I of herpes simplex virus. J Gen Virol 72:939–947

Ross LJN, Milne B, Schat KA (1985) The restriction enzyme analysis of MDV DNA and homology between strains. In: Calnek BW, Spencer JL (eds) International Symposium on Marek's Disease, Proceedings. The American Association of Avian Pathologists, Inc., Kennett Square, pp 51–67

Silva RF, Barnett JC (1991) Restriction endonuclease analysis of Marek's disease virus DNA: differentiation of viral strains and determination of passage history. Avian Diseases 35:487–495

Silva RF, Lee LF (1984) Monoclonal antibody-mediated immunoprecipitation of proteins from cells infected with Marek's disease virus or turkey herpesvirus. Virology 136:307–320

Silva RF, Witter RL (1985) Genomic expansion of Marek's disease virus DNA is associated with serial in vitro passage. J Virol 54:690–696

Silva RF, Witter RL (1992) A non-defective recombinant Marek's disease virus with two copies of inserted foreign DNA. In: Proceedings of the XIX World's Poultry Congress. Ponsen & Looijen, Wageningen, The Netherlands, pp 140–143

Telford EAR, Watson MS, McBride K, Davison AJ (1992) The DNA sequence of equine herpesvirus-1. Virology 189:304–316

Tsushima Y, Jang HK, Izumiya Y, Cai JS, Kato K, Miyazawa T, Kai C, Takahashi E, Mikami T (1999) Gene arrangement and RNA transcription of the BamHI fragments K and M2 within the non-oncogenic Marek's disease virus serotype 2 unique long genome region. Virus Res 60(1):101–110
Tulman ER, Afonso CL, Lu Z, Zsak L, Rock DL, Kutish GF (2000) The genome of a very virulent Marek's disease virus. J Virol 74:7980–7988

Protein-Coding Content of the Sequence of Marek's Disease Virus Serotype 1

B. LUPIANI, L.F. LEE, and S.M. REDDY

Avian Disease and Oncology Laboratory, Agricultural Research Service, 3606 East Mount Hope Road, East Lansing, MI 48823, USA

1 Introduction

MDV is classified as a member of the family *Herpesviridae* (Roizman et al. 1992). Because of its lymphotropism, MDV was originally classified as a γ-herpesvirus (Roizman et al. 1981). However, later on, based on its genomic organization and significant sequence similarity to other α-herpesviruses, such as herpes simplex virus (HSV) and varicella-zoster virus (VZV), it was classified in the subfamily *Alphaherpesvirinae* (Buckmaster et al. 1988; Roizman et al. 1992). Like other α-herpesviruses, MDV establishes a life-long latent infection, but differs from other members of the subfamily in that it goes latent not in sensory nerve ganglia but in activated T cells.

The highly cell-associated nature of MDV has made it very difficult to study the different stages in MDV replication. However, due to the collinearity of the MDV genome with other α-herpesvirus and the significant level of homology among the viral proteins, one could speculate that at least the lytic phase takes place in a way similar to that described for other members of the subfamily.

MDV is thought to enter the cell by fusion of the viral envelope with the cell membrane. Although the composition of MDV tegument is not known, MDV genome codes for a number of proteins present in the tegument of other α-herpesviruses such as the virion host shut-off protein (UL41 or MDV054) and VP16 (UL48 or MDV061). These proteins, among others, are thought to migrate to the nucleus where UL41 shuts off the cellular protein synthesis and VP16 transactivates MDV immediate-early genes (α genes) such as ICP4 (RSI or MDV084/MDV100) and ICP27 (UL54 or MDV068). These immediate-early genes in turn regulate the expression of early (β genes) and late genes (γ genes). Some of the early genes are enzymes required for viral DNA replication (see Sect. 2.2) and late genes include most of the structural genes such as capsid, envelope, and tegument proteins. It is expected that with the availability of new techniques such as real-time PCR (polymerase chain reaction) and DNA microarray, the kinetics of expression of the different viral proteins could be determined and this will help us understand the molecular mechanisms of MDV replication.

2 MDV Proteins

The DNA sequence of two strains of MDV, GA (virulent MDV) and Md5 (very virulent MDV), has been completed (Lee et al. 2000; Tulman et al. 2000). The organization of MDV unique long region (UL) is collinear to that of HSV and, in the unique short region (US), there are deletions and changes in gene order. On the other hand, the repeat long regions (TRL and IRL) and repeat short regions (TRS and IRS) are about 30 and 100% longer than their counterparts in HSV-1, respectively. Translation of MDV genomic sequence revealed over 100 potential open reading frames (ORFs) (LEE et al. 2000; Tulman et al. 2000) (Table 1). Among these, 57 ORFs in the UL region (UL1–UL55), 1 in the TRL/IRL regions (LORF1 or ICP0), 1 in the TRS/IRS regions (ICP4), and 7 in the US region (US1, US2, US3, US6, US7, US8, and US10) have homologues in the HSV genome. Throughout this review, we have used the nomenclature given for the ORFs of the GA strain of MDV (LEE et al. 2000) and indicated in parenthesis that of Md5 (Tulman et al. 2000). In cases where the putative ORFs have not been described for GA, we used the nomenclature given for the ORFs of the Md5 strain of MDV and vice versa. According to the GA nomenclature, ORFs that have counterparts in HSV-1 are named after them: UL for those in the UL region and US for those in the US region, followed by a number. MDV unique ORFs are referred to as LORFs (long ORFs), R-LORFs (repeat-long ORFs), or SORFs (short ORFs) depending on whether the start codon is initiated within the UL, RL, or US regions, respectively. In Table 1 we have indicated the percentage of identity between MDV serotype 1 and 2 proteins and the highest percentage identity with other non-MDV proteins. Although most of these ORFs have not been characterized yet, we have indicated their potential function based on their homology to proteins of other herpesviruses. It is interesting to note that most of the MDV-encoded proteins present the highest degree of homology with their counterparts in equine herpesviruses (EHV-1 and EHV-4). In the following section we will describe the current knowledge on MDV proteins and their function.

2.1 Transcriptional Regulators

MDV, like all other herpesviruses, codes for a number of proteins that play an important role in the transcriptional regulation of viral genes. In this section, we will cover the current knowledge on MDV-encoded transcriptional regulators.

2.1.1 ICP4

The MDV protein homologue of HSV-1 ICP4 and VZV ORF 62 is encoded by the RS1 (MDV084/MDV100) gene and was first described by ANDERSON et al. (1992). This ICP4 gene codes for a 1415 amino acids (aa) long protein, a size similar to that of other α-herpesviruses. However, when the sequence upstream of the gene was available it was evident that there was a much larger ORF of 2323 aa in this region. A similar large ICP4 has been found in MDV serotype 3 (ZELNIK et al. 1996), indicating that ICP4 is much larger in MDV than in other α-herpesviruses. Northern blot analysis of MDV-infected cells indicated the presence of mRNA species much larger than that expected for the 1415 aa long protein, although the size was not indicated (ANDERSON et al. 1992). Studies directed to detect ICP4 proteins in MDV-infected cells are controversial. MORGAN et al. (1996b) described the development of a polyclonal sera against the carboxy end of the short ICP4. Immunoprecipitation analysis of MDV-infected cells showed the presence of 3

Table 1. Open reading frames (*ORF*) encoded by the genome of GA and Md5 strains of MDV

GA		Md5		Location	Function/ description	MDV 2		Best match non-MDV		
ORF	Start–stop (aa)[a]	ORF	Start–stop (aa)[a]			I/L (%I)[b]	Acc. no (aa)[c]	Organism	I/L (%I)[b]	Acc. no (aa)[c]
–	–	MDV001	1517–2173 (219)	TRL	?	–	–	–	–	–
R-LORF1	1052–333 (239)	MDV002	2478–1885 (198)	TRL	ICP0	–	–	HSV-2	30/105 (28)	CAB06705 (826)
R-LORF2	2075–2138 (143)	MDV003	3555–3492 (134)	TRL	vIL-8 CXC chemokine (exon 3)	?	?	Mouse (MIP2)	30/65 (46)	033166 (100)
	1764–1899		3316–3181	TRL	vIL-8 CXC chemokine (exon 2)					
	1437–1665		3082–2881	TRL	vIL-8 CXC chemokine (exon 1)					
R-LORF3	1722–2030 (102)	R-LORF3	3139–3444 (102)	TRL	?	–	–	–	–	–
R-LORF4	2711–2334 (125)	R-LORF4	4179–3754 (142)	TRL	?	–	–	–	–	–
R-LORF5	2686–3033 (115)	–	–	TRL	?	–	–	–	–	–
R-LORF6	4817–4200 (205)	R-LORF6	6281–5667 (205)	TRL	?	–	–	–	–	–
MDV004	4414–4821 (136)	MDV004	5878–6285 (136)	TRL	23kDa nuclear protein	–	–	–	–	–
R-LORF7	5289–4270 (339)	MDV005	6753–5737 (339)	TRL	Meq (jun/fos homologue)	–	–	–	–	–
R-LORF8	5615–6025 (137)	R-LORF8	7077–7481 (135)	TRL	?	–	–	–	–	–
R-LORF9	8604–8284 (107)	R-LORF9	10051–9731 (107)	TRL	?	–	–	–	–	–
R-LORF10	9009–9308 (100)	R-LORF10	10455–10739 (95)	TRL	?	–	–	–	–	–
MDV006c	9254–9062 (71)	MDV006c	10719–10508 (71)	TRL	14kDa lytic phase protein, C-terminal exon	–	–	–	–	–
MDV006b	10373–10307 (22)	MDV006b	11821–11755 (22)	TRL	14kDa lytic phase protein, alternate b N-terminal exon	–	–	–	–	–

MDV006a	11703–11661 (14)	MDV006a	13147–13105 (14)	TRL	14kDa lytic phase protein, alternate a N-terminal exon	–	–	–	–	–
R-LORF11	10152–10460 (103)	R-LORF11	11600–11908 (103)	TRL	?	–	–	–	–	–
R-LORF12	11948–11601 (115)	MDV007	13392–13048 (115)	TRL	?	–	–	–	–	–
R-LORF13	12280–12594 (104)	R-LORF13	13724–14035 (104)	TRL	EBNA1	–	–	HVP	25/61 (40)	AAA66373 (476)
R-LORF14	12389–12853 (155)	MDV008	13833–14297 (155)	TRL/UL	pp24	38/91 (41)	BAA82891 (106)	–	–	–
LORF1	12894–11893 (333)	MDV009	14338–13340 (333)	TRL/UL	Hypothetical protein	–	–	–	–	–
LORF2	13091–13186 (756)	MDV010	14535–14630 (756)	UL	Phospholipase (exon 1)	401/763 (52)	BAA82891 (759)	AAdV	123/564 (21)	AAC71680 (721)
	13257–15431		14701–16872		Phospholipase (exon 2)					
–	–	MDV011	17431–17685 (85)	UL	?	–	–	–	–	–
LORF3	16351–17544 (198)	MDV012	17828–18979 (384)	UL	Hypothetical protein MDV-2	178/384 (46)	BAA82893 (363)	–	–	
UL1	17737–18221 (195)	MDV013	19172–19756 (195)	UL	Glycoprotein L (gL)	102/169 (60)	BAA82894 (190)	SVZV	38/195 (26)	AAD41747 (175)
LORF4	18173–17748 (142)	LORF4	19608–19183 (142)	UL	?	–	–	–	–	
UL2	18206–19144 (313)	MDV014	19641–20579 (313)	UL	Uracil DNA glycosylase	204/321 (63)	BAA82895 (323)	BHV-1	129/228 (56)	CAA06133 (301)
UL3	19172–19855 (228)	MDV015	20607–21290 (228)	UL	Nuclear phosphoprotein	153/215 (71)	BAA82896 (215)	EHV-4	124/212 (58)	AAC59579 (211)
UL4	21180–20377 (268)	MDV016	22615–21812 (268)	UL	Late nuclear protein	144/288 (53)	BAA82898 (270)	EHV-1	58/138 (42)	P28943 (225)
UL5	23810–21237 (858)	MDV017	25245–22672 (858)	UL	Helicase/primase	683/849 (80)	BAA82899 (858)	EHV-4	521/857 (60)	AAC59575 (880)

Table 1. (*Contd.*)

GA		Md5		Location	Function/ description	MDV 2		Best match non-MDV		
ORF	Start–stop (aa)[a]	ORF	Start–stop (aa)[a]			I/L (%I)[b]	Acc. no (aa)[c]	Organism	I/L (%I)[b]	Acc. no (aa)[c]
UL6	23882–26035 (718)	MDV018	25313–27478 (722)	UL	Capsid protein	493/724 (68)	BAA82900 (788)	EHV-1	323/695 (46)	P28944 (753)
UL7	25875–26789 (305)	MDV019	27318–28232 (305)	UL	Hypothetical protein	179/303 (59)	BAA82901 (301)	DEV	131/281 (46)	AAC03544 (312)
UL8	29132–26829 (768)	MDV020	30578–28272 (769)	UL	DNA Helicase/ Primase associated protein	483/766 (63)	BAA82902 (764)	EHV-4	263/757 (34)	AAC59572 (751)
UL9	31671–29149 (841)	MDV021	33117–30595 (841)	UL	Origin binding protein	605/877 (68)	BAA82903 (875)	EHV-1	428/851 (50)	P28947 (887)
UL10	31770–33041 (424)	MDV022	33216–34487 (424)	UL	Glycoprotein M (gM)	329/424 (77)	BAA82904 (424)	EHV-4	140/418 (33)	AAC59570 (450)
UL11	33352–33101 (84)	MDV023	34798–34547 (84)	UL	Myristylated tegument protein	43/80 (53)	BAA82905 (81)	HSV-2	12/28 (42)	CAB06771 (96)
UL12	34902–33331 (524)	MDV024	36348–34777 (524)	UL	Deoxyribo-nuclease	287/507 (56)	BAA82906 (525)	EHV-4	183/495 (36)	AAC59568 (565)
UL13	36434–34896 (513)	MDV025	37880–36342 (513)	UL	Serine/threonine protein kinase (tegument)	337/512 (65)	BAA82907 (500)	EHV-1	187/490 (38)	P28966 (594)
UL14	36898–36155 (248)	MDV026	38329–37601 (243)	UL	Minor tegument protein	128/237 (54)	BAA82908 (222)	VZV	53/148 (35)	P09295 (199)
UL15	36913–37956 (737)	MDV027	38360–39403 (737)	UL	DNA packing protein (virion) (exon 1)	589/736 (80)	BAA82909 (748)	EHV-1	446/731 (61)	P28969 (734)
	41457–42626		42885–44051	UL	DNA packing protein (virion) (exon 2)					
UL16	39078–37999 (360)	MDV028	40525–39446 (360)	UL	Tegument protein	233/352 (66)	BAA82910 (351)	EHV-4	120/324 (37)	AAC59563 (369)
UL17	41313–39085 (743)	MDV029	42741–40555 (729)	UL	Tegument protein, DNA packaging	415/721 (57)	BAA82911 (722)	EHV-4	251/740 (33)	AAC59562 (706)
UL18	43704–42748 (319)	MDV030	45132–44167 (319)	UL	Nucleocapsid protein (VP23)	268/320 (83)	BAA82912 (321)	EHV-1	142/307 (46)	P28921 (314)

UL19	48005–43833 (1391)	MDV031	49439–45261 (1393)	UL	Major capsid protein (VP5)	1202/1393 (86)	BAA82913 (1392)	EHV-1	792/1365 (58)	P28920 (1376)
UL20	49011–48310 (234)	MDV032	50447–49746 (234)	UL	Transmembrane protein, virus egress	128/228 (56)	BAA82914 (234)	EHV-1	62/200 (31)	P28971 (239)
UL21	49272–50909 (546)	MDV033	50708–52345 (546)	UL	Tegument protein	315/546 (57)	BAA82915 (532)	EHV-4	167/563 (29)	AAC59558 (529)
UL22	53499–51061 (813)	MDV034	54936–52498 (813)	UL	Glycoprotein H (gH)	465/809 (57)	BAA82917 (838)	SVZV	156/605 (25)	AAB04139 (852)
UL23	54740–53685 (352)	MDV036	56177–55122 (352)	UL	Thymidine kinase	261/350 (74)	BAA82918 (352)	EHV-4	122/340 (35)	AAC59555 (352)
UL24	54695–55666 (323)	MDV035	56132–57043 (304)	UL	Membrane protein	154/311 (49)	BAA82919 (316)	BHV-1	80/185 (43)	AAA61544 (293)
UL25	55701–57449 (583)	MDV037	57141–58889 (583)	UL	DNA packaging	430/581 (74)	BAA82920 (594)	EHV-1	278/590 (47)	P28928 (587)
UL26	57493–59406 (636)	MDV038	58933–60921 (663)	UL	VP24 capsid maturational protease, scaffold protein	385/664 (57)	BAA82921 (639)	EHV-1	228/702 (32)	P28936 (646)
UL26.5	58447–59406 (320)	MDV039	59887–60921 (345)	UL	Minor capsid scaffold protein	150/345 (43)	BAA82922 (333)	HSV-1	92/372 (24)	CAA32319 (329)
LORF5	59557–59913 (119)	LORF5	60996–61352 (119)	UL	?	–	–	–	–	–
UL27	62213–59619 (865)	MDV040	63652–61058 (865)	UL	Glycoprotein B (gB)	715/865 (82)	BAA82923 (865)	PHV-1	445/851 (52)	CAA92272 (881)
UL28	64669–62291 (793)	MDV041	66108–63730 (793)	UL	DNA packaging and cleavage	563/791 (71)	BAA83752 (787)	EHV-1	363/771 (47)	P28973 (766)
UL29	68436–64864 (1191)	MDV042	69875–66303 (1191)	UL	ssDNA binding protein	961/1188 (80)	BAA83753 (1190)	EHV-1	568/1224 (46)	P28932 (1209)
UL30	68705–72358 (1218)	MDV043	70144–73803 (1220)	UL	DNA polymerase large catalytic subunit	908/1220 (74)	BAA78719 (1190)	EHV-4	671/1242 ()	AAC59546 (1220)

Table 1. (*Contd.*)

GA		Md5		Location	Function/ description	MDV 2		Best match non-MDV		
ORF	Start–stop (aa)[a]	ORF	Start–stop (aa)[a]			I/L (%I)[b]	Acc. no (aa)[c]	Organism	I/L (%I)[b]	Acc. no (aa)[c]
UL31	73187–72288 (300)	MDV044	74632–73733 (300)	UL	Nuclear phosphoprotein	231/304 (75)	BAA82927 (304)	EHV-4	160/261 (61)	AAC59544 (326)
UL32	75129–73207 (641)	MDV046	76574–74652 (641)	UL	Cleavage/ packaging of DNA (virion)	441/643 (66)	BAA82928 (626)	VZV	271/645 (42)	P09782 (585)
UL33	75128–75487 (120)	MDV045	76573–76974 (134)	UL	Cleavage/ packaging of DNA	87/134 (64)	BAA82929 (131)	EHV-1	57/127 (42)	P28953 (162)
UL34	75623–76450 (276)	MDV047	77070–77900 (277)	UL	Membrane associated phosphoprotein	173/271 (63)	BAA82930 (279)	HSV-1	93/188 (49)	P10718 (275)
UL35	76537–76926 (130)	MDV048	77987–78379 (131)	UL	VP26 (capsid)	89/127 (70)	BAA82931 (127)	BHV-1	37/94 (39)	CAA06098 (124)
UL36	86967–76993 (3325)	MDV049	88471–78446 (3342)	UL	Large tegument protein	1716/3185 (54)	BAA82932 (3064)	EHV-4	973/3171 (31)	AAC59539 (3534)
LORF6	86615–87079 (155)	LORF6	88119–88583 (155)	UL	?	–	–	–	–	–
UL37	90322–87185 (1046)	MDV050	91826–88689 (1046)	UL	Tegument protein	656/1046 (62)	BAA82933 (1040)	EHV-4	322/1047 (30)	AAC595381 (1021)
LORF7	90033–90392 (120)	–	–	UL	?	–	–	–	–	–
UL38	90695–92104 (470)	MDV051	92197–93606 (470)	UL	VP19C (capsid)	295/472 (62)	BAA82934 (470)	EHV-1	174/427 (40)	P28935 (465)
UL39	92332–94797 (822)	MDV052	93831–96296 (822)	UL	Ribonucleotide reductase large subunit	603/792 (76)	BAA82935 (79)	PRV	405/736 (55)	P50643 (835)
UL40	94853–95881 (343)	MDV053	97380–9635 (343)	UL	Ribonucleotide reductase small subunit	256/323 (79)	BAA82936 (353)	EHV-1	200/320 (62)	P28847 (321)
UL41	97257–95935 (441)	MDV054	98756–97434 (441)	UL	Virion host shut off protein (tegument)	289/439 (65)	BAA82937 (423)	EHV-1	185/507 (36)	P28957 (497)

UL42	97914–99020 (369)	MDV055	99413–100519 (369)	UL	DNA polymerase auxiliary subunit	272/365 (74)	BAA32578 (369)	HSV-1	85/279 (30)	P10226 (488)
UL43	99183–100442 (420)	MDV056	100682–101941 (420)	UL	Probable membrane protein	229/422 (54)	BAA82939 (414)	EHV-1	83/411 (20)	P28959 (401)
UL44	100665–102167 (501)	MDV057	102164–103666 (501)	UL	Glycoprotein C (gC)	364/501 (72)	BAA82940 (478)	EHV-1	121/467 (25)	P12889 (468)
LORF8	103493–102870 (208)	LORF8	104992–104369 (208)	UL	Hypothetical 23.0kDa protein	–	–	–	–	–
UL45	103033–103665 (211)	MDV058	104532–105164 (211)	UL	Envelope transmembrane protein, cell fusion	149/211 (70)	BAA82941 (210)	EHV-1	44/179 (24)	P36323 (227)
UL46	105507–103804 (568)	MDV059	107006–105303 (568)	UL	VP11/VP12 tegument phospho proteins	231/559 (41)	BAA82942 (581)	EHV-4	125/377 (33)	AAC59528 (743)
UL47	108075–105652 (808)	MDV060	109574–107151 (808)	UL	VP13/VP14 tegument phospho proteins	408/844 (48)	BAA82943 (810)	EHV-4	124/508 (24)	AAC59527 (864)
UL48	109597–108317 (427)	MDV061	111096–109816 (427)	UL	VP16 (IEG transactivator) (tegument)	248/409 (60)	BAA32584 (424)	EHV-4	174/416 (41)	Q00028 (454)
UL49	110454–109708 (249)	MDV062	111953–111207 (249)	UL	VP22 phosphoprotein (tegument)	130/252 (51)	BAA82945 (241)	VZV	71/237 (29)	P09272 (302)
UL49.5	110890–110600 (95)	MDV064	112386–112102 (95)	UL	Envelope protein	67/95 (70)	BAA82946 (95)	ILTV	19/59 (32)	CAA74677 (117)
UL50	110867–112174 (436)	MDV063	112366–113673 (436)	UL	dUTPase	254/389 (65)	BAA82947 (395)	EHV-1	68/161 (42)	P28892 (326)

Table 1. (*Contd.*)

GA		Md5		Location	Function/ description	MDV 2		Best match non-MDV		
ORF	Start–stop (aa)[a]	ORF	Start–stop (aa)[a]			I/L (%I)[b]	Acc. no (aa)[c]	Organism	I/L (%I)[b]	Acc. no (aa)[c]
UL51	113014–112268 (249)	MDV065	114513–113767 (249)	UL	Virion phosphoprotein	126/181 (69)	BAA82948 (248)	EHV-1	81/201 (40)	P28961 (245)
UL52	113016–116240 (1075)	MDV066	114515–117736 (1074)	UL	Helicase/primase complex	717/1072 (66)	BAA82949 (1071)	EHV-4	420/1092 (38)	P28964 (1081)
UL53	116222–117283 (354)	MDV067	117718–118779 (354)	UL	Glycoprotein K (gK)	226/351 (64)	BAA33003 (354)	EHV-1	107/348 (30)	P28933 (343)
UL54	117434–118852 (473)	MDV068	118929–120347 (473)	UL	ICP27	260/485 (53)	BAA33004 (476)	PRV	118/362 (32)	Q85232 (361)
LORF9	119794–118988 (269)	MDV069	121289–120483 (269)	UL	Hypothetical protein	118/192 (61)	BAA33005 (222)	–	–	–
UL55	119962–120459 (166)	MDV070	121457–121954 (166)	UL	?	92/162 (56)	BAA33006 (167)	EHV-4	47/161(29)	AAC59557 (200)
LORF10	121399–120821 (193)	MDV071	122897–122316 (194)	UL	Hypothetical gene 2 protein	–	–	EHV-4	31/102 (30)	AAC59545 (256)
LORF11	124742–122034 (903)	MDV072	126241–123533 (903)	UL	?	489/900 (54)	BAA82954 (966)	–	–	–
LORF12	125390–124947 (148)	LORF12	126823–126446 (126)	UL	?	–	–	–	–	–
R-LORF14a	126288–125419 (290)	MDV073	127787–126918 (290)	UL/IRL	pp38	59/137 (43)	BAA82955 (222)	–	–	–
R-LORF13a	126397–126026 (124)	R-LORF13a	127896–127525 (124)	UL/IRL	?	–	–	–	–	–
R-LORF 12	126729–127073 (115)	MDV074	128228–128572 (115)	IRL	Hypothetical protein	–	–	–	–	–
R-LORF11	128525–128217 (103)	R-LORF11	130020–129712 (103)	IRL	?	–	–	–	–	–
MDV075a	126974–127016 (14)	MDV075a	128473–128515 (14)	IRL	14kDa lytic phase protein, alternate a N-terminal exon	–	–	–	–	–
MDV075b	128304–128370 (22)	MDV075b	129799–129865 (22)	IRL	14kDa lytic phase protein, alternate b N-terminal exon	–	–	–	–	–

MDV075c	129404–129614 (71)	MDV075c	130901–131112 (71)	IRL	14kDa lytic phase protein, C-terminal exon	–	–	–	–	–
R-LORF10	129667–129368 (95)	R-LORF10	131165–130881 (95)	IRL	?	–	–	–	–	–
R-LORF9	130072–130392 (107)	R-LORF9	131569–131889 (107)	IRL	?	–	–	–	–	–
R-LORF8	133061–132651 (137)	R-LORF8	134543–134139 (135)	IRL	Nuclear localization signal	–	–	–	–	–
R-LORF7	133387–134406 (339)	MDV076	134867–135883 (339)	IRL	Meq (jun/fos homologue)	–	–	–	–	–
MDV077	134262–133855 (136)	MDV077	135742–135335 (136)	IRL	23kDa nuclear protein	–	–	–	–	–
R-LORF6	133859–134476 (205)	R-LORF6	135339–135953 (205)	IRL	?	–	–	–	–	–
R-LORF5	135990–135646 (115)	–	–	IRL	?	–	–	–	–	–
R-LORF4	135965–136339 (125)	R-LORF4	137441–137866 (142)	IRL	?	–	–	–	–	–
R-LORF3	136854–136649 (102)	R-LORF3	138481–138176 (102)	IRL	?	–	–	–	–	–
R-LORF2	136538–136601 (143)	MDV078	138065–138128 (134)	IRL	vIL-8 CXC chemokine (exon 1)	–	–	Mouse (MIP2)	30/65 (46)	033166 (100)
	136777–136912		138304–138439	IRL	vIL-8 CXC chemokine (exon 2)					
	137011–137239		138538–138739	IRL	vIL-8 CXC chemokine (exon 3)					
R-LORF1	137624–138343 (239)	MDV079	139142–139735 (198)	IRL	ICP0	–	–	–	–	–

Table 1. (*Contd.*)

GA		Md5		Location	Function/ description	MDV 2		Best match non-MDV		
ORF	Start–stop (aa)[a]	ORF	Start–stop (aa)[a]			I/L (%I)[b]	Acc. no (aa)[c]	Organism	I/L (%I)[b]	Acc. no (aa)[c]
–	–	MDV080	140103–139447 (219)	IRL	?	–	–	–	–	–
–	–	MDV081	141149–141634 (162)	IRS	*a*-sequence	–	–	–	–	–
–	–	MDV082	143447–143124 (108)	IRS	?	–	–	–	–	–
–	–	MDV083	150610–150945 (112)	IRS	Antisense RNA	–	–	–	–	–
RS1	148438–141470 (2323)	MDV084	143807–150769 (2321)	IRS	ICP4	?	?	EHV-1	389/1350 (28)	P17473 (1487)
MDV085	149069–148767 (101)	MDV085	151396–151001 (132)	IRS	?	–	–	–	–	–
MDV086	149660–149932 (91)	MDV086	151988–152248 (87)	IRS	Cytoplasmic protein	–	–	–	–	–
SORF1	151064–150798 (89)	–	–	IRS	?	–	–	–	–	–
SORF2	151254–151790 (179)	MDV087	153443–153979 (179)	IRS/US	SORF2 (motif from FPV early protein FPV250)	–	–	FPV	40/88 (45)	P14362 (140)
US1	151960–152496 (179)	MDV088	154149–154685 (179)	US	ICP22	85/172 (49)	BAA32006 (173)	FHV-1	48/143 (33)	BAA07697 (334)
US10	152789–153427 (213)	MDV089	154978–155616 (213)	US	Capsid/tegument associated phosphoprotein	124/214 (57)	BAA32007 (231)	HSV-1	54/163 (33)	BAA44956 (213)
SORF3	154596–153544 (351)	MDV090	156785–155733 (351)	US	SORF3 hypothetical protein	181/345 (52)	BAA32008 (322)	–	–	–
US2	155635–154826 (270)	MDV091	157824–157015 (270)	US	?	191/265 (72)	BAA32009 (271)	EHV-1	67/161 (41)	P28964 (418)
US3	155747–156952 (402)	MDV092	157936–159141 (402)	US	Serine/threonine protein kinase	231/390 (59)	BAA32010 (391)	CHV	133/355 (37)	AAB93491 (400)
SORF4	157065–157505 (147)	MDV093	159254–159694 (147)	US	SORF4 (hypothetical protein)	–	–	–	–	–

US6	157676–158884 (403)	MDV094	159865–161073 (403)	US	Glycoprotein D (gD)	186/341 (54)	BAA32011 (385)	RHV-1	50/151 (33)	AAD46119 (191)
US7	158994–160058 (355)	MDV095	161183–162247 (355)	US	Glycoprotein I (gI)	170/349 (48)	BAA32012 (355)	HSV-1	56/220 (25)	BAA44952 (384)
US8	160200–161690 (497)	MDV096	162389–163879 (497)	US	Glycoprotein E (gE)	207/431 (48)	BAA32013 (488)	SHV-1	112/431 (25)	AAD51327 (578)
–	–	MDV097	165001–164555 (149)	US/TRS	SORF2-like	–	–	FPV	19/39 (48)	P14362 (140)
MDV098	163093–162821 (91)	MDV098	166456–166196 (87)	TRS	Cytoplasmic protein	–	–	–	–	–
MDV099	163684–163986 (101)	MDV099	167048–167443 (132)	TRS	?	–	–	–	–	–
RS1	164315–171283 (2323)	MDV100	167675–174637 (2321)	TRS	ICP4	?	?	EHV-1	389/1350 (28)	P17473 (1487)
–	–	MDV101	167834–167499 (112)	TRS	Antisense RNA	–	–	–	–	–
–	–	MDV102	174997–175320 (108)	TRS	?	–	–	–	–	–
–	–	MDV103	177295–176810 (162)	TRS	*a*-sequence	–	–	–	–	–

AAdV, avian adenovirus; BHV-1, bovine herpesvirus 1; CHV, canine herpesvirus; DEV, duck enteritis virus; EHV-1, equine herpesvirus 1; EHV-4, equine herpesvirus 4; FHV-1, feline herpesvirus 1; FPV, fowlpox virus; HSV-1, herpes simplex virus 1; HSV-2, herpes simplex virus 22; HVP, herpesvirus papio; ILTV, infectious laryngotracheitis virus; PHV-1, Phocine herpesvirus type 1; PRV, pseudorabies virus; RHV-1, Rangiferine herpesvirus 1; SHV-1, simian herpesvirus; SVZV, simian varicella-zoster virus; VZV, varicella-zoster virus;

[a] Nucleotide start and stop in the respective strain sequences. Size of the protein is indicated in number of amino acids in parenthesis;

[b] Identity of the two proteins in respect to total length checked. The percentage of identity is indicated in parenthesis;

[c] Accession number of the protein in the database. Size of the protein in amino acids is indicated in parenthesis;

The nomenclature for the ORFs of the GA strain is that described by Lee et al. (2000). The nomenclature for the ORFs of the Md5 strain is MDV followed by a number depending on their location in the genome (G. Kutish and D. Rock, personal communication). In cases where an ORF is present but has not been described, the nomenclature used is that of the other strain.

protein species of 210, 140, and 80kDa, while by Western blot analysis only the 140 and 80kDa species were detected. On the other hand, Xing et al. (1999) described a monoclonal antibody generated against a baculovirus-expressed ICP4 that recognized only a 155kDa protein in MDV-infected cells by Western blot.

Although the exact role of ICP4 in MDV replication has not been determined, there are speculations that it plays a role similar to the ICP4 protein of other α-herpesviruses, e.g., transcriptional transactivator. There are reports that transfection of MDCC-MSB-1 (MSB-1), an MDV-induced T cell line, with the short form of MDV ICP4, resulted in an increased transcription and translation of the MDV-specific phosphoproteins pp41, pp38, and pp24, and it also increased the transcription of endogenous ICP4. These results together indicate that ICP4 can upregulate its own promoter as well as that of MDV pp38, pp24, and pp41 (Pratt et al. 1994; Endoh 1996). In addition, a truncated form of ICP4 containing the first 1440 aa of the long form of ICP4 decreased the expression of these phosphoproteins. This could be due to the removal of the ICP4 transactivator domain leaving the DNA-binding domain intact, or by removal of the DNA-binding domain leaving the region responsible for binding of transcription factors intact so these would be removed from the pool (Pratt et al. 1994). In addition, Endoh (1996) showed that the expression of ICP4 was not enough to reactivate the virus and produce a lytic infection. MDV ICP4 has also been associated with maintenance of transformation (Xie et al. 1996). In this report, MSB-1 cells transfected with ICP4 antisense RNA reduced the relative colony formation by MSB-1 cells in soft agar, indicating that ICP4 plays a role in the maintenance of transformation.

2.1.2 ICP27

MDV UL54 (MDV068) codes for a 473 aa protein which presents significant homology to ICP27 of other α-herpesviruses (Ren et al. 1994). The region with higher homology is the carboxy end which contains a conserved zinc finger domain $C_{(442)}$-X_4-$C_{(447)}$-X_{13}-$H_{(461)}$-$C_{(467)}$. Western blot analysis of MDV-infected cells using polyclonal serum raised against MDV ICP27 identified the presence of a 55-kDa protein. In addition, immunofluorescence of MDV-infected cells indicated that ICP27 localizes to the nuclei. Immunoprecipitation analysis and phosphatase treatment of MDV-infected cells radiolabeled with ^{32}P indicated that MDV ICP27 is a phosphoprotein. Like HSV-1 ICP27, MDV ICP27 presents both intrinsic *trans*-activation and *trans*-repression activities. Transient expression assays indicated that MDV ICP27 *trans*-activates MDV pp14 and pp38 bidirectional early promoters, as well as a heterologous RSV-LTR U3 promoter. On the other hand, ICP27 strongly repressed MDV thymidine kinase (TK) early promoter, but had no effect on immediate-early (ICP27 and ICP4) and late (gB) promoters. In contrast to HSV-1 ICP27, MDV ICP27 did not show a cooperative activity with ICP4 (Ren et al. 1996).

2.1.3 VP16

MDV UL48 (MDV061) codes for a 427 aa protein, with an estimated molecular weight of 48kDa (YANAGIDA et al. 1993; BOUSSAHA 1996). This protein is encoded as a bicistronic message of 2.5kb. VP16 is a component of the viral tegument and is thought to function as a transcriptional activator of immediate-early genes. Western blot analysis of MDV-infected cells using antibody raised against HSV VP16, identified a polypeptide of 48kDa, suggesting that there are one or more epitopes conserved between the VP16 protein of both viruses. The VZV homologue to HSV VP16, ORF10, shares with MDV VP16 the absence of the acidic transactivator domain at the carboxy terminal end. However, VZV ORF10 and MDV VP16 are able to transactivate homologous as well as HSV immediate-early gene promoters (MORIUCHI et al. 1993; BOUSSAHA et al. 1996). The transcriptional activation domain in MDV and VZV is located at the amino terminus of the protein, while in HSV-1 is at the carboxy end. In HSV, VP16 is essential for replication (WEINHEIMER et al. 1992), whereas in VZV it was shown to be dispensable for replication (COHEN and SEIDEL 1994). The role of VP16 in MDV replication remains unknown.

2.2 MDV-Encoded Enzymes

Like in other herpesviruses, MDV genome codes for a number of enzymes required for viral DNA synthesis and virus replication. The enzymes identified in the MDV genome are: DNA polymerase complex (UL30 or MDV043, and UL32 or MDV046), helicase-primase complex (UL5 or MDV017, UL8 or MDV020, and UL52 or MDV066), phospholipase (LORF2 or MDV010), uracil DNA glycosylase (UL2 or MDV014), deoxyribonuclease (UL12 or MDV024), serine/threonine protein kinases (UL13 or MDV025, and US3 or MDV092), thymidine kinase (UL23 or MDV036), ribonucleotide reductase complex (UL39 or MDV052, and UL40 or MDV053), and dUTPase (UL50 or MDV063). This section will describe the current knowledge on MDV-encoded enzymes.

2.2.1 DNA Polymerase

Like in other herpesviruses, MDV DNA polymerase is sensitive to drugs such as phosphonoacetate (PA) (LEE et al. 1976; NAZERIAN and LEE 1976). MDV DNA polymerase is coded by UL30 (MDV043). Although it was previously reported that MDV DNA polymerase is 1180 aa (SUI et al. 1995), the new sequence data available indicate that it is 1220 aa in Md5 and 1218 aa in GA. Analysis of the domains in MDV DNA polymerase indicates that the middle region of the protein has three domains responsible for the 3′-5′ exonuclease activity similar to those of HSV-1 (HAFFEY et al. 1990). The carboxy terminal region of the protein has six domains with size and structural profile similar to those of the catalytic subunit of the Klenow fragment (JOYCE et al. 1987) and of herpesvirus DNA polymerases

(Blanco et al. 1991). This catalytic subunit region contains short runs of positively charged side chains (lysine and arginine), which are necessary for binding to the negatively charged polyphosphate backbone of the DNA. An RNase H domain, described in the DNA polymerase of other herpesviruses (Crute and Lehman 1989), has also been found in the amino terminal region of the MDV DNA polymerase, just upstream of the 3′-5′ exonuclease domain. Immunoprecipitation analysis of MDV-infected cells using a rabbit serum generated against a trpE-MDV DNA polymerase fusion protein revealed a 135-kDa protein not present in uninfected cells. The size of this protein is in agreement to that estimated for the UL30 ORF.

2.2.2 Thymidine Kinase

Thymidine kinases (TK) catalyze the conversion of thymidine deoxyribonucleosides to the monophosphate forms as part of the "salvage pathway" of nucleotide synthesis (Roizman 1978). MDV, like other herpesviruses, present a novel TK activity that can be differentiated from the host cell enzyme (Schat et al. 1984). MDV TK is encoded by UL23 (MDV036). The protein is composed of 352 aa (Scott et al. 1989), presents the highest homology with the TK of equine herpesvirus 4, and shows no overall homology to chicken cytoplasmic TK (Kwoh and Engler 1984). Although the overall homology among TK of herpesviruses is lower than that of other proteins, there are regions important for TK function that are highly conserved. For example, aa 19–36 are conserved in proteins that use ATP and GTP in catalysis (Kit 1985; Otsuka and Kit 1984; Gentry 1985) and are thought to form part of a nucleotide phosphate-binding pocket. Similarly, aa 129–145 are highly conserved and are thought to be involved in thymidine binding (Kit 1985). Like other herpesviruses, MDV TK is sensitive to acyclovir (Ross 1985); however, its role in pathogenicity has not been determined.

2.2.3 Ribonucleotide Reductase

All known α-herpesviruses encode a ribonucleotide reductase (RR). RR is an enzyme which catalyzes the reduction of ribonucleoside diphosphates to the corresponding deoxyribonucleotides and is thus instrumental in increasing the deoxyribonucleoside triphosphate (dNTP) pools to the levels required for optimal viral DNA replication in infected cells. The importance of this viral enzyme in infectivity, virulence, and in reactivation of the virus from latency in animal models has been demonstrated in several studies (Cameron et al. 1988; Heinemann and Cohen 1994).

MDV ribonucleotide reductase large subunit (RR1) is encoded by UL39 (MDV052). RR1 is 822 aa long and has an estimated molecular weight of 92.8kDa. Like RR1 in EHV-1, VZV and pseudorabies virus (PRV), MDV RR1 is about 350 aa shorter than its counterpart in HSV-1 (Lee et al. 2000; Nikas et al. 1986).

MDV ribonucleotide reductase small subunit (RR2) is encoded by UL40 (MDV053). RR2 is 343 aa long with an estimated molecular weight of 39.1kDa. The carboxy end of MDV RR1 contains a consensus sequence (YSGSVSNDL) which is conserved between MDV and HSV. This sequence has been shown to be important for RR activity in HSV (Dutia et al. 1986; Cohen et al. 1986).

A monoclonal antibody has been developed that recognizes MDV RR (L.F. Lee et al., unpublished data). In infected cells, RR is present in the cytoplasm and presents a granular distribution. At present, the role of RR in MDV pathogenicity has not been determined.

2.2.4 US3 Protein Kinase

Phosphorylation of viral proteins by protein kinases (PK) has a role in the establishment of transformation by oncogenic viruses as well as in regulation of virus replication and latency (Leader and Katan 1988).

US3 (MDV092) of MDV codes for a 402 aa polypeptide with significant homology to other α-herpesvirus serine/threonine kinases (Sakaguchi et al. 1993). Mouse serum produced against a baculovirus-expressed MDV US3 PK identified two proteins of 44 and 45kDa in chicken embryo fibroblasts infected with a vaccine strain of MDV serotype 1, but not in the tumor cell line MSB-1 or in cells from an MDV-induced tumor. These results indicate that US3 PK is only produced during productive infection. In the same study it was shown that the baculovirus-expressed US3 PK was able to phosphorylate protamine and therefore presents PK activity. In addition, they also showed that the substitution of US3 gene with the *lac*Z gene had no effect on virus replication in tissue culture or in the vaccine-induced immunity in chickens.

2.2.5 UL13 Protein Kinase

MDV UL13 (MDV025) codes for a 513 aa polypeptide with an estimated molecular weight of 58.9kDa (Reddy et al. 1999). Sequence analysis of UL13 ORF indicates that it shares significant homology with a serine/threonine protein kinase present in other herpesviruses. Protein kinases share a number of conserved motifs, some of which are present in MDV UL13 (Wierenga and Hol 1983).

Rabbit polyclonal serum raised against a trpE-MDV UL13 fusion protein was able to identify a 60-kDa protein in cells infected with the three MDV serotypes (Reddy et al. 1999). In addition, immunoprecipitates of MDV serotype 1-infected cells obtained with the trpE-MDV UL13 polyclonal serum showed protein kinase activity, indicating that MDV UL13 codes for a functional protein kinase.

2.3 Structural Proteins

MDV genome encodes a number of proteins that are structural components of the virion. These structural proteins can be classified into three groups depending on their location in the virion: nucleocapsid proteins, tegument proteins, and envelope proteins. In this section we will cover the current knowledge on MDV structural proteins.

2.3.1 Nucleocapsid Proteins

MDV codes for a number of nucleocapsid proteins (see Table 1); however, only the major capsid protein has been characterized.

MDV major capsid protein (VP5) is encoded by UL19 (MDV031). VP5 has an estimated molecular weight of 150kDa and is composed of 1393 (Md5) or 1391 (GA) aa. MDV VP5 shows a high degree of homology with its counterpart in other α-herpesviruses including MDV serotype 2 (Table 1). Immunoprecipitation analysis of MDV (serotypes 1, 2, and 3) infected cells using a rabbit polyclonal serum, raised against part of MDV VP5, identified a protein of approximately 150kDa. This size is in agreement with the size estimated from the deduced amino acid sequence (Reddy et al. 2000). In addition, treatment of infected cells with PA, a viral DNA synthesis inhibitor, abolished the production of VP5, indicating that MDV VP5 is a late gene.

2.3.2 Tegument Proteins

In addition to VP16 (see Sect. 2.1.2), only two other MDV-encoded tegument proteins have been previously described, UL47 and UL49.

2.3.2.1 UL47

MDV UL47 (MDV060) is one of the MDV-encoded proteins with less identity to its counterparts in other herpesviruses (Yanagida et al. 1993). The protein is 808 aa long and has an estimated molecular weight of 91.9kDa. It has been shown that the UL47 protein of HSV-1 and EHV-4 are phosphorylated and glycosylated (Lemaster and Roizman 1980; Whittaker et al. 1991). However, it has not been determined whether such modifications are also present in MDV UL47. The role of UL47 in MDV replication and infectivity remains to be determined.

2.3.2.2 UL49

MDV UL49 (MDV062) codes for a protein of 249 aa and has an estimated molecular weight of 27.6kDa. MDV UL49 has low homology with its counterpart in other herpesviruses, but a significant homology is detected in the central region of the protein. When compared to the UL49 protein of other herpesviruses, MDV UL49 presents a truncation at the amino end (Yanadiga et al. 1993). The role of MDV UL49 is still unknown.

2.3.3 Glycoproteins

In herpesviruses, membrane glycoproteins (gB, gC, gD, gH, and gL) are involved in the initial infection process and cell-to-cell spread. MDV is highly cell associated, and spreading from infected to uninfected cells is thought to take place by intracellular bridge formation (Kaleta and Neumann 1977). In this section we will cover the current knowledge on MDV glycoproteins.

2.3.3.1 gB

gB is highly conserved among herpesviruses. MDV gB or B-antigen complex is encoded by UL27 (MDV040). The estimated size of the protein is 865 aa with an estimated molecular weight of 95.5kDa (Ross et al. 1989). MDV gB has a potential signal peptide at aa 1–21 which shares low homology to that of other herpesviruses. On the other hand, the large hydrophobic region at the carboxy end (aa 683–770) corresponds to the transmembrane region and has high homology with the same region in HSV-1 (58%). The extracellular region of MDV gB (aa 22–682) contains several antigenic epitopes and regions relevant to penetration, rate of entry, and neutralization. MDV gB has nine potential *N*-linked glycosylation sites, and one of them, although not favorable, is conserved among several α-herpesviruses. The region preceding the transmembrane domain was found to have high homology with HSV, PRV, and VZV, as was the region immediately after the transmembrane domain. In addition, MDV gB has ten cysteine residues that are conserved among all the α-herpesviruses (Ross et al. 1989).

Immunoprecipitation analysis of MDV-infected cells using polyclonal serum, raised against specific gB peptides (Ross et al. 1989), or a monoclonal antibody (IAN86) (Sithole et al. 1988) identified 3 polypeptides of approximately 100, 60, and 49kDa. Kinetic studies and tunicamycin inhibition of MDV glycosylation showed that gB was initially synthesized as a glycosylated precursor of 100kDa and then proteolytically cleaved to the 60 and 49kDa peptides during intracellular transport (Chen and Velicer 1992). Later studies by Yoshida et al. (1994b) showed that the cleavage takes place at the predicted consensus sequence R-X-R-R (aa 431–434) and that the three arginine residues at the cleavage site are essential for recognition by cellular proteases. In addition, they showed that proteolytic cleavage of gB is not essential for its intracellular transport and surface expression.

gB has been shown to be a major component of the virion envelope (Chubb and Churchill 1968). It has also been shown that gB is highly immunogenic and produces neutralizing antibodies in infected birds (Ikuta et al. 1984; Isfort et al. 1986a). Nazerian et al. (1992) described a fowlpox recombinant virus expressing MDV gB. When used as a vaccine, this recombinant fowlpox virus was able to protect chickens from developing MD after challenged with a virulent strain of MDV.

2.3.3.2 gC

MDV gC is coded by UL42 (MDV057) and is also known as A-antigen. gC has a heterogenous molecular weight of 57–65kDa with the backbone protein having a molecular weight of 53kDa. The protein is mainly extracellular, it is detected mainly in culture supernatants and sera of infected birds, and, to a lesser extent, associates with the plasma membrane of infected cells. Approximately 95% of MDV serotype 1 gC is secreted while the remaining 5% stays bound to the plasma membrane of infected cells (Isfort et al. 1986b). gC is also the major viral antigen produced during infection (Van Zaane et al. 1982). It has been shown that gC production is lost during serial passage of MDV isolates in tissue culture (Churchill et al. 1969; Ikuta et al. 1983). These findings led to the speculation that the production of gC was related to pathogenicity and oncogenicity. However, this hypothesis was rejected since oncogenic and nononcogenic field viruses were shown to induce the production of antibodies against gC (Biggs and Milne 1972), and a clone of the oncogenic virus JM was shown not to produce gC (Purchase et al. 1971). In addition, it was later reported that the loss of gC expression after serial passage in tissue culture was not due to alteration of the gC gene or its promoter, but possibly due to a viral factor required for efficient transcription of gC (Wilson et al. 1994). Recently, Morgan et al. (1996a) described the construction of an MDV (strain RB1B) mutant where a *lac*Z/gpt cassette was introduced in the gC gene, disrupting the coding region for gC. This recombinant virus seemed to be attenuated with reduced infectivity, oncogenicity, and horizontal transmission. However, until a revertant virus is generated the real effects of the deletion of gC cannot be confirmed.

2.3.3.3 gD

US6 (MDV094) codes for the gD of MDV (Ross et al. 1991; Brunouskis and Velicer 1995). gD is composed of 403 aa and has an estimated molecular weight of 42.6kDa and 53kDa when glycosylated. Like other glycoproteins, gD has a signal peptide at the amino end and a hydrophobic region close to the carboxy end that could function as a transmembrane region. MDV gD contains four potential *N*-linked glycosylation sites and seven cysteine residues, six of which are conserved among the gD of many herpesviruses indicating the possibility of a common tertiary structure.

It has been reported that gD is expressed at very low levels in tissue culture (Ono et al. 1996). On the other hand, Tan and Velicer (1996) could not detect expression of gD in cell culture. Immunohistochemistry analysis of the feather follicle of infected birds showed that gD is expressed, but at a lower level than other viral antigens such as pp38 or gB (Niikura et al. 1999). Although gD of HSV-1 is essential for fusion and penetration in cell culture (Roizman and Sears 1991), there is evidence that in PRV (Hanssens et al. 1995) and MDV (Isfort et al. 1994; Anderson et al. 1998) gD is not essential for cell-to-cell spread. In HSV-1, gD is able to induce neutralizing antibodies (Eisenberg et al. 1985). On the other hand, a fowlpox recombinant virus expressing MDV gD protein failed to induce protective immunity in chickens (Nazerian et al. 1996). It has also been hypothesized that in MDV, gD functions to produce cell-free virus in the feather follicle and therefore is required for horizontal transmission. However, Anderson et al. (1998) showed that an oncogenic MDV serotype 1 mutant virus, lacking the gene for gD, was able to replicate well in tissue culture and chickens, produced tumors, and was capable of horizontal transmission. In addition, Isfort et al. (1994) showed that the gD gene is a common target for retrovirus integration resulting in the disruption of the gD ORF.

2.3.3.4 gH

MDV UL22 (MDV034) codes for gH (Scott et al. 1993; Wu et al. 1999). MDV gH is composed of 813 aa and has an estimated molecular weight of 91kDa. MDV gH precursor has four hydrophobic helices located at aa 1–18, 619–641, 660–682, and 770–792. The amino terminal helix appears to be the signal peptide and the carboxy terminal helix may be the transmembrane domain. gH has a small cytoplasmic domain (aa 793–813) and nine potential *N*-linked glycosylation sites located in regions predicted to be in the extracellular domain. Monoclonal antibodies and rabbit antiserum directed against a GST-gH fusion protein specifically reacted in MDV-infected cells and the fluorescence mainly localized to the cytoplasm.

As in other herpesviruses, MDV gH has been shown to interact with MDV gL (Wu 1998). It was shown that co-expression of gH and gL was essential for subcellular transportation and cell surface expression in a baculovirus expression system. In addition, using a fowlpox expression system, Wu (1998) showed that gH acts as a membrane anchor for gL and that in the absence of gH, gL was secreted into the supernatant. This study also showed that aa 451–659 of gH are essential for gH-gL interaction.

2.3.3.5 gI

MDV gI is encoded by US7 (MDV095). gI is composed of 355 aa and has an estimated molecular weight of 37kDa (Brunovskis and Velicer 1995). MDV gI has a signal peptide of about 18 aa (aa 1–18) and a potential transmembrane domain at positions 269–289. gI has five potential *N*-linked glycosylation sites and several cysteine residues are highly conserved when compared to the gI sequence of other herpesviruses (Brunovskis and Velicer 1995). In other herpesvirus systems, gI has been shown to interact with gE forming a complex that acts as a receptor for the Fc domain of IgG (Johnson and Feenstra 1987; Zuckermann et al. 1988; Whealy et al. 1993; Yao et al. 1993; Whitbeck et al. 1996). In HSV, the gI-gE heterodimer seems to be important for cell-to-cell spread of the virus during infection (Dingwell 1994). Immunoprecipitation analysis of MDV-infected cells using polyclonal sera specific for MDV gI or gE resulted in the precipitation of both gE and gI (Brunovskis and Velicer 1992; S.M. Reddy, unpublished data). These results indicate that these two proteins, like in other herpesviruses, interact with each other; however, their role in MDV infectivity and pathogenicity remains unknown.

2.3.3.6 gL

MDV gL is encoded by UL1 (MDV013). gL is composed of 195 aa and has an estimated molecular weight of 18kDa (Yoshida et al. 1994a). Immunoprecipitation analysis of MDV-infected cells using a polyclonal rabbit serum raised against a trpE-gL fusion protein revealed a polypeptide of 25kDa. Endo F treatment of this immunoprecipitated polypeptide resulted in a product of 15kDa consistent with the estimated size for gL and the presence of two potential *N*-linked glycosylation sites (Yoshida et al. 1994a). Like other α-herpesviruses, MDV gL has been shown to interact with gH, and this interaction is essential for the proper cellular location of both proteins (Wu 1998).

2.3.3.7 gK

MDV UL53 (MDV067) encodes the gK (Ren et al. 1994). MDV gK consists of 354 aa and has an estimated molecular weight of 39.5kDa. Like the gK of other herpesviruses, MDV gK has a signal peptide at the amino end (aa 1–30), four potential *N*-linked glycosylation sites, and four potential transmembrane domains (aa 29–41, 217–241, 255–274, and 319–335). The expression and role of MDV gK in tissue culture or infected chickens has not been studied.

2.3.3.8 gp82

MDV membrane glycoprotein gp82, is encoded by UL32 (MDV046), is 641 aa long and has an estimated molecular weigh of 71.5kDa. Hydrophobic analysis of the deduced amino acid sequence of gp82 indicated that it lacks a signal peptide but has four potential transmembrane spanning domains (Lee et al. 1996). In addition, endoglycosydase treatment of immunoprecipitated proteins indicated that gp82 is O-linked glycosylated. Monoclonal antibodies developed against trpE-gp82 fusion proteins indicated that gp82 localizes to the cell membrane of infected cells (Wu et al. 1997). This is in agreement with the results obtained for EHV-1 gp300 but contradicts the results obtained with HSV-1 UL32 (Whittaker et al. 1992) that localizes mainly to the cytoplasm and to a lesser extent to the nucleus. Although the function of MDV gp82 has not yet been determined, it has been shown that antibodies against both EHV-1 gp300 and MDV gp82 were able to restrict plaque size in infected cells (Whittaker et al. 1992; Lee et al. 1996). These results suggest that these proteins may be involved in the cell-to-cell fusion process.

2.4 MDV Unique Proteins

MDV genome codes for a number of proteins with no homology to any other herpesvirus protein or even to any other protein described to date. It has been hypothesized that some of these proteins, like Meq and pp38, play a role in MDV transformation. Others, like vIL-8, may play a role in the initial stages of viral infectivity. In this section, we will cover the current knowledge on MDV unique proteins.

2.4.1 Meq

Meq is encoded by R-LORF7 (MDV005/MDV076) and is the most extensively characterized MDV protein. The estimated size of Meq is 339 aa with a predicted molecular weight of 40kDa (JONES et al. 1992). Meq is consistently expressed in lymphoblastoid tumor cells and has significant homology with leucine zipper class of nuclear oncogenes (JONES et al. 1992). Meq consists of an amino terminal proline-glutamine-rich region, a basic region, a leucine zipper, as well as a proline-rich *trans*-activator domain at the carboxy terminus (QIAN et al. 1995; LIU et al. 1997). These characteristics suggest that Meq may be related to the *fos/jun* family of transcriptional activators and like them, Meq localizes to the nucleus and also the nucleolus (LIU et al. 1997). Overexpression of Meq in a rat fibroblast cell line resulted in cell transformation and protected the transformed cells from apoptosis, suggesting that it may be involved in oncogenesis (LIU et al. 1998). The role of Meq in MDV oncogenesis remains to be determined. A more detailed description of Meq characteristics and functions will be discussed elsewhere in this book.

2.4.2 pp38/pp24

MDV phosphorylated proteins, pp24 and pp38, are encoded from the opposite junction regions between the UL and TRL and between the UL and IRL, respectively, of the MDV serotype 1 genome, and therefore share the amino terminal 65 aa (ZHU et al. 1994).

MDV pp38 is encoded by R-LORF14a (MDV073). The estimated size of the protein is 290 aa with an estimated molecular weight of 31kDa. This protein is present in both MDV serotypes 1 and 2 and is encoded by an unspliced message of 1.9kb. pp38 localizes primarily in the cytoplasm of MDV-infected cells and in MD tumor cell lines (CUI et al. 1991). It is highly conserved among strains of MDV, although some variation has been observed. pp38 is expressed as an early viral protein and has been shown to be phosphorylated at serine residues (NAKAJIMA et al. 1987). This protein is relatively rich in acidic residues, with glutamic acid and aspartic acid composing 15% of the amino acid residues. As kinases have been strongly implicated in oncogenesis, and since pp38 appears to share the promoter-enhancer of the transformation-related 1.8kb gene family (BRADLEY et al. 1989), it has been speculated that pp38 may be involved in tumorigenicity (CUI et al. 1991).

MDV pp24 is encoded by R-LORF14 (MDV008). The estimated size of pp24 is 155 aa with an estimated molecular weight of 16.7kDa. pp24 is present in both MDV serotypes 1 and 2 and is encoded by an unspliced message. Sequence analysis of pp24 did not show any significant homology with any other known proteins. pp24 has several potential phosphorylation sites for casein kinase II and protein kinase C. Like MDV pp38, pp24 is expressed in MD tumor cell lines as well as in productively infected cells (ZHU et al. 1994).

LEE et al. (1983) described the development of a monoclonal antibody (H19) specific for MDV serotype 1 pp38. Immunoprecipitation analysis of MDV-infected cells using H19 antibody identified 3 polypeptides of approximately 41, 38, and 24kDa (SILVA and LEE 1984). It has been suggested that the 41 and 38kDa polypeptides are two differently phosphorylated forms of pp38, while the 24kDa polypeptide is the pp24 protein coded by R-LORF14 (LI et al. 1994; LEE et al. 2000). On the other hand, Western blot analysis using H19 identified only the 41 and 38kDa polypeptides, suggesting that the antibody recognizes an epitope specific for pp38. Epitope mapping of pp38 showed that H19 binds to aa 107, based on sequence comparison between CVI988, a serotype 1 vaccine strain, and GA isolates of MDV (Cui et al. 1999). On the other hand, western blot analysis of MDV-infected cells using M21 antibody (Zhu et al. 1994) identified two polypeptides of approximately 38 and 24kDa. These results indicated that the M21 antibody reacts with the amino terminal region common to pp38 and pp24.

2.4.3 pp14

MDV pp14 is encoded by MDV006/ MDV075. pp14 is coded by two small alternately spliced messages, both having a common carboxy terminal end (Hong et al. 1995). The carboxy terminal end of these two protein forms is 71 aa whereas the amino terminal end is either 22 or 14 aa long. The estimated molecular weight of these two protein forms are 10.3 or 9.4kDa. Both forms of pp14 have potential casein kinase and histone kinase consensus phosphorylation sites. Immunoprecipitation analysis of MDV-infected cells using polyclonal serum against these two protein forms identified a single14kDa polypeptide (Hong and Coussens 1994). These two protein forms were found to be highly phosphorylated and were expressed in the cytoplasmic fractions. Sequence analysis of both pp14 forms failed to identify any significant homology with any known protein sequence. We cannot speculate the function of these two protein forms, however they are expressed in MDV transformed cell lines suggesting that they may be associated with MDV tumorigenicity.

2.4.4 vIL-8

MDV vIL-8 is encoded by R-LORF2 (MDV003/MDV078) and was initially discovered as a gene aberrantly fused to the nuclear localization signal of the *meq* gene (Peng et al. 1995). The estimated size of the protein is 134 aa with a calculated molecular weight of 14.9kDa (Liu et al. 1999). This protein is MDV serotype 1 specific and is encoded by a spliced message of 0.7kb. This polypeptide is coded by three exons and has the characteristics of a CXC chemokine. Exon I codes for the signal peptide of the molecule. Exons II and III contain four cysteine residues, which are positionally conserved in all CXC chemokines. Functional analysis of baculovirus-expressed vIL-8 showed that vIL-8 is a chemoattractant for chicken peripheral blood mononuclear cells (H.-J. Kung, personal communication). While there are reports of other herpesviruses encoding CC chemokine or their receptors, this is the first example of CXC chemokine being coded by a herpesvirus. Epstein-Barr virus (EBV) induces the production and release of cellular IL-8 after infection (McColl et al. 1997). Since IL-8 is reported to be chemoattractant in vitro for T cells, the ability of EBV to induce IL-8 production by neutrophils may enhance its ability to infect T lymphocytes via increased recruitment to the site of infection. On the other hand, MDV codes for its own IL-8 homologue. Therefore, it has been speculated that the function of MDV vIL-8 could be to recruit target cells for infection or to serve as a decoy to cellular IL-8-mediated immune responses (Liu et al. 1999).

2.4.5 SORF2

SORF2 (MDV087) is one of the MDV-encoded unique proteins. In most of the MDV serotype 1 strains tested until now, the gene for SORF2 is located in the US region; however, in Md5, part of SORF2 is located at the IRS/TRS junction region. This results in the generation of two proteins, SORF2 and SORF2-like (MDV097), with a common amino end and a different carboxy end as seen in pp38 and pp24. SORF2 codes for a 179-aa-long protein and its sequence has been shown to be highly conserved (Brunovskis and Velicer 1995; Liu 1999). On the other hand, SORF2-like protein is 145-aa long and shares with SORF2 the first 119 aa. SORF-2 has an estimated molecular weight of 21kDa while that of SORF2-like is about 15kDa. Immunofluorescence staining of MDV-infected cells indicated that SORF2 is expressed both in the cytoplasm and nucleus early in infection; however, it localizes to the cytoplasm once plaques start appearing (Liu 1999). MDV SORF2 presents significant homology to the fowlpox virus FPV250 ORF and to ORF4 of fowl adenovirus (Brunovskis and Velicer 1995). In vitro studies showed that SORF2 interacts with the chicken growth hormone (GH), (Liu 1999). Interestingly, GH has been associated with MD resistance that is heavily influenced by the major histocompatibility complex (MHC) (H.-C. Liu and H.H. Cheng, personal communication). However, the significance of this interaction remains to be determined. Deletion experiments showed that the SORF2 is not essential for virus growth *in vitro* and *in vivo* (Parcells et al. 1994, 1995) or for tumor formation (Parcells et al. 1995). Therefore, the role of SORF2 in MDV pathogenesis remains unknown.

2.4.6 vLIP

MDV vLIP (viral lipase) is encoded by LORF2 (MDV010). The estimated size of the protein is 756 aa with a calculated molecular weight of 84kDa. This protein is encoded by a spliced message. The first exon is very short and codes for a 32-aa signal peptide, whereas the second exon codes for a 724-aa protein. This is the first example of a lipase being encoded by a herpesvirus (H.-J. Kung, personal communication). However, the vaccinia virus genome codes for a lipase which has been shown to be essential for cell-to-cell transmission and membrane biogenesis (Baek et al. 1997). Although the role of MDV vLIP is still unknown, integration of a reticuloendotheliosis virus LTR into the vLIP gene did not affect the virus replication in tissue culture, suggesting that this gene is not essential for virus replication.

2.5 Other Proteins

MDV genome encodes a large number of proteins that have not been mention in the previous sections. Of these, only the origin binding protein has been characterized.

2.5.1 Origin Binding Protein

In herpesviruses, DNA replication starts with the binding of the origin binding protein (OBP) to the viral DNA origin of replication. In turn, other replication-related proteins bind to the origin to form a functional replication complex. MDV OBP is encoded by UL9 (MDV021) (Wu et al. 1996). MDV OBP is 841 aa long and has an estimated molecular weight of 93.4kDa. MDV OBP shares with its counterparts in other α-herpesviruses a high degree of homology, including several structural motifs. For example, the amino end of MDV OBP presents six conserved helicase motifs, and their spatial arrangement is also conserved. There is also an ATP-binding domain highly conserved among α-herpesviruses as well as a leucine zipper motif thought to be involved in dimerization of the OBP. The carboxy end of MDV OBP also presents a conserved pseudo-leucine zipper motif probably important for DNA binding activity. It is interesting to note that a polyclonal serum raised against HSV-1 OBP identified a 85-kDa protein in MDV-infected cells.

3 Genetics of MDV

The role of individual proteins in virus replication is studied by deletion of individual genes from the viral genome. MDV genetics has lagged behind other herpesviruses due to the difficulty in introducing site-specific mutations into the genome of oncogenic strains of MDV. Initial studies with purified MDV DNA showed that it is infectious in cell culture (Wilson and Coussens 1991), indicating that cotransfection with a selectable marker could be used to generate MDV recombinants. Parcells et al. (1995) described for the first time the insertion of *lac*Z gene into the US region of a MDV serotype 1 oncogenic virus, resulting in the disruption of six genes. In this study, cells in culture were cotransfected with DNA isolated from MDV-infected cells and a plasmid containing the *lac*Z gene flanked by MDV specific sequences. Recombinants were selected by detecting the expression of β-galactosidase in cells containing the virus, and were subsequently purified by repeated rounds of plaque purification. The resulting recombinant virus had insertionally inactivated the first six genes in the US region (SORF1, SORF2, US1, US10, SORF3, and US2). Later on, Anderson et al. (1998) and Dienglewick and Parcells (1999) described two mutants in which the gD (US6) and the US2 genes had been replaced by the *lac*Z and green fluorescent protein, respectively. In both cases, expression of a foreign marker gene did not impair the oncogenic properties of the virus.

A more versatile method to generate recombinant MDV, which does not require plaque purification or expression of selectable marker, was recently described (S.M. Reddy et al., unpublished results). An overlapping cosmid library composed of five cosmids spanning the entire genome of a very virulent oncogenic strain (Md5) of MDV serotype 1 was developed. Transfection of these overlapping cosmids into cells in culture resulted in the generation of infectious virus. Transfection of cosmids, in which individual genes have been deleted or inserted, resulted in MDV with the expected mutation in the genome. The availability of this new technique will be very useful in generating MDV deletion mutants that will help in understanding the role of individual MDV proteins in virus replication, latency, and oncogenicity.

4 Conclusions

Prior sequence information from other herpesviruses has greatly facilitated the analysis of the MDV genome. Now, with the availability of the complete MDV genome sequence and the ability to generate knockout mutants, the unifying and diverging features of MDV compared to other herpesviruses will be unveiled.

MDV proteins could be classified into three different classes. Class I would include the proteins homologous to those of other α-herpesviruses that are involved in virus replication and assembly. Class II would include proteins present in all MDV serotypes (vLIP, SORF3) that may have a common function in the biology of these viruses. Class III would include proteins unique to the MDV serotype 1 and may be involved in the oncogenic properties of this virus (Meq, vIL-8, pp38, pp24). Interestingly, some of these MDV unique proteins (vLIP and SORF2) have homologues among other avian DNA viruses, which may have resulted from ancient recombination events (Lee et al. 2000). A better understanding of the function of MDV unique proteins will reveal the molecular mechanisms involved in MDV pathogenesis.

Acknowledgments. We are grateful to Drs. Gerald Kutish and Daniel Rock for providing Md5 sequence information prior to publication.

References

Anderson AS, Francesconi A, Morgan RW (1992) Complete nucleotide sequence of the Marek's disease virus ICP4 gene. Virology 189(2):657–667

Anderson AS, Parcells MS, Morgan RW (1998) The glycoprotein D (US6) homolog is not essential for oncogenicity or horizontal transmission of Marek's disease virus. J Virol 72(3):2548–2553

Baek SH, Kwak JY, Lee SH, Lee T, Ryu SH, Uhlinger DJ, Lambeth JD (1997) Lipase activities of p37, the major envelope protein of vaccinia virus. J Biol Chem 272(51):32042–32049

Biggs PM, Milne BS (1972) In: Biggs PMM, deThé, Payne LN (eds) Oncogenesis and herpesviruses. IARC Scientific publication No. 2, International Agency for Research on Cancer, Lyon, pp 88–94

Blanco L, Bernad A, Blasco MA, Salas M (1991) A general structure for DNA-dependent DNA polymerases. Gene 100:27–38

Boussaha M (1996) Identification and molecular characterization of Marek's disease virus (MDV) homolog of the herpes simplex virus type 1 (HSV-1) VP16 gene. Ph.D. Dissertation. Michigan State University, East Lansing, MI, USA

Boussaha M, Sun W, Pitchyangkura R, Triezenberg S, Coussens P (1996) Marek's disease virus UL48 (VP16) contains multiple functional domains and transactivates both homologous and heterologous immediate early gene promoters. In: Silva RF, Cheng HH, Coussens PM, Lee LF, Velicer LF (eds) Current research on Marek's disease. Proc 5th International Symposium on Marek's disease, East Lansing, MI, pp 182–188

Bradley G, Hayashi M, Lancz G, Tanaka A, Nonoyama M (1989) Structure of the Marek's disease virus BamHI-H gene family: genes of putative importance for tumor induction. J Virol 63(6):2534–2542

Brunovskis P, Velicer LF (1992) Analysis of Marek's disease virus glycoproteins D, I and E. In: Proceedings 19th world's poultry congress, vol. 1. 4th International Symposium on Marek's disease. Posen and Looijen, Wageningen, The Netherlands

Brunovskis P, Velicer LF (1995) The Marek's disease virus (MDV) unique short region: alphaherpesvirus-homologous, fowlpox virus-homologous, and MDV-specific genes. Virology 206(1):324–338

Buckmaster AE, Scott SD, Sanderson MJ, Boursnell ME, Ross NL, Binns MM (1988) Gene sequence and mapping data from Marek's disease virus and herpesvirus of turkeys: implications for herpesvirus classification. J Gen Virol 69:2033–2042

Cameron JM, McDougall I, Marsden HS, Preston VG, Ryan DM, Subak-Sharpe JH (1988) Ribonucleotide reductase encoded by herpes simplex virus is a determinant of the pathogenicity of the virus in mice and a valid antiviral target. J Gen Virol 69:2607–2612

Chen X, Velicer LF (1992) Expression of the Marek's disease virus homolog of herpes simplex virus glycoprotein B in *Escherichia coli* and its identification as B antigen. J Virol 66(7):4390–4398

Chubb RC, Churchill AE (1968) Precipitating antibodies associated with Marek's disease. Vet Rec 82: 4–7

Churchill AE, Chubb RC, Baxendale W (1969) The attenuation, with loss of oncogenicity, of the herpes-type virus of Marek's disease (strain HPRS-16) on passage in cell culture. J Gen Virol 4(4):557–564

Cohen EA, Gaudreau P, Brazeau P, Langelier Y (1986) Specific inhibition of herpesvirus ribonucleotide reductase by a nonapeptide derived from the carboxy terminus of subunit 2. Nature 321(6068): 441–443

Cohen JI, Seidel K (1994) Varicella-zoster virus (VZV) open reading frame 10 protein, the homolog of the essential herpes simplex virus protein VP16, is dispensable for VZV replication in vitro. J Virol 68(12):7850–7858

Crute JJ, Lehman IR (1989) Herpes simplex-1 DNA polymerase. Identification of an intrinsic 5′-3′ exonuclease with ribonuclease H activity. J Biol Chem 264(32):19266–19270

Cui ZZ, Lee LF, Liu JL, Kung HJ (1991) Structural analysis and transcriptional mapping of the Marek's disease virus gene encoding pp38, an antigen associated with transformed cells. J Virol 65(12):6509–6515

Cui Z, Qin A, Lee LF, Wu P, Kung HJ (1999) Construction and characterization of a H19 epitope point mutant of MDV CVI988/rispens strain. Acta Virol 43(2–3):169–173

Dienglewicz RL, Parcells MS (1999) Establishment of a lymphoblastoid cell line using a mutant MDV containing a green fluorescent protein expression cassette. Acta Virol 43(2–3):106–112

Dingwell KS, Brunetti CR, Hendricks RL, Tang Q, Tang M, Rainbow AJ, Johnson DC (1994) Herpes simplex virus glycoproteins E and I facilitate cell-to-cell spread in vivo and across junctions of cultured cells. J Virol 68(2):834–845

Dutia BM, Frame MC, Subak-Sharpe JH, Clark WN, Marsden HS (1986) Specific inhibition of herpesvirus ribonucleotide reductase by synthetic peptides. Nature 321(6068):439–441

Eisenberg RJ, Cerini CP, Heilman CJ, Joseph AD, Dietzschold B, Golub E, Long D, Ponce DL, Cohen GH (1985) Synthetic glycoprotein D-related peptides protect mice against herpes simplex virus challenge. J Virol 56:1014–1017

Endoh D (1996) Enhancement of gene expression by Marek's disease virus homologue of the herpes simplex virus-1 ICP4. Jpn J Vet Res 44(2):136–137

Gentry GA (1985) Locating a nucleotide binding site in the thymidine kinase of vaccinia virus and of herpes simplex virus by scoring triply aligned protein sequences. Proc Nat Acad Sci USA 80:1565–1569

Haffey ML, Novotny J, Bruccoleri RE, Carroll RD, Stevens JT, Matthews JT (1990) Structure-function studies of the herpes simplex virus type 1 DNA polymerase. J Virol 64(10):5008–5018

Hanssens FP, Nauwynck HJ, Mettenlieter TC (1995) Role of glycoprotein gD in the adhesion of pseudorabies virus infected cells and subsequent cell-associated virus spread. Arch Virol 140(10):1855–1862

Heineman TC, Cohen JI (1994) Deletion of the varicella-zoster virus large subunit of ribonucleotide reductase impairs growth of virus in vitro. J Virol 68(5):3317–3323

Hong Y, Coussens PM (1994) Identification of an immediate-early gene in the Marek's disease virus long internal repeat region which encodes a unique 14-kilodalton polypeptide. J Virol 68(6):3593–3603

Hong Y, Frame M, Coussens PM (1995) A 14-kDa immediate-early phosphoprotein is specifically expressed in cells infected with oncogenic Marek's disease virus strains and their attenuated derivatives. Virology 206(1):695–700

Ikuta K, Ueda S, Kato S, Hirai K (1983) Monoclonal antibodies reactive with the surface and secreted glycoproteins of Marek's disease virus and herpesvirus of turkeys. J Gen Virol 64:2597–2610

Ikuta K, Ueda S, Kato S, Hirai K (1984) Identification with monoclonal antibodies of glycoproteins of Marek's disease virus and herpesvirus of turkeys related to virus neutralization. J Virol 49(3): 1014–1017

Isfort R, Jones D, Kost R, Witter R, Kung HJ (1992) Retrovirus insertion into herpesvirus in vitro and in vivo. Proc Natl Acad Sci USA 89(3):991–995

Isfort RJ, Qian Z, Jones D, Silva RF, Witter R, Kung HJ (1994) Integration of multiple chicken retroviruses into multiple chicken herpesviruses: herpesviral gD as a common target of integration. Virology 203(1):125–133

Isfort RJ, Sithole I, Kung H-J, Velicer LF (1986a) Molecular characterization of Marek's disease herpesvirus B antigen. J Virol 59(2):411–419

Isfort RJ, Stringer RA, Kung HJ, Velicer LF (1986b) Synthesis, processing, and secretion of the Marek's disease herpesvirus A antigen glycoprotein. J Virol 57(2):464–474

Johnson DC, Feenstra V (1987) Identification of a novel herpes simplex virus type 1-induced glycoprotein which complexes with gE and binds immunoglobulin. J Virol 61(7):2208–2216

Jones D, Lee L, Liu JL, Kung HJ, Tillotson JK (1992) Marek's disease virus encodes a basic-leucine zipper gene resembling the fos/jun oncogenes that is highly expressed in lymphoblastoid tumors. Proc Natl Acad Sci USA 89(9):4042–4046

Joyce CM, Steitz TA (1987) DNA polymerase I: from crystal structure to function via genetics. Trend Biochem Sci 12:288–292

Kaleta EF, Neumann U (1977) Investigations on the mode of transmission of the herpesvirus of turkeys in vitro. Avian Pathol 6:33–39

Kit S (1985) Thymidine kinase. Microbiol Sci 2:369–375

Kwoh TJ, Engler JA (1984) The nucleotide sequence of the chicken thymidine kinase gene and the relationship of its predicted polypeptide to that of the vaccinia virus thymidine kinase. Nucleic Acids Res 12(9):3959–3971

Leader DP, Katan M (1988) Viral aspects of protein phosphorylation. J Gen Virol 69:1441–1464

Lee LF, Liu X, Witter RL (1983) Monoclonal antibodies with specificity for three different serotypes of Marek's disease viruses in chickens. J Immunol 130(2):1003–1006

Lee LF, Nazerian K, Leinbach SS, Reno JM, Boezi JA (1976) Effect of phosphonoacetate on Marek's disease virus replication. J Natl Cancer Inst 56(4):823–827

Lee LF, Wu P, Sui D (1996) Identification and transcriptional analysis of Marek's disease virus gene encoding membrane glycoprotein gp82. In: Silva RF, Cheng HH, Coussens PM, Lee LF, Velicer LF (eds) Current research on Marek's disease. Proc. 5th International Symposium on Marek's disease, East Lansing, MI, pp 245–250

Lee LF, Wu P, Sui D, Ren D, Kamil J, Kung H-J, Witter RL (2000) The complete unique long sequence and the overall genomic organization of the GA strain of Marek's disease virus. Proc Nat Acad Sci USA (in press)

Lemaster S, Roizman B (1980) Herpes simplex virus phosphoproteins. II. Characterization of the virion protein kinase and of the polypeptides phosphorylated in the virion. J Virol 35(3):798–811

Li D, Green PF, Skinner MA, Jiang C, Ross N (1994) Use of recombinant pp38 antigen of Marek's disease virus to identify serotype 1-specific antibodies in chicken sera by western blotting. J Virol Methods 50(1–3):185–195

Liu H-C (1999) Identification of a host protein interacting with the Marek's Disease virus SORF2 protein. Ph.D. Dissertation. Michigan State University, East Lansing, MI, USA

Liu JL, Lee LF, Ye Y, Qian Z, Kung HJ (1997) Nucleolar and nuclear localization properties of a herpesvirus bZIP oncoprotein, MEQ. J Virol 71(4):3188–3196

Liu JL, Lin SF, Xia L, Brunovskis P, Li D, Davidson I, Lee LF, Kung HJ (1999) MEQ and V-IL8: cellular genes in disguise?. Acta Virol 43(2–3):94–101

Liu JL, Ye Y, Lee LF, Kung HJ (1998) Transforming potential of the herpesvirus oncoprotein MEQ: morphological transformation, serum-independent growth, and inhibition of apoptosis. J Virol 72(1):388–395

McColl SR, Roberge CJ, Larochelle B, Gosselin J (1997) EBV induces the production and release of IL-8 and macrophage inflammatory protein-1 alpha in human neutrophils. J Immunol 159(12):6164–6168

Morgan R, Anderson A, Kent J, Parcells M (1996a) Characterization of Marek's disease virus RB1B-based mutants having disrupted glycoprotein C or glycoprotein D homolog genes. In: Silva RF, Cheng HH, Coussens PM, Lee LF, Velicer LF (eds) Current research on Marek's disease. Proc. 5th International Symposium on Marek's disease, East Lansing, MI, pp 207–212

Morgan R, Xie Q, Cantello J (1996b) The Marek's disease virus ICP4 gene: update on sense and antisense RNAs and characterization of the gene product. In: Silva RF, Cheng HH, Coussens PM, Lee LF, Velicer LF (eds) Current research on Marek's disease. Proc. 5th International Symposium on Marek's disease, East Lansing, MI, pp 160–163

Moriuchi H, Moriuchi M, Straus SE, Cohen JI (1993) Varicella-zoster virus open reading frame 10 protein, the herpes simplex virus VP16 homolog, transactivates herpesvirus immediate-early gene promoters. J Virol 67(5):2739–2746

Nakajima K, Ikuta K, Naito M, Ueda S, Kato S, Hirai K (1987) Analysis of Marek's disease virus serotype 1-specific phosphorylated polypeptides in virus-infected cells and Marek's disease lymphoblastoid cells. J Gen Virol 68:1379–1389

Nazerian K, Lee LF (1976) Selective inhibition by phosphonoacetic acid of MDV DNA replication in a lymphoblastoid cell line. Virology 74(1):188–193

Nazerian K, Lee LF, Yanagida N, Ogawa R (1992) Protection against Marek's disease by a fowlpox virus recombinant expressing the glycoprotein B of Marek's disease virus. J Virol 66(3):1409–1413

Nazerian K, Witter RL, Lee LF, Yanagida N (1996) Protection and synergism by recombinant fowl pox vaccines expressing genes from Marek's disease virus. Avian Dis 40:368–376

Niikura M, Witter RL, Jang HK, Ono M, Mikami T, Silva RF (1999) MDV glycoprotein D is expressed in the feather follicle epithelium of infected chickens. Acta Virol 43(2–3):159–163

Nikas I, McLauchlan J, Davison AJ, Taylor WR, Clements JB (1986) Structural features of ribonucleotide reductase. Proteins 1(4):376–384

Ono M, Jang HK, Maeda K, Kawaguchi Y, Tohya Y, Niikura M, Mikami T (1996) Detection of Marek's disease virus serotype 1 (MDV1) glycoprotein D in MDV1-infected chick embryo fibroblasts. J Vet Med Sci 58(8):777–780

Otsuka H, Kit S (1984) Nucleotide sequence of the marmoset herpesvirus thymidine kinase gene and predicted amino acid sequence of thymidine kinase polypeptide. Virology 135(2):316–330

Parcells MS, Anderson AS, Cantello JL, Morgan RW (1994) Characterization of Marek's disease virus insertion and deletion mutants that lack US1 (ICP22 homolog), US10, and/or US2 and neighboring short-component open reading frames. J Virol 68(12):8239–8253

Parcells MS, Anderson AS, Morgan TW (1995) Retention of oncogenicity by a Marek's disease virus mutant lacking six unique short region genes. J Virol 69(12):7888–7898

Peng Q, Zeng M, Bhuiyan ZA, Ubukata E, Tanaka A, Nonoyama M, Shirazi Y (1995) Isolation and characterization of Marek's disease virus (MDV) cDNAs mapping to the BamHI-I2, BamHI-Q2, and BamHI-L fragments of the MDV genome from lymphoblastoid cells transformed and persistently infected with MDV. Virology 213(2):590–599

Pratt WD, Cantello J, Morgan RW, Schat KA (1994) Enhanced expression of the Marek's disease virus-specific phosphoproteins after stable transfection of MSB-1 cells with the Marek's disease virus homologue of ICP4. Virology 201(1):132–136

Purchase HG, Burmester BR, Cunningham CH (1971) Responses of cell cultures from various avian species to Marek's disease virus and herpesvirus of turkeys. Am J Vet Res 32:1821–1823

Qian Z, Brunovskis P, Rauscher F 3rd, Lee L, Kung HJ (1995) Transactivation activity of Meq, a Marek's disease herpesvirus bZIP protein persistently expressed in latently infected transformed T cells. J Virol 69(7):4037–4044

Reddy SM, Sui D, Wu P, Lee L (1999) Identification and structural analysis of a MDV gene encoding a protein kinase. Acta Virol 43(2–3):174–180

Reddy SM, Sui D, Wu P, Qin A, Lee L (2000) Identification and characterization of a Marek's disease virus gene encoding major capsid protein. 6th International Symposium on Marek's disease, Montreal, Canada

Ren D, Lee LF, Coussens PM (1994) Identification and characterization of Marek's disease virus genes homologous to ICP27 and glycoprotein K of herpes simplex virus-1. Virology 204(1):242–250

Ren D, Lee LF, Coussens PM (1996) Regulatory function of Marek's disease virus ICP27 gene product. In: Silva RF, Cheng HH, Coussens PM, Lee LF, Velicer LF (eds) Current research on Marek's disease. Proc. 5th International Symposium on Marek's disease, East Lansing, MI, pp 170–175

Roizman B (1978) The Herpesviruses. In: Nayak DP (ed) The molecular biology of animal viruses. New York & Basel, Marcel Dekker, Vol. 2, pp 769–848

Roizman B (1989) The herpesviruses. In: Nayak DP (ed) The molecular biology of animal viruses. New York and Basel, Marcel Dekker, Vol. 2. pp 769–848

Roizman B, Carmichael LE, Deinhardt F, de The G, Nahmias AJ, Plowright W, Rapp F, Sheldrick P, Takahashi M, Wolf K (1981) Herpesviridae: definition, provisional nomenclature, and taxonomy. Intervirol 16:201–217

Roizman B, Desrosiers RC, Fleckenstein B, Lopez C, Minson AC, Studdert MJ (1992) The Family Herpesviridae: an update. Arch Virol 123:425–449, pp 849–895

Roizman B, Sears AE (1991) Herpes simplex viruses and their replication. In: Fields BN, Knipe DM (eds) Fundamental virology (2nd ed). Raven Press, New York

Ross LJ (1985) Molecular biology of the virus. In: Payne LN (ed) Marek's disease. The Hague, Martinus Nijhoff, pp 113–150

Ross LJ, Binns MM, Pastorek J (1991) DNA sequence and organization of genes in a 5.5kbp EcoRI fragment mapping in the short unique segment of Marek's disease virus (strain RB1B). J Gen Virol 72:949–954

Ross LJ, Sanderson M, Scott SD, Binns MM, Doel T, Milne B (1989) Nucleotide sequence and characterization of the Marek's disease virus homologue of glycoprotein B of herpes simplex virus. J Gen Virol 70:1789–1804

Sakaguchi M, Urakawa T, Hirayama Y, Miki N, Yamamoto M, Zhu GS, Hirai K (1993) Marek's disease virus protein kinase gene identified within the short unique region of the viral genome is not essential for viral replication in cell culture and vaccine-induced immunity in chickens. Virology 195(1):140–148

Schat KA, Schinazi RF, Calnek BW (1984) Cell-specific antiviral activity of 1-(2-fluoro-2-deoxy-beta-D-arabinofuranosyl)-5-iodocytosine (FIAC) against Marek's disease herpesvirus and turkey herpesvirus. Antiviral Res 4(5):259–270

Scott SD, Ross NL, Binns MM (1989) Nucleotide and predicted amino acid sequences of the Marek's disease virus and turkey herpesvirus thymidine kinase genes; comparison with thymidine kinase genes of other herpesviruses. J Gen Virol 70:3055–3065

Scott SD, Smith GD, Ross NL, Binns MM (1993) Identification and sequence analysis of the homologues of herpes simplex virus type 1 glycoprotein H in Marek's disease virus and the herpesvirus of turkey. J Gen Virol 74:1185–1190

Silva RF, Lee LF (1984) Monoclonal antibody-mediated immunoprecipitation of proteins from cells infected with Marek's disease virus or turkey herpesvirus. Virology 136(2):307–320

Sithole I, Lee LF, Velicer LF (1988) Synthesis and processing of the Marek's disease herpesvirus B antigen glycoprotein complex. J Virol 62(11):4270–4279

Sui D, Wu P, Kung H-J, Lee LF (1995) Identification and characterization of a Marek's disease virus gene encoding DNA polymerase. Virus Res 36(2–3):269–278

Tan X, Velicer LF (1996) Marek's disease virus gD expression is downregulated at the transcription level in cell culture. In: Silva RF, Cheng HH, Coussens PM, Lee LF, Velicer LF (eds) Current research on Marek's disease. Proc. 5th International Symposium on Marek's disease, East Lansing, MI, pp 213–218

Tulman ER, Afonso CL, Lu Z, Zsak L, Rock DL, Kutish GF (2000) The genome of a very virulent Marek's disease virus. J Virol 74(17):7980–7988

Van Zaane D, Brinkhof JM, Westenbrink F, Gielkens AL (1982) Molecular-biological characterization of Marek's disease virus. I. Identification of virus-specific polypeptides in infected cells. Virology 121(1):116–132

Weinheimer SP, Boyd BA, Durham SK, Resnick JL, O'Boyle DR II (1992) Deletion of the VP16 open reading frame of herpes simplex virus type 1. J Virol 66(1):258–269

Whealy ME, Card JP, Robbins AK, Dubin JR, Rziha HJ, Enquist LW (1993) Specific pseudorabies virus infection of the rat visual system requires both gI and gp63 glycoproteins. J Virol 67(7):3786–3797

Whitbeck JC, Knapp AC, Enquist LW, Lawrence WC, Bello LJ (1996) Synthesis, processing, and oligomerization of bovine herpesvirus 1 gE and gI membrane proteins. J Virol 70(11):7878–7884

Wierenga RK, Hol WG (1983) Predicted nucleotide-binding properties of p 21 protein and its cancer-associated variant. Nature 302(5911):842–844

Wilson MR, Coussens PM (1991) Purification and characterization of infectious Marek's disease virus genomes using pulsed field electrophoresis. Virology 185:673–680

Wilson MR, Southwick RA, Pulaski JT, Tieber VL, Hong Y, Coussens PM (1994) Molecular analysis of the glycoprotein C-negative phenotype of attenuated Marek's disease virus. Virology 199(2):393–402

Whittaker GR, Bonass WA, Elton DM, Halliburton IW, Killington RA, Meredith DM (1992) Glycoprotein 300 is encoded by gene 28 of equine herpesvirus type 1: a new family of herpesvirus membrane proteins?. J Gen Virol 73:2933–2940

Whittaker GR, Riggio MP, Halliburton IW, Killington RA, Allen GP, Meredith DM (1991) Antigenic and protein sequence homology between VP13/14, a herpes simplex virus type 1 tegument protein, and gp10, a glycoprotein of equine herpesvirus 1 and 4. J Virol 65(5):2320–2326

Wu P, Lee LF, Reed WM (1997) Serological characteristics of a membrane glycoprotein gp82 of Marek's disease virus. Avian Dis 41(4):824–831

Wu P (1998) Functional analysis of glycoproteins H and L complex of Marek's disease virus. Ph.D. Dissertation. Michigan State University, East Lansing, MI, USA

Wu P, Reed WM, Yoshida S, Sui D, Lee LF (1999) Identification and characterization of glycoprotein H of MDV-1 GA strain. Acta Virol 43(2–3):152–158

Wu TF, Sun W, Boussaha M, Southwick R, Coussens PM (1996) Cloning and sequence analysis of a Marek's disease virus origin binding protein (OBP) reveals strict conservation of structural motifs among OBPs of divergent alphaherpesviruses. Virus Genes 13(2):143–157

Xie Q, Anderson AS, Morgan RW (1996) Marek's disease virus (MDV) ICP4, pp38, and meq genes are involved in the maintenance of transformation of MDCC-MSB-1 MDV-transformed lymphoblastoid cells. J Virol 70(2):1125–1131

Xing Z, Xie Q, Morgan RW, Schat KA (1999) A monoclonal antibody to ICP4 of MDV recognizing ICP4 of serotype 1 and 3 MDV strains. Acta Virol 43(2–3):113–120

Yanagida N, Yoshida S, Nazerian K, Lee LF (1993) Nucleotide and predicted amino acid sequences of Marek's disease virus homologues of herpes simplex virus major tegument proteins. J Gen Virol 74:1837–1845

Yao Z, Jackson W, Forghani B, Grose C (1993) Varicella-zoster virus glycoprotein gpI/gpIV receptor: expression, complex formation, and antigenicity within the vaccinia virus-T7 RNA polymerase transfection system. J Virol 7(1):305–314

Yoshida S, Lee LF, Yanagida N, Nazerian K (1994a) Identification and characterization of a Marek's disease virus gene homologous to glycoprotein L of herpes simplex virus. Virology 204(1):414–419

Yoshida S, Lee LF, Yanagida N, Nazerian K (1994b) Mutational analysis of the proteolytic cleavage site of glycoprotein B (gB) of Marek's disease virus. Gene 150(2):303–306

Zelník V, Kopácek J, Rejholcavá O, Kabát P, Pastorek J (1996) ICP4 homologues of Marek's disease virus and herpesvirus of turkeys are larger than their alphaherpesvirus counterparts. In: Silva RF, Cheng HH, Coussens PM, Lee LF, Velicer LF (eds) Current research on Marek's disease. Proc. 5th International Symposium on Marek's disease, East Lansing, MI, pp 164–169

Zhu GS, Iwata A, Gong M, Ueda S, Hirai K (1994) Marek's disease virus type 1-specific phosphorylated proteins pp38 and pp24 with common amino acid termini are encoded from the opposite junction regions between the long unique and inverted repeat sequences of viral genome. Virology 200(2): 816–820

Zuckermann FA, Mettenleiter TC, Schreurs C, Sugg N, Ben-Porat T (1988) Complex between glycoproteins gI and gp63 of pseudorabies virus: its effect on virus replication. J Virol 62(12):4622–4626

A Complete Genomic DNA Sequence of Marek's Disease Virus Type 2, Strain HPRS24

Y. Izumiya[1], H.-K. Jang[2], M. Ono[3], and T. Mikami[4]

[1] Department of Veterinary Microbiology, Faculty of Agriculture, The University of Tokyo, 1-1-1 Yayoi, Bunkyo-ku, Tokyo 113-8657, Japan

[2] Department of Microbiology and Immunology, Louisiana State University Health Sciences Center, 1501 Kings' Highway Shreveport, LA 71130-3932, USA

[3] Biomedical Research Laboratories, Sankyo Co. Ltd., 1-2-58 Hiromachi, Shinagawa-ku, Tokyo 140-0005, Japan

[4] The National Research Center for Protozoan Disease, Obihiro University of Agriculture and Veterinary Medicine, Nishi 2 Inada-cho, Obihiro 080-8555, Japan

1 Introduction

Marek's disease (MD) is a contagious lymphoproliferative disorder of chickens, and has been a major cause of poultry mortality in many countries since the 1960s. The causative agent of MD, a highly cell-associated herpesvirus called Marek's disease virus (MDV), was isolated (Bankowski et al. 1969; Churchill and Biggs 1967, 1968; Nazerian et al. 1968; Solomon et al. 1968; Witter et al. 1969) and live vaccines were developed by serial passages in cultured cells (Churchill et al. 1969a,b) or by use of apathogenic herpesvirus isolated from turkeys (Kawamura et al. 1969; Okazaki et al. 1970; Witter et al. 1970). At present, vaccines derived from all three serotypes offer different levels of protection against the disease, either alone or in bivalent and trivalent combinations (Calnek and Witter 1997).

MDV was at first biologically classified as a gamma-herpesvirus based on its lymphotropism and other biological natures (Roizman et al. 1981), but recent molecular biological aspects of the MDV genome are not consistent with this classification. MDV can induce lymphomas similar to some gamma-herpesviruses, but any other alpha-herpesviruses cannot induce tumors in natural hosts. However, molecular biological properties of MDV are more closely related to alpha-herpesviruses than to gamma-herpesviruses. Therefore, we cannot begin to elucidate the oncogenicity by comparative studies with oncogenic gamma-herpesviruses such as Epstein-Barr virus. One effective approach to this subject is a comparative study with other MDV serotypes, because the serotyping is also connected with oncogenesis. Based on this idea, we present a summary of the complete DNA genome sequence of nononcogenic MDV serotype 2 (MDV2), and compare the genetic contents with those of oncogenic MDV and with other members of the *Alphaherpesvirinae*.

The MDV2 genome is 164,270bp in size and has a base composition of 53.6% G+C. The MDV2 genome contains 68 open reading frames (ORFs) in the unique long (UL) region, 12 ORFs in the unique short (US) region, 9 ORFs in each of the internal and terminal repeat long (IRL and TRL) regions, and 2 ORFs in each of the internal and terminal repeat short (IRS and TRS) regions. Most MDV2 genes identified exhibit sequence homology with counterparts in herpes simplex virus type 1 (HSV-1) (McGeoch et al. 1988). Therefore, to avoid confusion, the nomenclature of the MDV2 genes are followed after their HSV-1 homologues. Putative MDV2 specific genes are named basically on their predicted amino acid numbers. According to the orientation of the MDV2 genome map, the arrangement of these ORFs sequentially numbered 1–102 is shown in Fig. 1.

2 Classification and Serotyping

MDV was divided into three serotypes on the basis of data obtained by indirect-immunofluorescence, agar-gel precipitin, and neutralization tests (von Bulow and

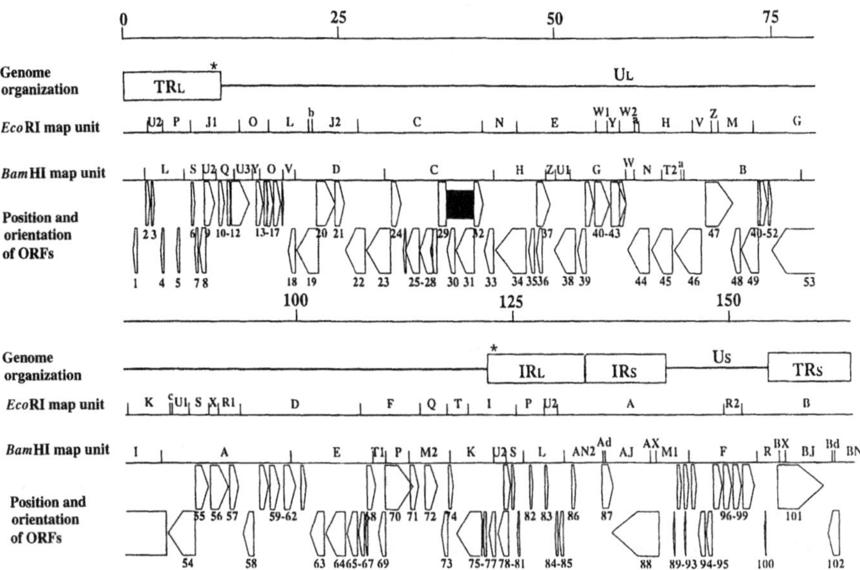

Fig. 1. Predicted gene organization and final restriction maps of the MDV2 genome. Restriction maps of the investigated *Bam*HI and *Eco*RI fragments are shown on successive lines. Individual *Bam*HI and *Eco*RI fragments are named alphabetically according to their sizes (Ono et al. 1992). The unique long (*UL*) and short (*US*) regions of the MDV2 genome are *boldly lined* and the inverted repeat regions are *boxed* and designated, TRL, IRL, IRS, and TRS. Location and orientations of the identified MDV2 ORFs are depicted by *open arrows* and sequentially numbered from *left to right* with respect to the orientation of the MDV2 genome map. ORFs 11–12, 29–32, and 75–76 indicated by *filled boxes*, are expressed as a spliced transcript, respectively. *Stars* on TRL and IRL regions indicate the putative MDV2 origin sequences of DNA replication. Exact positions and putative properties of these MDV2 ORFs are listed in Table 2. The scale is in kilobase (kb)

Biggs 1975a,b; von Bulow et al. 1975). Later, the presence of the three serotypes was confirmed by serotype-specific monoclonal antibodies (Lee et al. 1983), specific lectines (Malkinson et al. 1986), migration behavior of serotype-specific polypeptides by two-dimensional gel electrophoreses (Van Zaane et al. 1982), and restriction enzyme analysis (Hirai et al. 1986; Ross et al. 1983). In addition, MDV serotype 1 (MDV1) has been further subdivided into several groups based on its variable pathogenicity or oncogenicity in chickens (Table 1). All virulent strains are included in MDV1, designated Gallid herpesvirus 2. MDV2, designated Gallid herpesvirus 3, consists of nononcogenic strains isolated from healthy birds (Lin

Table 1. Some commonly studied isolates

Virus type	Serotype	Oncogenicity	Representative isolates
MDV1	1	Very high	RB1B, Md5, Md11
		High	GA, JM, HPRS16
		Moderate	HPRS17, Conn-A
		Low	Cu-2, CV1988
MDV2	2	None	HPRS24, SB-1, HN-1
HVT	3	None	FC126, HPRS26, WTHV-1

et al. 1990a,b; Schat and Calnek 1978). MDV serotype 3, which is known as the herpesvirus of turkeys (HVT) and designated Meleagrid herpesvirus 1, consists of apathogenic strains.

3 Genomic Structure of MDV

Herpesviruses are grouped into six types based on their genomic structures (Roizman and Baines 1991). Genomes of group A herpesviruses have a single set of repeated sequences at the termini, which are in the same orientation. Both genomes of B and C herpesviruses have numerous repeats of the same set of sequences at both termini in the same orientation; besides, those of group C herpesviruses have a variable number of tandem repeats of a different sequence internally. Genomes of group D herpesviruses have two segments, and the short sequence is composed of a unique region inserted between a single set of inverted repeats. Group E genomes, exemplified by HSV-1, are composed of linked long (L) and short (S) components. The L and S components are composed of UL and US sequences, which are bracketed by inverted repeats (Fig. 1). Genomes of F herpesviruses have neither tandem nor inverted repeats. Alpha-herpesviruses are classified as group D or E herpesviruses.

It was at first demonstrated by electron microscopy that genomic structures of MDV1 and HVT were similar to those of group E herpesvirus (Cebrian et al. 1982). Later, these electron microscopic observations of the MDV1 and HVT genomes were supported by restriction enzyme analyses, which also established a collinear relation between the MDV1 and HVT genomes (Fukuchi et al. 1984; Igarashi et al. 1987). Sequencing of random fragments of the MDV1 and HVT genomes and mapping data of these fragments revealed that the MDV1 and HVT genomes are similar to alpha-herpesviruses in gene arrangement and DNA homology (Buckmaster et al. 1988). These findings suggested the need to reclassify MDV from *Gammaherpesviridae* to *Alphaherpesviridae* (Roizman and Baines 1991).

The MDV2 genome was considered to differ from those of MDV1 and HVT because restriction enzyme digestion patterns of the MDV2 genome differ entirely from those of MDV1 and HVT (Hirai et al. 1981; Ross et al. 1983; Silva and Barnett 1991).

Later, Ono et al. (1992) developed a genomic library of restriction enzyme fragments of MDV2 and demonstrated a collinear relationship among the DNAs of three MDV serotypes in both the UL and US regions. Furthermore, two pairs of these fragments in this library hybridized each other, suggesting that MDV2 DNA has two pairs of repeats and that its structure is similar to those of group E herpesviruses. These libraries have also contributed to sequence analysis of the MDV2 genome.

4 MDV2 Genomic Sequence Analyses

The physical maps of restriction enzyme fragments (*Bam*HI and *Eco*RI) of MDV2 DNA were constructed and reported in our previous study (Ono et al. 1992). To complete the DNA sequence, we newly cloned *Eco*RI-A and -B fragments into cosmid vector using the charomid cloning kit (Nippon Gene). For sequencing, a set of overlapping subclones was prepared by exonuclease III digestion using a commercially available kit (Strategene, La Jolla, Calif., USA). DNA sequencing was performed with double-stranded plasmid templates, and their sequences were determined on both strands as described previously (Jang et al. 1998). The sequences at the ends of the genomes that were not generated in these libraries were obtained from the *Eco*RI junction fragment B (Fig. 1). From our previous *Bam*HI and *Eco*RI libraries, we could not confirm whether the *Eco*RI-P and U2 fragments are located in the IRL or TRL region. Therefore, we sequenced three clones, independently isolated, and confirmed their locations. The final map revealing the prototypical orientation of the viral genome generated with the *Bam*HI and *Eco*RI fragments is shown in Fig. 1. As the precise ends of the molecule are presently not known, we have chosen to number the sequence from the start of the putative direct repeat (DR1) that was found in the MDV1 genome by Kishi et al. (1991). By analogy with the *a*-sequence of other herpesviruses, this is the closest feature to the end of the genome (Mocarski and Roizman 1982; Tamashiro et al. 1984). Homology searches from the GenBank database were performed using the UWGCG sequence analysis program packages BESTFIT and PILEUP. Pairwise percentage identities were calculated by dividing the number of identical residues by the total number of residues in the aligned region of homology including gaps as described previously (Jang et al. 1998).

5 Unique Short Region of MDV2

The MDV2 US region comprises 12,109bp with an overall G+C content of 49.1%, which is somewhat greater than 41% for the MDV1 US region (Brunovskis and Velicer 1995) and 43.5% for the HVT US region (Zelnik et al. 1993), and contains 12 potential ORFs likely to encode for proteins (Fig. 2). Seven of them exhibited homologies to HSV-1 US1 (ICP22), US2, US3 (protein kinase), US6 (gD), US7 (gI), US8 (gE), and US10 genes (Table 2). These ORFs are conserved in a similar arrangement with those of HSV-1, except for US10, which is transposed in the US regions in all three serotypes. Compared to other MDV serotypes, the MDV2 US-encoded proteins showed 46–70% and 33–59% identities with their MDV1 and HVT equivalents at the amino acid level, respectively (Table 2) (Jang et al. 1998).

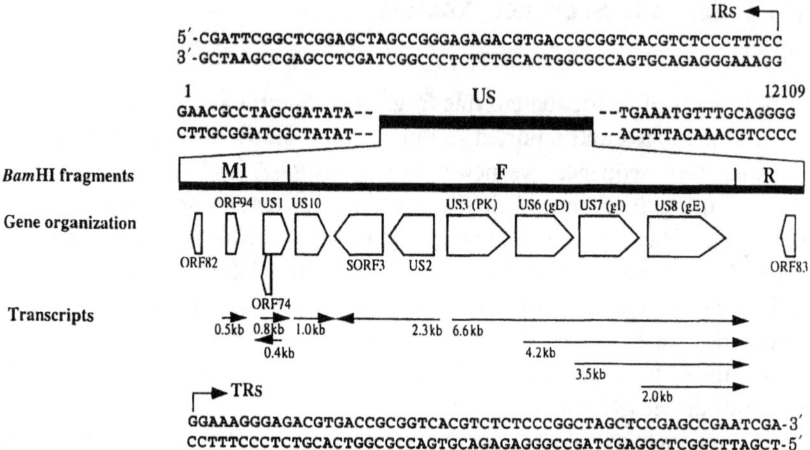

Fig. 2. Location of the IRS/US and US/TRS junctions in the MDV2 genome and organization of the MDV2 US genes. The IRS and TRS inverted repeat sequences are in *bold type*, and extremities of the IRS and TRS repeats are indicated by *arrows*. The entire US region is indicated by the *bold line* along with the *Bam*HI fragments M1, F and R. The approximate length and orientation of the identified genes are indicated by *open arrows*. The locations and 5′ to 3′ orientations of viral transcripts are indicated by *black arrows*

5.1 Junctions Between US and Repeat Sequences

Ono et al. (1992) showed that two *Bam*HI fragments, M1 and R, positioned at junctions between the US and inverted repeat regions in the MDV2 genome have partial homology with each other. This result was practically confirmed by the sequence analyses of these fragments (Fig. 2) (Jang et al. 1998). From these results, MDV2 IRS/US and US/TRS junctions are localized 153bp upstream of the ORF82 stop codon and 138bp downstream of the ORF83 start codon, respectively.

5.2 MDV2 Genes in US Region

5.2.1 Characterization of US Gene-Encoded Glycoproteins

The MDV2 US region contains three genes with the potential to encode glycoproteins (Tables 2 and 3). The MDV2 gD, gI, and gE genes contain ORFs capable of encoding 385, 355, and 488 amino acids, respectively. These proteins possess N-and C-terminal hydrophobic domains consistent with signal and anchor regions, and also contain conserved two, three, and four potential N-linked glycosylation sites in putative extracellular domains, respectively (Jang et al. 1996a,b, 1997). The predicted molecular weights of the corresponding gene products range from 38.4 to 54.3kDa, and MDV2 gI and gE are confirmed using recombinant proteins to be expressed in infected cells in chickens and conserved common antigenic epitope(s) beyond serotypes (Jang et al. 1996b, 1997). Analysis of MDV2 gD has lagged behind these proteins, while the MDV1 gD protein is well characterized and known to be expressed in infected cells in vitro and in the feather follicle epithelium of infected chickens (ONO et al. 1995, 1996; NIIKURA et al. 1999). The protein is also known to be nonessential for virus replication, horizontal transmission, and oncogenesis (ANDERSON et al. 1998).

5.2.2 Characterization of US Gene-Encoded Other Virion Proteins

The location and properties of the identified US genes are summarized in Table 2. The deduced products of MDV2 ORF82, 94, 74, and 83 exhibited no similarity with any known herpesviral proteins, but other gene products revealed significant similarities with US products of other herpesviral proteins, such as US1, US2, US3, US10, SORF3, gD, gI, and gE. The MDV2 US1 gene is capable of encoding a 173-amino-acid protein. The product revealed significant homologies with the US1 gene products of other alpha-herpesviruses, but these homologies are mainly restricted to the central portion of the protein. The product of the US10 gene is predicted to consist of 231 amino acids with a zinc-finger motif in the C-terminal part (residues 144–159) similar to the product of the infectious laryngotracheitis virus (ILTV) US10 gene (WILD et al. 1996). The SORF3 protein consists of 322 amino acids and exhibits homology to the putative MDV-specific SORF3 proteins of MDV1 and HVT. Multiple alignments among the MDVs indicated that the most conserved region of the SORF3 products is located in the N-terminal half of the proteins (JANG et al. 1998). Surprisingly, the corresponding gene in ILTV was recently found as two copies located in the repeat region (WILD et al. 1996). Thus, it is thought to be specific to avian herpesviruses. However, further work is required to determine the properties and functional features of this protein. The overall homology of the deduced 271 amino acids of the MDV2 US2 protein to its homologues in other alpha-herpesviruses is rather low. However, it is noteworthy that the N-terminal region of US2 is strongly hydrophobic and well conserved among MDVs and other alpha-herpesviruses. The predicted US3 product comprises 391 amino acids and has a consensus pattern of a serine/threonine protein kinase motif. Multiple alignments showed that the serine/threonine motif residues are highly conserved among MDVs. Later, ZELNIK et al. (1995) reported that US1, US10, and SORF3 genes of HVT were nonessential for virus replication.

5.3 Transcription Map of US Region

Transcriptional mapping of the identified genes in the MDV2 US region was performed by Northern blot analyses. As shown in Fig. 2, except for the ORF82 and ORF83 genes, all others identified in the US region were confirmed to be transcribed either as gene-specific and/or 3′-coterminal transcripts in the virus-infected cells (JANG et al. 1998).

Table 2. Features of MDV2 gene

Position of ORF			Protein		Identity to alpha-herpesvirus homologues[e]					MDV2 proposed nomenclature and properties of HSV-1 counterparts
No.[a]	Start[b]	Stop[c]	Codons	Mol. mass[d]	MDV1	EHV-1	BHV-1	VZV	HSV-1	
1 (r)*	1342	812	176	19.2	–[f]	–	–	–	–	Unknown
2*	2833	3378	181	19.9	–	–	–	–	–	Unknown
3*	3939	4289	116	12.7	–	–	–	–	–	Unknown
4 (r)*	5273	4806	155	16.8	–	–	–	–	–	Unknown
5 (r)*	6763	6455	102	11.6	–	–	–	–	–	Unknown
6*	8359	8751	130	14.5	–	–	–	–	–	Unknown
7 (r)*	8942	8547	131	14.8	–	–	–	–	–	Unknown
8*	9924	10907	327	36.6	–	–	–	–	–	Unknown
9 (r)*	10016	9225	263	28.5	–	–	–	–	–	Unknown
10*	11621	12148	175	18.8	47.1	–	–	–	–	pp24
11*	12385	12480								Probable membrane glycoprotein
12*	12557	14740	759	83.0	52.4	–	–	–	–	ORF759
13*	15606	16697	437	48.8	NS	–	–	–	–	ORF437 (unknown)
14	16799	17371	190	21.0	60.7	30.1	29.6	32.1	28.2	UL1 (glycoprotein L)
15	17238	18209	323	35.7	63.2	53.9	56.6	53.3	52.2	UL2 (Urasil-DNA glycosylase
16	18284	18931	215	24.1	71.8	58.5	53.8	53.3	54.4	UL3
17	18949	19218	89	9.9	NS	36.1	35.8	25.0	–	UL3.5
18 (r)	20124	19312	270	28.9	NS	41.1	44.5	34.8	32.8	UL4
19 (r)	22738	20162	858	97.0	NS	61.4	58.1	56.6	57.5	UL5 (DNA helicase/primase complex)
20	22662	25028	788	88.6	NS	46.2	47.7	48.7	44.6	UL6 (Possible virion protein)
21	24826	25732	301	34.3	NS	35.3	34.7	35.6	31.6	UL7
22 (r)	28080	25786	764	84.8	NS	34.0	33.6	36.8	28.1	UL8 (DNA helicase/primase complex)
23 (r)	30730	28103	875	98.1	71.9	49.6	48.8	46.0	49.0	UL9 (Ori-binding protein)
24	30824	32098	424	46.9	NS	33.2	30.5	33.2	27.7	UL10 (glycoprotein M)
25 (r)	32406	32158	81	9.1	NS	40.5	40.6	41.6	40.9	UL11 (Myristylated virion protein)
26 (r)	33962	32385	525	59.3	NS	37.4	40.1	40.1	38.9	UL12 (Deoxyribonuclease)
27 (r)	35455	33953	500	56.5	NS	43.9	37.9	37.8	38.4	UL13 (Virion protein kinase)
28 (r)	35895	35227	222	25.3	NS	31.4	34.2	34.9	33.0	UL14
30 (r)	38084	37029	351	39.6	NS	32.7	33.7	30.6	31.4	UL16
31 (r)	40310	38142	722	78.9	NS	35.0	34.1	33.0	31.4	UL17

29	35940	36983	748	82.5	NS	63.5	60.0	58.0	60.4	UL15 (Possible DNA packaging protein)
32	40431	41633	321	34.5	NS	47.5	47.6	49.6	43.9	UL18 (Capsid protein)
33 (r)	42712	41747	1392	153.9	NS	57.3	56.4	54.9	52.0	UL19 (Major capsid protein)
34 (r)	47019	42841	234	26.6	NS	35.4	26.0	28.0	25.0	UL20 (Multiply hydrophobic protein)
35 (r)	48016	47312	273	29.6	NS	–	–	–	–	ORF273 (unknown)
36 (r)*	49143	48322	532	58.8	NS	35.2	36.4	29.3	38.2	UL21
37	48125	49723	838	92.4	57.5	24.0	25.5	26.1	23.5	UL22 (glycoprotein H)
38 (r)	52407	49891	352	39.8	73.9	30.5	28.1	27.8	29.1	UL23 (Thymidine kinase)
39 (r)	53578	52520	316	35.8	NS	39.8	42.8	35.6	41.7	UL24
40	53497	54447	594	65.0	NS	58.3	58.1	42.7	53.3	UL25 (Virion protein)
41	54571	56355	639	69.3	NS	64.0	65.4	56.9	60.8	UL26 (Protease)
42	56373	58292	333	35.9	NS	48.1	57.1	16.4	31.3	UL26.5 (Capsid assembly protein)
43	57291	58292	865	97.2	82.3	50.9	50.8	52.4	50.0	UL27 (glycoprotein B)
44 (r)	61013	58416	811	91.5	NS	47.7	45.6	45.2	48.0	UL28 (Probable virion protein)
45 (r)	63449	61014	1190	130.3	NS	59.4	47.8	46.8	56.3	UL29 (ssDNA binding protein)
46 (r)	67347	63775	1190	133.5	76.5	59.8	57.7	54.7	54.4	UL30 (DNA polymerase)
47	67597	71169	304	34.2	NS	60.0	60.0	59.5	51.6	UL31
48 (r)	72007	71093	626	69.3	NS	45.7	42.7	46.5	51.5	UL32 (Probable virion protein)
49 (r)	73904	72024	131	14.7	NS	48.3	51.4	40.8	45.2	UL33
50	73903	74298	279	30.7	NS	52.7	48.6	54.3	50.0	UL34 (Possible virion protein)
51	74366	75205	127	14.0	NS	47.2	38.7	35.6	35.6	UL35 (Capsid protein)
52	75291	75674	3064	333.1	NS	33.9	34.0	30.3	30.5	UL36 (Large tegument protein)
53 (r)	84936	75742	1040	115.0	NS	29.9	30.1	29.3	28.3	UL37 (Tegument protein)
54 (r)	88246	85124	470	51.8	NS	46.1	43.3	39.9	39.9	UL38 (Capsid protein)
55	88616	90028	797	88.6	NS	52.3	54.0	53.6	50.6	UL39 (Ribonucleotide reductase)
56	90207	92600	353	39.8	NS	62.7	62.7	59.3	61.3	UL40 (Ribonucleotide reductase)
57	92627	93688	423	47.9	NS	48.1	42.6	36.3	41.5	UL41 (Host shut-off virion protein)
58 (r)	95021	93750	369	40.0	NS	31.1	30.7	27.0	29.8	UL42 (DNA pol processivity factor)
59	95685	96794	414	44.0	NS	18.4	23.1	41.7	27.5	UL43 (Multiply hydrophobic protein)
60	96901	98145	478	53.1	76.6	30.7	32.9	27.8	27.8	UL44 (glycoprotein C)
61	98355	99791	210	23.0	71.0	24.4	–	–	17.6	UL45 (Virion protein)
62	100415	101047	581	64.2	49.6	31.0	33.0	30.4	27.7	UL46 (Tegument protein)
63 (r)	102912	101167	810	89.9	54.0	28.1	28.2	26.4	29.8	UL47 (Tegument protein)
64 (r)	105524	103092								

Table 2. (*Contd.*)

Position of ORF			Protein		Identity to alpha-herpesvirus homologues[e]					MDV2 proposed nomenclature and properties of HSV-1 counterparts
No.[a]	Start[b]	Stop[c]	Codons	Mol. mass[d]	MDV1	EHV-1	BHV-1	VZV	HSV-1	
65 (r)	107051	105777	424	47.2	62.7	42.3	42.3	45.4	36.9	UL48 (Tegument protein: VP16)
66 (r)	107824	107099	241	26.9	54.2	42.5	42.0	49.0	39.5	UL49 (Tegument protein)
67 (r)	108249	107962	95	10.0	70.5	37.3	32.5	46.2	41.7	UL49.5 (Possible membrane protein)
68	108274	109413	395	43.8	NS	35.5	42.2	36.5	31.8	UL50 (duTPase)
69 (r)	110348	109608	248	26.4	NS	45.5	45.2	45.5	40.3	UL51 (Virion protein)
70	110368	113583	1071	118.6	NS	49.6	45.6	48.7	50.6	UL52 (DNA helicase/primase complex)
71	113562	114626	354	39.2	64.7	31.9	29.0	29.4	26.9	UL53 (glycoprotein K)
72	114791	116221	476	54.2	54.8	36.6	33.6	34.3	37.5	UL54 Transactivator (ICP27)
73 (r)*	117133	116465	222	25.2	NS	20.9	–	–	–	ORF-I (unknown)
74	117310	117813	167	18.2	NS	36.9	–	29.0	36.3	UL55
75 (r)*	120710	118062	873	109.0	NS	–	–	–	–	ORF873 (Probable glycoprotein)
76 (r)*	121088	120834	84	9.5	44.0	–	–	–	–	ORF873s
77 (r)*	121962	121297	221	23.3	45.6	–	–	–	–	pp38
78*	123567	124358	263	28.5	–[f]	–	–	–	–	Unknown
79 (r)*	123659	122676	327	36.6	–	–	–	–	–	Unknown
80*	124641	125036	131	14.8	–	–	–	–	–	Unknown
81 (r)*	125224	124832	130	14.5	–[f]	–	–	–	–	Unknown
82*	126820	127128	102	11.6	–	–	–	–	–	Unknown
83 (r)*	128310	128777	155	16.8	–	–	–	–	–	Unknown
84 (r)*	129644	129294	116	12.7	–	–	–	–	–	Unknown
85 (r)*	130750	130205	181	19.9	–	–	–	–	–	Unknown
86*	132241	132771	176	19.2	–	–	–	–	–	Unknown
87*	134225	135304	359	39.5	–	–	–	–	–	Unknown
88 (r)	141456	135355	2033	217.1	47.3	32.6	29.6	36.5	28.8	Major IE transactivatior (ICP4)
89 (r)*	143081	142833	82	9.1	–	–	–	–	–	Unknown
90*	143523	143807	94	10.0	–	–	–	–	–	Unknown
91 (r)*	144475	144251	74	8.7	–	–	–	–	–	Unknown
92	144277	144798	173	19.4	46.0	31.0	–	25.0	20.0	US1 (Transactivator: ICP22)
93	145023	145718	231	25.5	58.0	24.0	–	20.0	29.0	US10 (Virion protein)
94*	146835	145870	322	37.3	57.0	–	–	–	–	SORF3 (unknown)
95 (r)	147842	147027	271	30.1	70.0	33.0	26.0	–	36.0	US2
96 (r)	147991	149166	391	43.7	59.0	34.0	32.0	31.0	36.0	US3 (Protein kinase)
97	149491	150648	385	42.8	55.0	23.0	23.0	–	24.0	US6 (glycoprotein D)
98	150721	151788	355	38.4	49.0	21.0	24.0	22.0	20.0	US7 (glycoprotein 1)

99	151915	153381	488	54.3	47.0	21.0	22.0	21.0	19.0	US8 (glycoprotein E)
100 (r)*	154651	154400	83	9.1	–	–	–	–	–	Unknown
101	156015	162116	2033	217.1	47.3	32.6	29.6	36.5	28.8	Major IE transactivator (ICP4)
102 (r)*	163236	162157	359	39.5	–	–	–	–	–	Unknown

* Different genetic contents among other alpha-herpesviruses are indicated by asterisks;
NS, no sequence data were available from the GenBank database;
[a] ORFs located on the reverse DNA strand are indicated by a parenthetical *r*;
[b] Location of the first base or its complement in first ATG, except for ORFs 12 and 32, whose first base is given in exon;
[c] Location of the third base or its complement in stop codon, except for ORFs 11 and 29, whose third last base is given in exon;
[d] The molecular masses of primary translation products were calculated from predicted amino acid sequences;
[e] Homologies were determined by pairwise comparison with the UWGCG program BESTFIT and expressed as a percentage;
[f] The absence of positional counterparts is indicated by dashes.

Table 3. Identity of the predicted amino acid glycoproteins among the three MDV serotypes[a]

Cording region	Genes	Identities with corresponding protein		
		MDV2 and MDVI	MDV2 and HVT	MDVI and HVT
US	gD (US6)	55	41	46
	gl (US7)	49	36	40
	gE (US8)	47	39	44
UL	ORF759	52	NS	NS
	gL (ULI)	61	NS	NS
	gM (ULIO)	NS	NS	NS
	gH (UL22)	57	50	56
	gB (UL27)	83	76	82
	gC (UL44)	73	65	71
	gK (UL53)	65	NS	NS
	ORF873	NS	39	NS

NS, no sequence data available;
[a] Values were obtained using the UWGCG program BESTFIT, and are expressed as percentage identity. References: gB (Kato et al. 1999b), gC (Kitazawa et al. 1993), gD (Jang et al. 1996a), gE (Jang et al. 1997), gH (Shimojima et al. 1997), gI (Jang et al. 1996b), gK (Tsushima et al. 1999), gM (Cai et al. 1999), ORF873 (Tsushima et al. 1999).

6 Unique Long Region of MDV2

The MDV2 UL region is 109,933bp in size with a G+C content of 50.3% and contains 68 potential ORFs capable of encoding protein. Of the 68 ORFs, 60 ORFs exhibited significant homologies to the UL genes of other alpha-herpesviruses, while the remaining 8 ORFs, mostly found around the UL and its flanking repeat junctional regions, had no apparent relation to any known herpesviral proteins, except for several counterparts of other MDV serotypes. The MDV2 candidate for the origin of DNA replication did not exist in the UL region as observed in the genomes of varicella-zoster virus (VZV) and bovine herpesvirus type 1 (BHV-1) (Davison and Scott 1986; Schwyzer and Ackermann 1996). The precise locations of the MDV2 UL ORFs and characteristics of the predicted amino acid sequences are summarized in Table 2. It is presumed that transcriptional initiation starts at the first ATG codon in each ORF because of the lack of contradicting experimental data. Functional properties of MDV2 UL proteins were deduced from the known alpha-herpesvirus gene functions, mainly in HSV-1 (Roizman and Sears 1996).

6.1 Junctions Between UL and Repeat Sequences: Characterization of MDV2 pp38 and pp24 Homologous Genes

The MDV1 pp24 and pp38 genes were identified to reside across the junction regions of TRL/UL and UL/IRL, respectively (Cui et al. 1991; Becker et al. 1994; Ross 1999). The N-terminal 58 amino acids in pp24 and pp38 proteins were identical because they were encoded from the inverted repeat regions (Zhu et al. 1994). Based on Southern blot analysis, the TRL/UL and UL/IRL junctions of MDV2 were previously localized to the *Bam*HI fragments K and Q, respectively (Ono et al. 1992). We identified a homologue of the MDV1 pp38 within the right-end part of the *Bam*HI-K fragment that encompassed the putative MDV2 origin of DNA replication (Ono et al. 1994). By sequencing the entire *Bam*HI-Q fragment and comparing the sequence to that of the *Bam*HI-K, we identified an additional homologue of MDV1 pp24 and the corresponding repeat of the MDV2 origin within the left-end part of *Bam*HI-Q. The results also allow us to determine the junctions of the UL region and its flanking repeat regions in the MDV2 genome. As in MDV1, the TRL/UL and UL/IRL junction regions of MDV2 were located within the pp24 and pp38 homologous genes, respectively, and possessed identical 68 amino acids in the N-terminus of their proteins (Fig. 3). However, the gene organization at the UL/IRL junctions of MDV2 and HVT is different since a pp38 homologue in the HVT genome lacks a large proportion of the N-terminal part of the protein and is located entirely within the IRL region (Smith et al. 1995). Apart from these observations, unexpected findings in the vicinity of the MDV2 UL/TRL junction were observed as follows: (1) the MDV2 pp24 homologue contains three different elements of tandemly repeated 12bp GC-rich sequences as shown in Fig. 3 (designated TR1, TR2, and TR3, respectively); and (2) a stretch of 81bp sequences encoding the middle 27 amino acids of the pp24 homologue is repeated within this protein twice and inversely repeated across the UL/IRL junction region of the pp38 homologue. These repeat elements are uniquely observed in MDV2 homologues but not in MDV1; however, the significance of these elements is unknown at present, since the structural and functional features of their gene products have not been determined. The function of the MDV1 pp38/pp24 complex remains unclear and its role in oncogenic transformation has not been confirmed. However, MDV2 pp24 is of particular interest since its predicted amino acid sequence showed 47.1% identity only to the MDV1 pp24 C-terminal sequence of 66 amino acids, which were located at residues 90–155 in the MDV1 protein and 110–175 in the MDV2 protein. The C-terminal amino acid sequences of both pp24 homologues of MDV1 and MDV2 contain putative hydrophobic anchor domains for membrane association, similar for both pp38 homologues of MDV1 and MDV2 (Ono et al. 1994; Zhu et al. 1994). Consistent with previous studies, the N-terminal sequences in both pp38 and pp24 proteins of MDV are specific to each serotype.

6.2 MDV2 Genes in UL Region

6.2.1 Characterization of UL Gene-Encoded Glycoproteins

The MDV2 UL region contains at least six genes with the potential to encode glycoproteins, gB (Kato et al. 1999b), gC (Kitazawa et al. 1993), gH (Shimojima et al. 1997a), gK (Tsushima et al. 1999), gL, and gM (Cai et al. 1999) (Table 3).

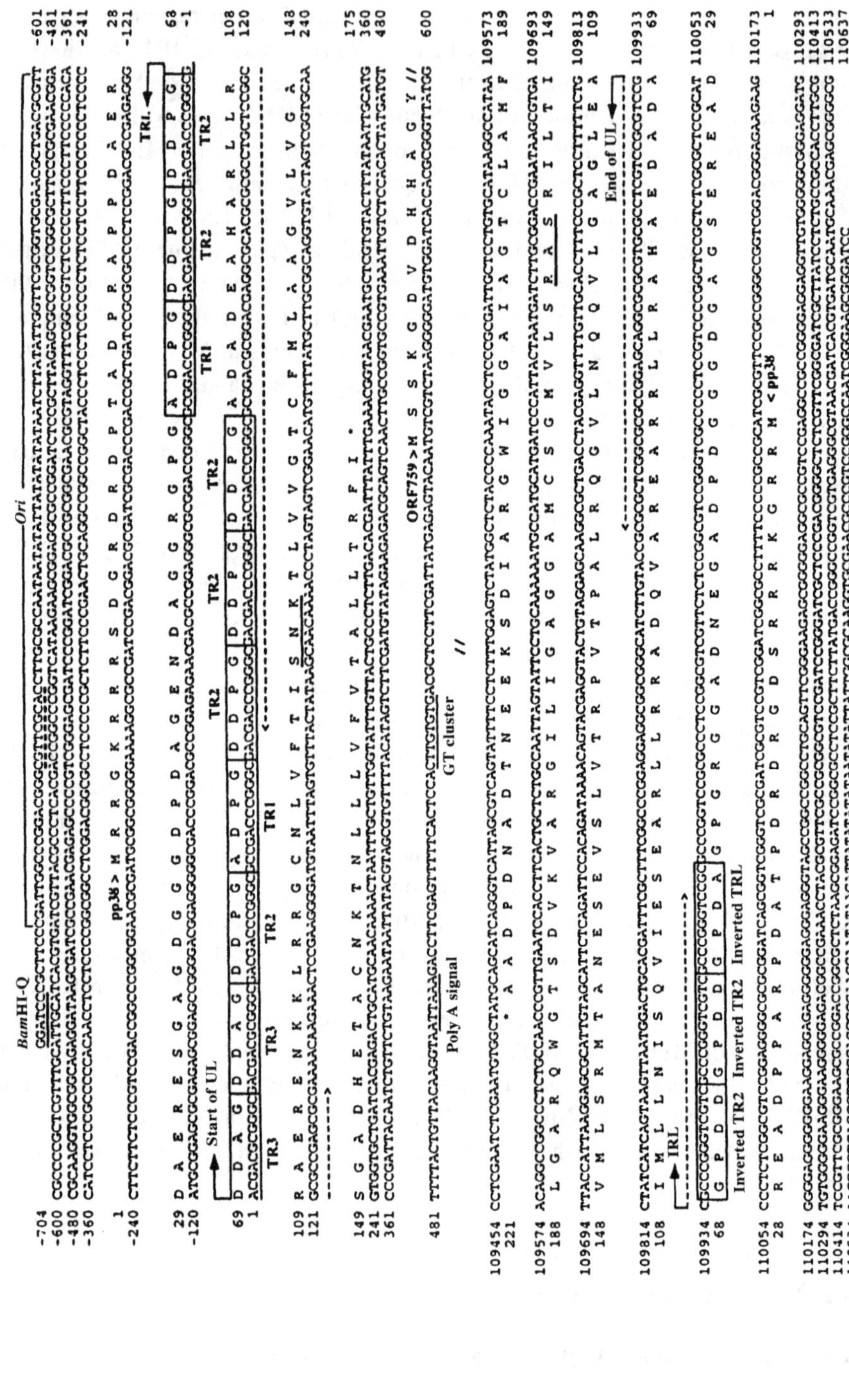

```
            BamHI-Q                 Ori
-704        GGATCCCGCTTCCCGATTGGCCCGGACGGGCGTTCGCACCTTGCGCCAATAATATATTATATATATATAATCTTATATTGGTTCGCGGTGCGAACGCTGACGCGTT -601
-600 CGCCCCGCTCGTTTGCATTGCATCACGTGATCGTTACGCCCTCACGACCGGCCCGGTCATAAGAAGCGGAGGCGCCGGATCTCCGCTTAGAGCGCCGGTCCGGCGCTTCCCGCGAACGGA -481
-480 CGCAAGGTGGCGGCAGAGGATAAGCGATCGCCGAACGAGAGCCCCGTCGGGAGCGATCCCGGATCGGACGCCGCGGCGAACGCGTAGGTTTCGGCCGTCTCCCCCTTCCCTTCCCCCACA -361
-360 CATCCTCCCGCCCCCACAACCTCCTCCCCGGCGGCCTCGGACGGCGCCTCCCCCGCTCTTCCCGAACTGCAGGCCGGCCGGCTACCCTCCCTCCCCCCTCTCCTCCTTCCCCCCCTCCCC -241

   1                              pp38 > M  R  R  G  K  R  R  R  S  D  G  R  D  R  D  P  T  A  D  P  R  A  P  P  D  A  E  R    28
-240 CTTCTTCTCCCGTCCGACCGGCCCGGCGGAACGCGATGCGGCGGGGAAAAGGCGCCGATCCGACGGACGCGATCGCGACCCGACCGCTGATCCGCGCGCCCCTCCGGACGCCGAGAGGG -121
                                                                                                                     TRL <-
  29 D  A  E  R  E  S  G  A  G  D  G  G  G  D  P  D  A  G  E  N  D  A  G  G  R  G  P  G [A  D  P  G][D  D  P  G][D  D  P  G]   68
-120 ATGCGGAGCGCGAGAGCGGAGCCGGGGACGGAGGGGGCGACCCGGACGCCGGAGAGAACGACGCCGGAGGGCGCGGACCGGGCGCGGACCCGGGCGACGACCCGGGCGACGACCCGGGCG   -1
                                                                                            TR1         TR2         TR2
 -> Start of UL                                  TR2         TR2         TR2
  69 [D  D  A  G][D  D  A  G][D  D  P  G][A  D  P  G][D  D  P  G][D  D  P  G][D  D  P  G] A  D  A  D  E  A  H  A  R  L  L  R   108
   1 ACGACGCGGGCGACGACGCGGGCGACGACCCGGGCGCCGACCCGGGCGACGACCCGGGCGACGACCCGGGCGACGACCCGGGCGCGGACGCGGACGAGGCGCACGCGCGCCTGCTCCGGC  120
      TR3         TR3         TR2         TR1      <----------------------------------------------------------------------

 109 R  A  E  R  E  N  K  K  L  R  R  G  C  N  L  V  F  T  I  S  N  K  T  L  V  V  G  T  C  F  M  L  A  A  G  V  L  V  G  A   148
 121 GCGCCGAGCGCGAAAACAAGAAACTCCGAAGGGGATGTAATTTAGTGTTTACTATAAGCAACAAAACCCTAGTAGTCGGAACATGTTTTATGCTTGCGGCAGGTGTACTAGTCGGTGCAA  240
     ----------->

 149 S  G  A  D  H  E  T  A  C  N  K  T  N  L  L  L  V  F  V  T  A  L  L  T  R  F  I  *                                          175
 241 GTGGTGCTGATCACGAGACTGCATGCAACAAAACTAATTTGCTGTTGGTATTTGTTACTGCCCTCTTGACACGATTTATTTGAAACGGTAACGAATGCTCGTGTACTTTATAATTGCATG  360
 361 CCCGATTACAATCTGTTCTGTAAGAATAATTATACGTAGCGTGTTTTACATAGTCTTCGATGTATAGAAGAGACGCAGTCAACTTGCCGGTGCCGTGAAATTGTCTCCACACTATGATGT  480

                                                                                        ORF759 > M  S  S  K  G  D  V  D  H  H  A  G  Y //
 481 TTTTACTGTTACAAGGTAATTAAAGACCTTCGAGTTTTTCACTCCACTTGTGTGACGCTCCTTCGATTATGAGAGTACAATGTCGTCTAAGGGGGATGTGGATCACCACGCGGGTTATGG  600
                       Poly A signal                GT cluster     //

109454 CCTCGAATCTCGAATGTGGCTATGCAGCATCAGGGTCATTAGCGTCAGTATTTTCCTCTTTGGAGTCTATGGCTCTACCCCAAATACCTCCCGCGATTGCTCCTGTGCATAAGGCCATAA 109573
   221                         *  A  A  D  P  D  N  A  D  T  N  E  E  K  S  D  I  A  R  G  W  I  G  G  A  I  A  G  T  C  L  A  M  F    189

109574 ACAGGCCGGCCCTCTGCCAACCCGTTGAATCCACCTTCACTGCTCTGCCAATTAGTATTCCTGCAAAAAATGCCATGCATGATCCCATTACTAATGATCTTGCGGACCGAATAAGCGTGA 109693
   188   L  G  A  R  Q  W  G  T  S  D  V  K  V  A  R  G  I  L  I  G  A  G  G  A  M  C  S  G  M  V  L  S  R  A  S  R  I  L  T  I    149

109694 TTACCATTAAGGAGCGCATTGTAGCATTCTCAGATTCCACAGATAAAACAGTACGAGGTACTGTAGGAGCAAGGCGCTGACCTACGAGGTTTTGTTGCACCTTTCCCGCTCCTTTTTCTG 109813
   148   V  M  L  S  R  M  T  A  N  E  S  E  V  S  L  V  T  R  P  V  T  P  A  L  R  Q  G  V  L  N  Q  Q  V  L  G  A  G  L  E  A    109
                                                                                                                     End of UL <-
                                                                                       <-----------------------------------------
109814 CTATCATCAGTAAGTTAATGGACTGCACGATTTCGCTTTCGGCCCGGAGGAGGCGGCGGGCATCTTGTACCGCGCGCTCGGCGCGCCGGAGCAGGCGCGCGTGCGCCTCGTCCGCGTCCG 109933
   108   I  M  L  L  N  I  S  Q  V  I  E  S  E  A  R  L  L  R  R  A  D  Q  V  A  R  E  A  R  R  L  L  R  A  H  A  E  D  A  D  A     69
       -> IRL
       ------------------------------->
109934 CGCCCGGGTCGTCGCCCGGGTCGTCGCCCGGGTCCGCGCCCGGTCCGCGCCCTCCGGCGTCGTTCTCTCCGGCGTCCGGGTCGCCCCCTCCGTCCCCGGCTCCGCTCTCGCGCTCCGCAT 110053
    68  [G  P  D  D][G  P  D  D][G  P  D  A] G  P  G  R  G  G  A  D  N  E  G  A  D  P  D  G  G  G  D  G  A  G  S  E  R  E  A  D     29
       Inverted TR2 Inverted TR2 Inverted TRL

110054 CCCTCTCGGCGTCCGGAGGGGCGCGCGGATCAGCGGTCGGGTCGCGATCGCGTCCGTCGGATCGGCGCCTTTTCCCCCGCCGCATCGCGTTCCGCCGGGCCGGTCGGACGGGAGAAGAAG 110173
    28   R  E  A  D  P  P  A  R  P  D  A  T  P  D  R  D  R  G  D  S  R  R  R  K  G  R  R  M  < pp38                                 1

110174 GGGGAGGGGGGGAAGGAGGAGAGGGGGGAGGGAGGGTAGCCGGCCGGCCTGCAGTTCGGGAAGAGCGGGGGAGGCGCCGTCCGAGGCCGCCGGGGAGGAGGTTGTGGGGGCGGGAGGATG 110293
110294 TGTGGGGGAAGGGAAGGGGGAGACGGCCGAAACCTACGCGTTCGCCGCGGCGTCCGATCCGGGATCGCTCCCGACGGGGCTCTCGTTCGGCGATCGCTTATCCTCTGCCGCCACCTTGCG 110413
110414 TCCGTTCGCGGGAAGCGCCGGACCGGCGCTCTAAGCGGAGATCCGGCGCCTCCGCTTCTTATGACCGGGCCGGTCGTGAGGGCGTAACGATCACGTGATGCAATGCAAACGAGCGGGGCG 110533
110534 AACGCGTCAGCGTTCGCACCGCGAACCAATATAAGATTATATATATAATATATTATTGGCGCAAGGTGCGAACGCCCGTCCGGGCCAATCGGGAAGCGGGATCC                 110637
                                          Ori                                                                          BamHI-K
```

Fig. 3. Nucleotide sequence and predicted amino acid sequence of the MDV2 TRL/UL and UL/IRL junctional regions. Two *Bam*HI sites of the K and Q fragments are *underlined*. Also, potential transcriptional termination sites are *underlined*. The putative MDV2 origin of DNA replication is indicated by *broken lines*. *Bold dashed lines* indicate the locations of origin-binding protein consensus recognition motifs. ORF orientations are shown as *arrowheads*. The predicted amino acid residues are given in the single-letter code above (MDV2 pp24 homologue) or below (MDV2 pp38 homologue) and *asterisks* represent stop codons. Three tandems repeat elements are *boxed* and indicated by TR1, TR2, and TR3. *Dashed lines with arrowheads* indicate other repeated sequences between the MDV2 pp38 and pp24 homologous protein. *Boldly underlined* amino acids containing serine residues indicate potential consensus phosphorylation sites. Continuous sequences are not plotted and are indicated as (//). Dotted four bases are corrected in this study

The overall amino acid sequence homology between the deduced MDV2 gL protein and other alpha-herpesviral gL proteins is not significant, except for the MDV1 counterpart. However, an extending region from 67 to 94 residues in MDV2 is highly conserved among all of the gL proteins, as observed in MDV1 (Yoshida et al. 1994). The MDV2 gL protein possesses two potential N-glycosylation sites (residues 36–38 and 73–75), which are positionally conserved with those of MDV1. From all of the above results, nine structural MDV glycoprotein homologues have been identified in the MDV2 genome so far, and these are composed of six UL-encoded glycoproteins and three US-encoded glycoproteins (Table 3). Recently, UL49.5 homologous genes of pseudorabies virus (PRV) and BHV-1 were identified to encode a viral envelope protein (Jöns et al. 1996; Liang et al. 1996). The MDV2 UL49.5 gene encodes a 95-amino-acid protein with 71% identity to that of the MDV1 homologue (Yanagida et al. 1993; Izumiya et al. 1998), and exhibits characteristics of a membrane protein but lacks consensus N-glycosylation sequences as described elsewhere (Barnett et al. 1992; Jöns et al. 1996). Furthermore, two potential MDV-unique glycoprotein genes are found closely in both end portions of the UL region in all three serotypes (Becker et al. 1994; Smith et al. 1995; Tsushima et al. 1999). MDV2 ORF759 and ORF873 proteins correspond to these glycoproteins (Table 2) and their transcriptional patterns are also similar to the MDV1 homologues (Tsushima et al. 1999). The MDV2 ORF759 comprises two exons near the left terminus of the MDV2 UL region that are spliced together, and encodes a novel glycoprotein of which no homologue has been identified so far in other alpha-herpesviruses except for MDV1 (Becker et al. 1994). Compared to MDV1, the MDV2 ORF759 protein showed 52% identity with the MDV1 counterpart (Becker et al. 1994). There is an obvious signal peptide stretch (residues 11–32) having 86.4% homology with that of the MDV1 counterpart (Becker et al. 1994). Based on hydropathic analysis according to the method of Kyte and Doolittle and its location, a hydrophobic segment (residues 501–525) was predicted to be the transmembrane anchor domain. The external N-terminus of the MDV2 ORF759 protein contains eight cysteine residues and five potential sites for N-linked glycosylation, which were highly conserved between MDV1 and MDV2 proteins (Fig. 4). On the opposite side of the MDV2 UL region, we also identified a gene presumably encoding a glycoprotein whose transcripts are also spliced (Tsushima et al. 1999). The MDV1 counterpart is located in a similar

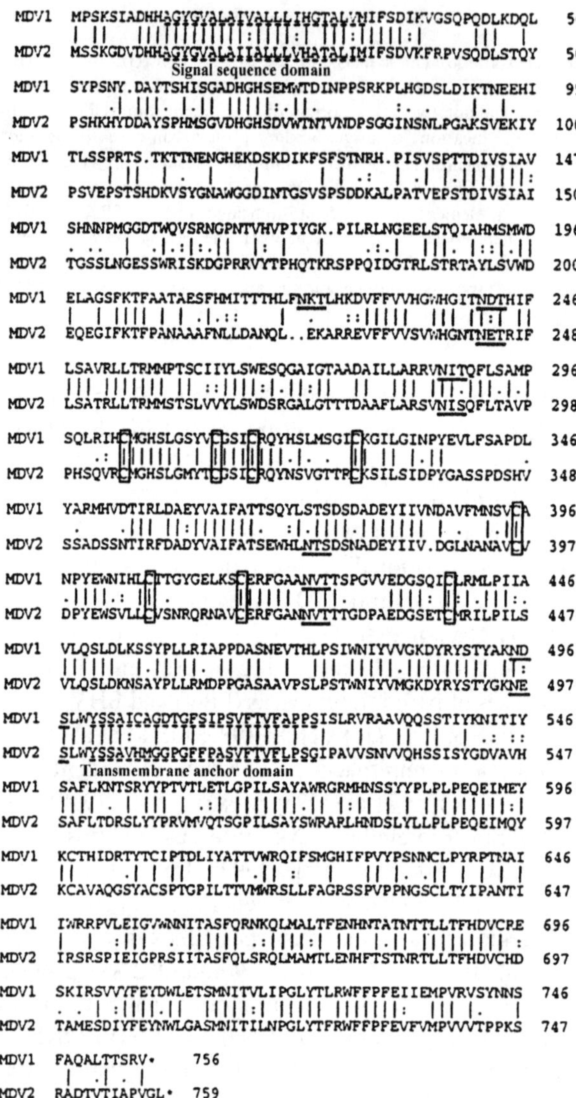

Fig. 4. Alignment of the predicted amino acid sequences of the putative MDV-unique glycoprotein genes localized in the UL left-end portion of MDV1 (Becker et al. 1994) or MDV2 (ORF759). Both the predicted signal sequence and transmembrane anchor domains are indicated by *dashed lines*. Conserved N-linked glycosylation sites are shown as *bold lines*. Eight cysteine residues are *boxed*, which may be involved in interchain disulfide bonding. The identity (or similarity) between two amino acid residues is indicated by a *vertical line* or *dot*, respectively

region of the MDV1 UL (Ross et al. 1993), but the HVT corresponding gene is somewhat different, which is located across the junction region of UL/IRL (Smith et al. 1995). In addition to these class I membrane glycoproteins, the MDV genome encodes several other potentially glycosylated proteins associated with the viral membrane, such as homologues of HSV-1 UL20, UL32, UL34, and UL43 (Wu et al. 1996; Izumiya et al. 1998, 1999b; Hatama et al. 1999). The MDV1 UL32 protein was recently characterized and detected in infected cells as a 82-kDa membrane glycoprotein (Wu et al. 1997). However, the properties and functional features of almost all these MDV2-encoded proteins, including the UL32 homologue, remain to be determined.

6.2.2 Characterization of UL Gene-Encoded Other Virion Proteins

As with other alpha-herpesviruses, homologues to the two exons of HSV-1 UL15 are also observed in the MDV2 genome and are expressed as a spliced transcript (Table 4), as seen in HSV-1 (Costa et al. 1985). The predicted protein of UL26.5 identical to the C-terminal part of the UL26 protein was identified in the MDV2 genome as well as in the genomes of other alpha-herpesviruses (Kato et al. 1999a).

MDV2 has several different genetic contents among other alpha-herpesviruses in the UL region (Table 2, asterisks). The UL3.5 gene is observed in the MDV2 genome, which is conserved in most mammalian (Telford et al. 1992; Dean and Cheung 1993; Schwyzer and Ackermann 1996) and avian alpha-herpesviruses (Fuch and Mettenleiter 1996), but not in HSV-1 and -2 (McGeoch et al. 1988; Dolan et al. 1998). Hatama et al. (1999) reported that the MDV2-specific ORF273 gene is presumably found inversely within the N-terminal part of the UL21 gene and is transcribed as a polycistronic transcript in virus-infected cells; however, its predicted protein has no significant homology to any known alpha-herpesvirus proteins. The UL45 and UL55 genes are conserved in the MDV2 genome, but not in BHV-1 (Schwyzer and Ackermann 1996). The UL45 gene is also absent in the VZV genome (Davison and Scott 1986). The MDV2 UL45 protein, comprised of 210 amino acids, has relatively high homology with that of the MDV1 homologue. The N-terminus of the HSV-1 UL45 protein represents a membrane-spanning or -associated region (Visalli and Brandt 1993). A similar hydrophobic stretch was found near the N-terminus (residues 40–70) of the potential MDV2 UL45 protein (Izumiya et al. 1998). Interestingly, the ORF-1 homologous gene, which is only conserved in the equine herpesvirus type 1 and PRV genomes so far (Telford et al. 1992; Baumeister et al. 1995), was found in the MDV2 genome (Tsushima et al. 1999). However, it is still unclear whether this counterpart is conserved in the MDV1 and HVT genomes or not. Besides these conserved genes, several MDV-specific genes are presumably identified within the UL region, although most of the MDV-specific genes were identified within the TRL and IRL sequences, including pp38, pp24, and several oncogenes (Kung et al.

Table 4. Splice donor and acceptor sequences

Gene	Donor sequence	Acceptor sequence
ORF759	CTGATTATG/[a]GTGAGTG (12472–12487)	TTTTAG/ATCTTTTCT (12551–12565)
UL15	AATACAAAC/GTGAGTA (36976–36991)	CTGCAG/AGCATTCGC (40426–40440)
ORF873s/873	AGTGAAAAT/GTGAGTT (120832–120817)	TTGCAG/AATGCCATA (120716–120702)

Slashes indicate splicing sites either at the 5′ end (donor) or 3′ end (acceptor) of the introns

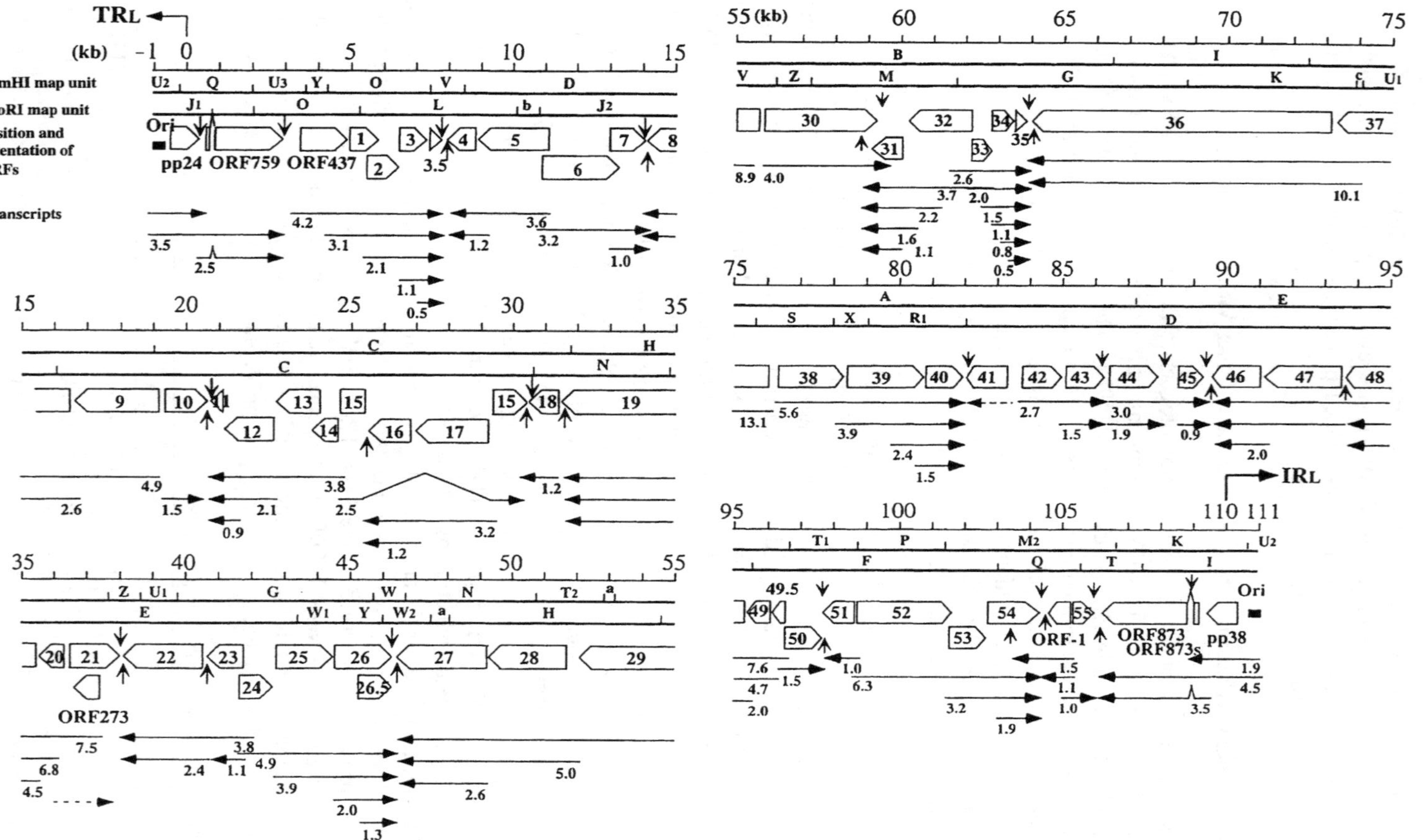

TRL
(kb)
BamHI map unit
EcoRI map unit
Position and orientation of ORFs
Transcripts
Ori
pp24 ORF759 ORF437
ORF273
IRL
ORF-1
ORF873
ORF873s
pp38

Fig. 5. Gene arrangement and schematic diagram of the transcription map in the MDV2 unique long (UL) genome region flanked by the inverted TRL and IRL sequences. Restriction enzyme maps of the investigated *Bam*HI and *Eco*RI fragments are shown in *bold on successive lines*. The locations and orientations of the identified MDV2 UL ORFs are indicated by *open arrows*. Numbers *1–55* in the MDV2 UL region correspond to the homologous gene of the of HSV-1 UL region. *Dashed lines with arrowhead* indicate the predicted transcripts only based on location of the polyadenylation sites. Splice sites are indicated by *triangles*. *Vertical arrows* indicate the proposed polyadenylation sites in the appropriate strand. The MDV2 origin candidates of DNA replication found in both TRL and IRL are marked with *filled squares*

1995; Calnek and Witter 1997). An apparently MDV-specific ORF was further identified in the MDV2 UL region and was located upstream from the 5′-terminal part of the UL1 homologue (Fig. 5). The MDV2 ORF437 encodes a unique polypeptide of 437 amino acids that shows no significant homology to any known herpesviral proteins. A matter of particular interest is the presence of a highly charged C-terminus in this protein. Of the C-terminal 72 amino acids of the ORF437 protein, 31 residues possess charged side chains, 11 of which are acidic and 20 basic in character (data not shown). MDV2 ORFs found in the left-end part of the UL region positionally corresponded to the ILTV-specific UL0 and UL[-1] genes (Ziemann et al. 1998). Interestingly, the genetic arrangements and transcriptional patterns of the MDV2 and ILTV genomes closely resemble each other, at least within both corresponding UL regions; however, there is no homology among the predicted proteins. As a result, the gene content within both UL termini of the three MDV serotypes is clearly distinct from those of other alpha-herpesviruses including ILTV. Similar cases have also been shown between the US terminal regions of the MDVs and other alpha-herpesviruses. Although the significance of these differences is not known at present, these regions may be considered to be preferred sites of recombination and acquisition of species-specific virus genes as speculated in our and other previous studies (Telford et al. 1992; Zelnik et al. 1993; Ono et al. 1994; Brunovski and Velicer 1995; Smith et al. 1995; Ziemann et al. 1998). A major focus of further studies will be whether these MDV-unique genes are related to the determinants of viral immunity, serotype-specific pathogenicity, and unique nature of MDVs.

6.3 Transcription Map of UL Region

Transcriptional mapping of the identified genes in the MDV2 UL region was performed by Northern blot analyses. A schematic diagram of MDV2 UL transcripts experimentally detected up to now is presented in Fig. 5. There are possibly 11 overlapping ORFs within the MDV2 UL region, which localize in two major clusters at UL5–UL14 and UL30–UL33, suggesting that transcriptional control elements may overlap within these protein-encoding regions. All of the putative polyadenylation sites possessing AATAAA (or minor AT/GTAAA variants) are also indicated in Fig. 5 (see vertical arrows). In many occasions, they are located at the distal portion of sets of genes transcribed into 3′-coterminal families of transcripts, as observed in other alpha-herpesviruses (Davison and Scott 1986; McGeoch et al. 1988; Telford et al. 1992; Schwyzer and Ackermann 1996; Fuchs and Mettenleiter 1996). Although we attempted to detect transcripts related to the MDV2 pp24, UL21, and UL41 homologous genes, no respective transcripts were found in the virus-infected cells (Izumiya et al. 1998; Hatama et al. 1999). Therefore, the transcript layouts are predicted based on the position of putative polyadenylation sites (Fig. 5, dashed arrows). Transcriptional mapping data from the UL20 to the pp38 homologous genes have been published in our previous studies (Cai et al. 1999; Hatama et al. 1999; Izumiya et al. 1998, 1999a,b; Kato et al. 1999a,b; Kitazawa et al. 1993; Ono et al. 1994; Shimojima et al. 1997a,b; Tsushima et al. 1999). As shown in Fig. 5, all of the identified genes, except the above three genes in the MDV2 UL region, were confirmed to be transcribed either as gene-specific and/or 3′-coterminal transcripts in virus-infected cells. Furthermore, at least three spliced transcripts were observed in the MDV2 UL region, and the exact positions of splicing sites are presented in Table 4.

7 Internal and Terminal Repeat Long Regions

MDV2 TRL and IRL regions consist of 11,825bp each and their G + C contents are 62.2%. There is a distinct origin of MDV2 with copies in both TRL and IRL (Figs. 1, 3). These two copies of origin contain an almost perfect palindromic structure with a central AT-rich region. Furthermore, both arms of the detected palindrome contain the sequence motif CGTTCGCAC (Fig. 3, dashed lines), which is identified as a recognition domain of the origin-binding protein encoded by the HSV-1 UL9 gene (Koff and Tegtmeyer 1988). Similar features were found and precisely described in other alpha-herpesvirus origin elements (Davison and Scott 1986; McGeoch et al. 1988; Telford et al. 1992; Schwyzer and Ackermann 1996). Although it has not been shown to function as an origin, the potential origin of DNA replication sequence of MDV2 HPRS24 strain exhibited 97.8% identity with the functional origin of a different strain of MDV2 (Camp et al. 1991).

Our complete sequence data also demonstrated that MDV2 possesses no positionally conserved ICP0 gene. The gene is not translocated anywhere; therefore, it is completely missing. The absence of an ICP0 homologous gene in the MDV2 genome seems possible since HSV-1 ICP0 was shown to be dispensable for virus replication (Cai et al. 1993; Stow and Stow 1986) and no ICP0 homologues have been identified in MDV1 and HVT (Calnek and Witter 1997). Recently, as with the MDVs, no ICP0 homologue has been identified in ILTV (Ziemann et al. 1998). This might indicate that avian herpesviruses have a somewhat different mechanism(s) for reactivation with mammalian alpha-herpesviruses. As shown in Fig. 1, there are nine potential ORFs in both the IRL and TRL regions, but gene arrangements in these regions are largely different with those found in other alpha-herpesvirus genomes (McGeoch et al. 1988; Davison and Scott 1986; Telford et al. 1992; Schwyzer and Ackermann 1996), including MDV1 (Ross 1999). Database searches revealed that these predicted proteins have no apparent relation to any known herpesviral proteins, indicating that these ORFs encoded MDV2-specific proteins.

From our previous comparison of the three serotypes of the MDV genomes by cross-hybridization (Ono et al. 1992), as well as the accumulating nucleotide sequence data for the MDV genomes as described above, it is known that the genomes of the three MDV serotypes are collinear in the UL and US regions but differ in the adjoining repeat regions. Therefore, it has been speculated that the candidate MDV oncogenes reside within the inverted repeat regions flanking UL and US (Kung et al. 1995). Until now, the MDV1 IRL genes were associated with the oncogenic nature of MDV1 (Ross 1999). Interestingly, sequence analyses practically demonstrated that the MDV2 genes found in both TRL and IRL regions showed no homology with those sequences found in the oncogenic MDV1 genome in those regions. However, it has already been demonstrated that MDV1 has complex transcription patterns in the repeat regions, characterized by the presence of alternatively spliced transcripts in both sense and antisense orientation (Ross 1999), but transcription patterns in these regions of MDV2 are still unknown. Transcriptional analyses in this region will be needed to fully understand the nononcogenic nature of MDV2.

Recently, the *meq* (Marek's disease virus *Eco*RI-Q) gene, which may play a role in oncogenesis, has been the focus of attention for a number of reasons. (1) It is consistently expressed in the vast majority of cells in transformed cell lines and in $CD4^+$ T cells from lymphomas (Ross et al. 1997); (2) Oligonucleotides and RNA complementary to the *meq* gene can prevent the growth of cell lines (Xie et al. 1996); (3) Overexpression of Meq protein in rat-2 fibroblasts by a recombinant murine retrovirus leads to cell transformation and gives anti-apoptotic activity to the cells (Liu et al. 1998); and (4) The *meq* gene encodes a transactivating bZIP protein homologous to those of the *fos* and *jun* oncogenes (Jones et al. 1992). Furthermore, the Meq protein interacts with the Jun protein and Meq/Jun heterodimers bind to an AP1-like sequence in the *meq* promoter region with higher affinity than Meq/Meq or Jun/Jun homodimers. The presence of an AP-1 site in the promoter of the *meq* gene suggests autoregulation of the *meq* gene expression. The binding of these heterodimers to AP-1 like sequences has been shown to upregulate the *meq* gene expression in vitro (Qian et al. 1995, 1996). In addition, our complete sequencing analyses of the nononcogenic MDV2 genome support the above conception, because the homologous *meq* gene is not conserved in the naturally nononcogenic MDV2 genome. Another difference between MDV1 and MDV2 is the increase in the 132-bp tandem repeats located in the IRL region of MDV1 as a result of virus attenuation by serial passages in fibroblast cells (Ross 1999); however, they were not observed in the counterpart region of the MDV2 genome in spite of at least 30 passages in the cells.

In future, as one possible strategy, we will attempt to substitute the MDV1 *meq* gene with putative nonessential genes of MDV2 and investigate the oncogenic potential of the recombinant MDV2 in chickens. The availability of the full genomic sequence of MDV2 will greatly enhance the capability of completing these future endeavors.

Previous works demonstrated that avian retroviruses such as reticuloendotheliosis virus, avian leukosis virus, or avian erythroblastosis virus could efficiently integrate into the MDV genome (ISFORT et al. 1992, 1994; JONES et al. 1996). The integrations are mainly observed in the MDV1 genomes. The great majority of these sites of virus integrations are not randomly distributed, but are rather clustered near the junctions of the TRL/UL, UL/IRL, IRS/US, and US/TRS and within the gD gene in the US region (ISFORT et al. 1994; ENDOH et al. 1998). Furthermore, JONES et al. (1996) suggested that such insertions could alter the expression of herpesvirus genes. Thus, transcripts originating from the reticuloendotheliosis virus long terminal repeat promoters were found to encode the adjoining MDV genes, SORF2, US1, and US10. However, the relationship between retrovirus integration and virus phenotype remains to be understood. While in this study we could not find retroviral integrations in this strain of the MDV2 genome, it may indicate that this strain did not have chances of coinfection with avian retroviruses in vitro and in vivo.

8 Identification of an *a*-Sequence of MDV2 that Contains Host Cell Telomere Repeat Sequences

In HSV-1, newly replicated DNA consists of head-to-tail concatemers that are cleaved to generate unit-length genomes bounded by the terminally reiterated *a*-sequence. The *a*-sequence is necessary and sufficient to direct cleavage and packaging of HSV-1 genomes (MOCARSKI and ROIZMAN 1981, 1982). Later, two sequence motifs designated pac-1 and pac-2 that are involved in the formation of unit-length genomes were first identified by their conservation at herpesvirus termini (DEISS et al. 1986). They are found within the HSV-1 *a*-sequence and at the termini of all herpesvirus genomes already sequenced. The pac-1 motifs consist of a 3- to 7-bp A- or T-rich region flanked on each side by 5–7 Cs (DEISS et al. 1986; TAMASHIRO et al. 1984) and the pac-2 motifs consist of 5- to 10-bp A-rich region that is often associated with the nearby CGCGGCG sequence (DEISS et al. 1986). Both pac-1 and pac-2 are located 30–35bp from the genome termini, generally at opposite ends of the viral genome except in the human cytomegalovirus (HCMV) (MCVOY et al. 1998). Consequently, for most herpesviruses, cleavage to release genomes from concatemers occurs between pac-1 and pac-2 (DEISS et al. 1986). In addition, recent detailed analysis of murine cytomegalovirus genome revealed the following: (1) Within pac-1, the poly C region was very important for cleavage and packaging, but the A-rich region was not; (2) Within pac-2, the A-rich region and adjacent sequences were essential for cleavage and packaging and the CGCGGCG sequence contributed to, but was not strictly essential for efficient cleavage and packaging; and (3) A second A-rich region was not important at all (MCVOY et al. 1998).

In MDV1, sequence analysis of the junction region between the L and S components revealed the existence of an *a*-like sequence, similar in location and structure, but not in sequence, to the *a*-sequence of HSV-1 (Kishi et al. 1991). Within the *a*-sequence of MDV1, a structure homologous to the internal direct repeats (DR2) of HSV-1 was found to contain 17 copies of the telomeric sequence, GGGTTA (Kishi et al. 1988). In addition, Southern blot analyses revealed that the telomeric sequence also localized at the L-S junction and termini of the HVT genome (Reilly and Silva 1993). It is striking that similar elements have been detected in other lymphotropic herpesviruses so far, including beta-herpesviruses [human herpesvirus 6 (HHV-6) (Thomson et al. 1994), human herpesvirus 7 (HHV-7) (Secchiero 1995)] and gamma-herpesvirus [equine herpesvirus 2 (EHV-2) (Telford et al. 1995)]. Except for the case of EHV-2, the telomere repeat sequence motifs existed proximally to the genomic termini and were adjacent to the pac-1 and pac-2 motifs (Thomson et al. 1994; Secchiero 1995). As shown in Fig. 6, telomere repeat sequence motifs were also identified within the MDV2 *a*-sequence located adjacently to the putative pac-1 and pac-2 motifs. Within the *a*-sequence of MDV2, 57 copies of the telomere repeat sequence were observed as structural homologues to DR2 of HSV-1 (Fig. 6). These elements consisted mainly of the TAACCC and four copies of the closely related TGACCC sequence. In addition, the MDV2 telomeric repeats contained other randomly distributed repeats differing from the TAACCC or TGACCC motifs, five copies of the TAAAGGCC sequence, and one copy of the TAACGGCC sequence (Fig. 6). DR1 elements identified in the MDV1 *a*-sequence (Kishi et al. 1991) were partially conserved in the MDV2 genome (Fig. 6). As unexpected findings in the MDV2 genome, we also found heterogeneity of the number of telomeric sequence in the TRS region (117 copies) and L-S junction region (57 copies). The same heterogeneity is also observed in HHV-7 and human herpesvirus 6B genomes (Megaw et al. 1998; Dominguez et al. 1999). This copy number variability was also present in different clones obtained with *Bam*HI digestion from total cellular DNA of MDV2-infected cells (data not shown). It may indicate that sequence heterogeneity naturally occurs in the telomere repeat regions. In HHV-6, it was shown that heterogeneity of the copy number of the elements was caused by extended viral passage in cultured cells (Lindquester and Pellett 1991). Therefore, sequence analysis of uncultured virus is required to understand more completely the structure of this region in wild-type virus. As it is different from other herpesviruses, the positions of pac-1 elements in both the MDV1 (Kishi et al. 1991) and the MDV2 genomes interchanged with those of pac-2 elements. Furthermore, compared with MDV2, pac-1 and pac-2 elements of MDV1 were not efficiently conserved (Kishi et al. 1991). Therefore, functional experiments will be required to identify whether these elements are actually *a*-sequences of MDVs or not.

Alternative functions for the telomere repeat sequences might include roles in viral replication at latent-phase, maintenance of the viral chromosome during latent infection in dividing cells (Thomson et al. 1994), and/or the site-specific integration of viral DNA into the telomeres of host cell chromosomes (Torelli et al. 1995; Delecluse et al. 1993). In situ hybridization analyses have already identified that

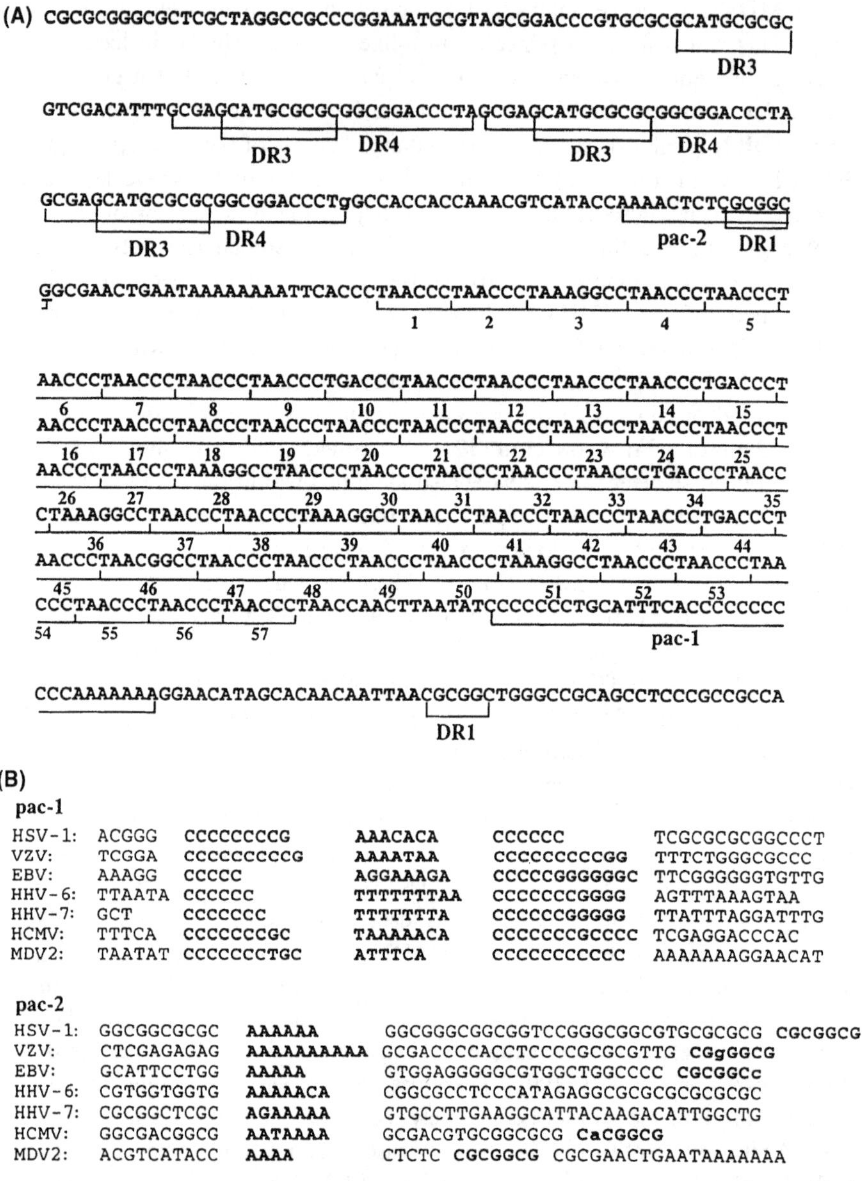

Fig. 6. **A** Nucleotide sequence of the junction region between the long and the short components in the MDV2 genome. Direct repeats are indicated by DR1, DR2, and DR3. The pac-1 and pac-2 elements are also indicated in this panel. Especially, the CGCGGCG motif of the pac-2 is *underlined*. Telomere repeat sequences are numbered from *1–57*. **B** Conserved sequences at herpesvirus termini. Alignments of the terminal sequence of HSV-1 (McGeoch et al. 1988), VZV (Davison and Wilkie 1981), EBV (Zimmermann and Hammershmidt 1995), HHV-6 (Thomson et al. 1994), HHV-7 (Secchiero et al. 1995), and HCMV (Tamashiro et al. 1984). The conserved components of pac-1 and pac-2 are shown in *boldface* and are set off by spaces. Bases with the CGCGGCG motifs of VZV and EBV that differ from the consensus are in *lowercase*

MDV1 DNA is integrated at multiple sites in the chromosomes of primary lymphoma cells in chickens (Delecluse et al. 1993). However, it is still unknown whether their sites of integration are the telomeres of host cell chromosomes or not, and the relationship between integrations and oncogenesis. In any case, the identification of telomere repeat sequence motifs in the genomes of several lymphotropic herpesviruses merits further investigation. It will be of particular interest to determine whether the telomere repeat sequence motifs are involved in replication, gene regulation, or biological properties of these viruses.

9 Internal and Terminal Repeat Short Regions: Identification of the MDV2 ICP4 Homologous Gene

In MDV1, the gene homologous to the HSV-1 ICP4 gene has been identified in the *Bam*HI fragment A (Anderson et al. 1992). The MDV1 ICP4 gene consists of 4245bp and has an overall G+C content of 52%. When we identified a protein-encoding region from the first initiation codon of the ORF, the MDV2 ICP4 gene consisted of 6099bp and had an overall G+C content of 66.7%. The MDV2 homologue, located in the TRS region, has the same size as the ICP4 gene, except for four nucleotides involving three amino acid changes with a gene located in the IRS region. The deduced 2033 amino acid MDV2 ICP4 protein exhibited 47.3% identity to that of MDV1. Database searches revealed that the MDV1 ICP4 protein exhibited significant similarities with the MDV2 ICP4 protein from the 810th methionine residue. Therefore, the precise location of the transcriptional initiation sites of both MDV ICP4 proteins needs to be determined.

A number of potential transcriptional regulatory sites lie adjacent to the MDV2 ICP4 gene. Degenerative versions of the putative TAATGARAT element, which is a sequence associated with HSV-1 immediate-early genes and is recognized by the VP16 protein, lie 285bp (TAAaGgAgT), 291bp (TAATGAtAa), and 431bp (TAATGgcAa) upstream from the first initiation codon. In addition, three SP-1 binding sites (GGGCGG) were also found 211, 232, and 367bp upstream from the first initiation codon. However, it remains to be determined whether these sequences are functionally related to the element.

The MDV2 ICP4 protein possesses a conserved serine-rich region (residues 947–979) as a putative site for phosphorylation (Xia et al. 1996). This region was also conserved in the MDV1 ICP4 protein (Anderson et al. 1992). Furthermore, similar to MDV1, the MDV2 ICP4 protein possessed two highly conserved regions, which are believed to be transactivation domains of early and late genes, as those of other alpha-herpesviruses. Therefore, this protein may share a similar function with those of other alpha-herpesviruses. The detailed transcriptional analyses of both sense and antisense copies of the MDV2 ICP4 gene are currently under investigation. However, further work is required to determine the properties and functional features of these proteins, especially their actual initiation sites.

10 Conclusive Remarks

On February 14, 2000, we determined a complete genomic DNA sequence of the MDV2 HPRS24 strain of 164,270bp. The sequence lays a good foundation for future studies of MDVs and related viruses. Furthermore, comparative studies with oncogenic MDV1 and other members of alpha-herpesviruses will be made possible, helping to understand its oncogenic nature and characterize the nature of MDVs.

We hope this work contributes to the progress of MDV molecular virology.

Acknowledgements. We should thank many people who supported the completion of this work. Without their help and encouragement, we would not have completed this work. First, we would like to express our appreciation to Prof. Eiji Takahashi, Prof. Chieko Kai, and Dr. Takayuki Miyazawa for their valuable help and advice. We are also deeply indebted to our coworkers, Dr. Jin-Shun Cai, Dr. Masahiro Niikura, Dr. Hiroto Kashiwase, Dr. Rika Iwanaga, Dr. Yasuto Murakami, Dr. Takehisa Kitazawa, Dr. Yukako Shimojima, Dr. Shinichi Hatama, Dr. Kentaro Kato, and Mr. Yoshinori Tsushima. This work was supported by grants from the Ministry of Education, Science, Sports and Culture, and the Ministry of Agriculture, Forestry and Fisheries of Japan. Y. Izumiya is supported by a research fellowships for young scientists of the Japan Society for Promotion of Science.

References

Anderson AS, Francesconi A, Morgan RW (1992) Complete nucleotide sequence of the Marek's disease virus ICP4 gene. Virology 189:657–667

Anderson AS, Parcells MS, Morgan RW (1998) The glycoprotein D (US6) homolog is not essential for oncogenicity or horizontal transmission of Marek's disease virus. J Virol 72:2548–2553

Bankowski RA, Moulton JE, Mikami T (1969) Characterization of Cal-1 strain of acute Marek's disease agent. Am J Vet Res 30:1667–1676

Barnett BC, Dolan A, Telford EAR, Davison AJ, McGeoch DJ (1992) A novel herpes simplex virus gene (UL49 A) encodes a putative membrane protein with counterparts in other herpesviruses. J Gen Virol 73:2167–2171

Baumeister J, Klupp BG, Mettenleiter TC (1995) Pseudorabies virus and equine herpesvirus 1 share a nonessential gene which is absent in other herpesviruses and located adjacent to a highly conserved gene cluster. J Virol 69:5560–5567

Becker Y, Asher Y, Tabor E, Davidson I, Malkinson M (1994) Open reading frames in a 4556 nucleotide sequence within MDV-1 BamHI-D DNA fragment: evidence for splicing of mRNA from a new viral glycoprotein gene. Virus Genes 8:55–69

Brunovskis P, Velicer LF (1995) The Marek's disease virus (MDV) unique short region: alphaherpesvirus-homologous, fowlpox virus-homologous, and MDV-specific genes. Virology 206:324–338

Buckmaster AE, Scott SD, Sanderson MJ, Boursnell ME, Ross NLJ, Binns MM (1988) Gene sequence and mapping data from Marek's disease virus and herpesvirus of turkeys: implications for herpesvirus classification. J Gen Virol 69:2033–2042

Cai JS, Jang HK, Izumiya Y, Tsushima Y, Kato K, Damiani AM, Miyazawa T, Kai C, Takahashi E, Mikami T (1999) Identification and structure of the Marek's disease virus serotype 2 glycoprotein M gene: comparison with glycoprotein M genes of *Herepesvindae* family. J Vet Med Sci 61:503–511

Cai W, Astor TL, Liptak LM, Cho C, Coen DM, Shaffer PA (1993) The herpes simplex virus type 1 regulatory protein ICP0 enhances virus replication during acute infection and reactivation from latency. J Virol 67:7501–7512

Calnek BW, Witter RL (1997) Marek's disease. In: Calnek BW (ed) Disease of Poultry, 10th ed. Iowa State University Press, Ames, pp 369–413

Camp HS, Coussens PM, Silva RF (1991) Cloning, sequencing, and functional analysis of a Marek's disease virus origin of DNA replication. J Virol 65:6320–6324

Cebrian J, Kaschka-Dierich C, Berthelot N, Sheldrick P (1982) Inverted repeat nucleotide sequences in the genomes of Marek's disease virus and the herpesvirus of the turkey. Proc Natl Acad Sci USA 79:555–558

Churchill AE, Biggs PM (1967) Agent of Marek's disease in tissue culture. Nature 215:528–530

Churchill AE, Biggs PM (1968) Herpes-type virus isolated in cell culture from tumors of chickens with Marek's disease. II. Studies in vivo. J Natl Cancer Inst 41:951–956

Churchill AE, Chubb RC, Baxendale W (1969a) The attenuation, with loss of oncogenicity, of the herpes-type virus of Marek's disease (strain HPRS-16) on passage in cell culture. J Gen Virol 4: 557–564

Churchill AE, Payne LN, Chubb RC (1969b) Immunization against Marek's disease using a live attenuated virus. Nature 221:744–747

Costa RH, Draper KG, Kelly TJ, Wagner EK (1985) An unusual spliced herpes simplex virus type 1 transcript with sequence homology to Epstein-Barr virus DNA. J Virol 54:317–328

Cui ZZ, Lee LF, Liu JL, Kung HJ (1991) Structural analysis and transcriptional mapping of the Marek's disease virus gene encoding pp38, an antigen associated with transformed cells. J Virol 65:6509–6515

Davison AJ, Scott JE (1986) The complete DNA sequence of varicella-zoster virus. J Gen Virol 67: 1759–1816

Davison AJ, Wilkie NM (1981) Nucleotide sequences of the joint between the L and S segments of herpes simplex virus types 1 and 2. J Gen Virol 55:315–331

Dean HJ, Cheung AK (1993) A 3′ coterminal gene cluster in pseudorabies virus contains herpes simplex virus UL1, UL2, and UL3 gene homologs and a unique UL3.5 open reading frame. J Virol 67: 5955–5961

Deiss LP, Chou J, Frenkel N (1986) Functional domains within the *a* sequence involved in the cleavage-packaging of herpes simplex virus DNA. J Virol 59:605–618

Delecluse HJ, Schüller S, Hammerschmidt W (1993) Latent Marek's disease virus can be activated from its chromosomally integrated state in herpesvirus-transformed lymphoma cells. Embo J 12: 3277–3286

Dolan A, Jamieson FE, Cunningham C, Barnett BC, McGeoch DJ (1998) The genome sequence of herpes simplex virus type 2. J Virol 72:2010–2021

Dominguez G, Dambaugh TR, Stamey FR, Dewhurst S, Inoue N, Pellett PE (1999) Human herpesvirus 6B genome sequence: coding content and comparison with human herpesvirus 6A. J Virol 73: 8040–8052

Endoh D, Ito M, Cho KO, Kon Y, Morimura T, Hayashi M, Kuwabara M (1998) Retroviral sequence located in border region of short unique region and short terminal repeat of Md5 strain of Marek's disease virus type 1. J Vet Med Sci 60:227–235

Fuchs W, Mettenleiter TC (1996) DNA sequence and transcriptional analysis of the UL1 to UL5 gene cluster of infectious laryngotracheitis virus. J Gen Virol 77:2221–2229

Fukuchi K, Sudo M, Lee YS, Tanaka A, Nonoyama M (1984) Structure of Marek's disease virus DNA: detailed restriction enzyme map. J Virol 51:102–109

Hatama S, Jang HK, Izumiya Y, Cai JS, Tsushima Y, Kato K, Miyazawa T, Kai C, Takahashi E, Mikami T (1999) Identification and DNA sequence analysis of the Marek's disease virus serotype 2 genes homologous to the herpes simplex virus type 1 UL20 and UL21. J Vet Med Sci 61:587–593

Hirai K, Ikuta K, Kato S (1981) Restriction endonuclease analysis of the genomes of virulent and avirulent Marek's disease viruses. Microbiol Immunol 25:671–681

Hirai K, Nakajima K, Ikuta K, Kirisawa R, Kawakami Y, Mikami T, Kato S (1986) Similarities and dissimilarities in the structure and expression of viral genomes of various virus strains immunologically related to Marek's disease virus. Arch Virol 89:113–130

Igarashi T, Takahashi M, Donovan J, Jessip J, Smith M, Hirai K, Tanaka A, Nonoyama M (1987) Restriction enzyme map of herpesvirus of turkey DNA and its collinear relationship with Marek's disease virus DNA. Virology 157:351–358

Isfort R, Jones D, Kost R, Witter R, Kung HJ (1992) Retrovirus insertion into herpesvirus in vitro and in vivo. Proc Natl Acad Sci USA 89:991–995

Isfort RJ, Qian Z, Jones D, Silva RF, Witter R, Kung HJ (1994) Integration of multiple chicken retroviruses into multiple chicken herpesviruses: herpesviral gD as a common target of integration. Virology 203:125–133

Izumiya Y, Jang HK, Cai JS, Nishimura Y, Nakamura K, Tsushima Y, Kato K, Miyazawa T, Kai C, Mikami T (1999a) Identification and sequence analysis of the Marek's disease virus serotype 2 gene homologous to the herpes simplex virus type 1 UL52 protein. J Vet Med Sci 61:683–687

Izumiya Y, Jang HK, Kashiwase H, Cai JS, Nishimura Y, Tsushima Y, Kato K, Miyazawa T, Kai C, Mikami T (1998) Identification and transcriptional analysis of the homologues of the herpes simplex virus type 1 UL41 to UL51 genes in the genome of nononcogenic Marek's disease virus serotype 2. J Gen Virol 79:1997–2001

Izumiya Y, Jang HK, Sugawara M, Ikeda Y, Miura R, Nishimura Y, Nakamura K, Miyazawa T, Kai C, Mikami T (1999b) Identification and transcriptional analysis of the homologues of the herpes simplex virus type 1 UL30 to UL40 genes in the genome of nononcogenic Marek's disease virus serotype 2. J Gen Virol 80:2417–2422

Jang HK, Niikura M, Song CS, Mikami T (1997) Characterization and expression of the Marek's disease virus serotype 2 glycoprotein E in recombinant baculovirus-infected cells: initial analysis of its DNA sequence and antigenic properties. Virus Res 48:111–123

Jang HK, Ono M, Kato Y, Tohya Y, Niikura M, Mikami T (1996a) Identification of a potential Marek's disease virus serotype 2 glycoprotein D gene with homology to herpes simplex virus glycoprotein D. Arch Virol 141:2207–2216

Jang HK, Ono M, Kim TJ, Cai JS, Tsushima Y, Niikura M, Mikami T (1996b) Marek's disease virus serotype 2 glycoprotein I gene: nucleotide sequence and expression by a recombinant baculovirus. J Vet Med Sci 58:1057–1066

Jang HK, Ono M, Kim TJ, Izumiya Y, Damiani AM, Matsumura T, Niikura M, Kai C, Mikami T (1998) The genetic organization and transcriptional analysis of the short unique region in the genome of nononcogenic Marek's disease virus serotype 2. Virus Res 58:137–147

Jones D, Brunovskis P, Witter R, Kung HJ (1996) Retroviral insertional activation in a herpesvirus: transcriptional activation of US genes by an integrated long terminal repeat in a Marek's disease virus clone. J Virol 70:2460–2467

Jones D, Lee L, Liu JL, Kung HJ, Tillotson JK (1992) Marek's disease virus encodes a basic-leucine zipper gene resembling the *fos/jun* oncogenes that is highly expressed in lymphoblastoid tumors. Proc Natl Acad Sci USA 89:4042–4046

Jöns A, Granzow H, Kuchling R, Mettenleiter TC (1996) The UL49.5 gene of pseudorabies virus codes for an O-glycosylated structural protein of the viral envelope. J Virol 70:1237–1241

Kato K, Jang HK, Izumiya Y, Cai JS, Tsushima Y, Miyazawa T, Kai C, Mikami T (1999a) Identification and sequence analysis of the Marek's disease virus serotype 2 homologous genes of the herpes simplex virus type 1 UL25, UL26 and UL26.5 genes. J Vet Med Sci 61:787–793

Kato K, Jang HK, Izumiya Y, Cai JS, Tsushima Y, Miyazawa T, Kai C, Mikami T (1999b) Identification of the Marek's disease virus serotype 2 genes homologous to the glycoprotein B (UL27), ICP18.5 (UL28) and major DNA-binding protein (UL29) genes of herpes simplex virus type 1. J Vet Med Sci 61:1161–1165

Kawamura H, King DJ, Anderson DP (1969) A herpesvirus isolated from kidney cell culture of normal turkeys. Avian Dis 13:853–863

Kishi M, Bradley G, Jessip J, Tanaka A, Nonoyama M (1991) Inverted repeat regions of Marek's disease virus DNA possess a structure similar to that of the *a* sequence of herpes simplex virus DNA and contain host cell telomere sequence. J Virol 65:2791–2797

Kishi M, Hrada H, Takahashi M, Tanaka A, Hayashi M, Nonoyama M, Josephs SF, Buchbinder A, Schachter F, Ablashi DV, Wong-Staal F, Salahuddin SZ, Gallo RC (1988) A repeat sequence, GGGTTA, is shared by DNA of human herpesvirus 6 and Marek's disease virus. J Virol 62: 4824–4827

Kitazawa T, Ono M, Maeda K, Kawaguchi Y, Kamiya N, Niikura M, Mikami T (1993) Nucleotide sequence of the glycoprotein C (gC) homologous gene of Marek's disease virus (MDV) serotype 2 and comparison of gC homologous genes among three serotypes of MDV. J Vet Med Sci 55:985–990

Koff A, Tegtmeyer P (1988) Characterization of major recognition sequences for a herpes simplex virus type 1 origin-binding protein. J Virol 62:4096–4103

Kung HJ, Tanaka A, Nonoyama M (1995) Two gene families of Marek's disease virus (MDV) with a potential role in tumor induction in chicken. Int J Oncol 6:997–1002

Lee LF, Liu X, Witter RL (1983) Monoclonal antibodies with specificity for three different serotypes of Marek's disease viruses in chickens. J Immunol 130:1003–1006

Liang X, Chow B, Raggo C, Babiuk LA (1996) Bovine herpesvirus 1 UL49.5 homolog gene encode a novel viral envelope protein that forms a disulfide-linked complex with a second virion structural protein. J Virol 70:1448–1454

Lin JA, Kitagawa H, Ono M, Iwanaga R, Kodama H, Mikami T (1990a) Isolation of serotype 2 Marek's disease virus from birds belonging to genus Gallus in Japan. Avian Dis 34:336–344

Lin JA, Kodama H, Itakura C, Onuma M, Mikami T (1990b) Evaluation of pathogenicity and protective efficacy of serotype 2 Marek's disease virus from birds belonging to genus Gallus in Japan. Jpn J Vet Sci 52:329–337

Lindquester GJ, Pellet PE (1991) Properties of the human herpesvirus 6 strain Z29 genome: G+C content, length, and presence of variable-length directly repeated terminal sequence elements. Virology 182:102–110

Liu JL, Ye Y, Lee LF, Kung HJ (1998) Transforming potential of the herpesvirus oncoprotein MEQ: morphological transformation, serum-independent growth, and inhibition of apoptosis. J Virol 72:388–395

Malkinson M, Orgad U, Becker Y (1986) Use of lectins to detect and differentiate subtypes of Marek's disease virus and turkey herpesvirus glycoproteins in tissue culture. J Virol Methods 13: 129–133

McGeoch DJ, Dalrymple MA, Davison AJ, Dolan A, Frame MC, McNab D, Perry LJ, Scott JE, Taylor P (1988) The complete DNA sequence of the long unique region in the genome of herpes simplex virus type 1. J Gen Virol 69:1531–1574

McVoy MA, Nixon DE, Adler SP, Mocarski ES (1998) Sequences within the herpesvirus-conserved *pac*1 and *pac*2 motifs are required for cleavage and packaging of the murine cytomegalovirus genome. J Virol 72:48–56

Megaw AG, Rapaport D, Avidor B, Frenkel N, Davison AJ (1998) The DNA sequence of the RK strain of human herpesvirus 7. Virology 244:119–132

Mocarski ES, Roizman B (1981) Site-specific inversion sequence of the herpes simplex virus genome: domain and structural features. Proc Natl Acad Sci USA 78:7047–7051

Mocarski ES, Roizman B (1982) Structure and role of the herpes simplex virus DNA termini in inversion, circularization, and generation of virion DNA. Cell 31:89–97

Nazerian K, Solomon JJ, Witter RL, Burmester BR (1968) Studies on the etiology of Marek's disease. II. Finding of a herpesvirus in cell culture. Proc Soc Exp Biol Med 127:177–182

Niikura M, Witter RL, Jang HK, Ono M, Mikami T, Silva RF (1999) MDV glycoprotein D is expressed in the feather follicle epithelium of infected chickens. Acta Virol 43:159–163

Okazaki W, Purchase HG, Burmester BR (1970) Protection against Marek's disease by vaccination with a herpesvirus of turkeys. Avian Dis 14:413–429

Ono M, Katsuragi-Iwanaga R, Kitazawa T, Kamiya N, Horimoto T, Niikura M, Kai C, Hirai K, Mikami T (1992) The restriction endonuclease map of Marek's disease virus (MDV) serotype 2 and collinear relationship among three serotypes of MDV. Virology 191:459–463

Ono M, Jang HK, Maeda K, Kawaguchi Y, Tohya Y, Niikura M, Mikami T (1996) Detection of Marek's disease virus serotype 1 (MDV1) glycoprotein D in MDV1-infected chick embryo fibroblasts. J Vet Med Sci 58:777–780

Ono M, Jang HK, Maeda K, Kawaguchi Y, Tohya Y, Niikura M, Mikami T (1995) Preparation of monoclonal antibodies against Marek's disease virus serotype 1 glycoprotein D expressed by a recombinant baculovirus. Virus Res 38:219–230

Ono M, Kawaguchi Y, Maeda K, Kamiya N, Tohya Y, Kai C, Niikura M, Mikami T (1994) Nucleotide sequence analysis of Marek's disease virus (MDV) serotype 2 homolog of MDV serotype 1 pp38, an antigen associated with transformed cells. Virology 201:142–146

Qian Z, Brunovskis P, Lee LF, Vogt PK, Kung HJ (1996) Novel DNA binding specificities of a putative herpesvirus bZIP oncoprotein. J Virol 70:7161–7170

Qian Z, Brunovskis P, Rauscher III F, Lee L, Kung HJ (1995) Transactivation activity of Meq, a Marek's disease herpesvirus bZIP protein persistently expressed in latently infected transformed T cells. J Virol 69:4037–4044

Reilly JD, Silva RF (1993) The number of copies of an *a*-like region in the serotype-3 Marek's disease virus DNA genome is variable. Virology 193:268–280

Roizman B, Baines J (1991) The diversity and unity of Herpesviridae. Comp Immunol Microbiol Infect Dis 14:63–79

Roizman B, Sears AE (1996) Herpes simplex viruses and their replication. In: Fiels BN, Knipe DM, Howley P, Chanock RM, Hirsch MS, Melnick JL, Monath TP, Roizman B (eds). Field of Virology 3rd ed. Lippincott-Raven, Philadelphia, pp 2231–2295

Roizman B, Carmichael LE, Deinhardt F, de-The G, Nahmias AJ, Plowright W, Rapp F, Sheldrick P, Takahashi M, Worf K (1981) Herpesviridae. Definition, provisional nomenclature, and taxonomy. The herpesvirus study group, the international committee on taxonomy of viruses. Intervirology 16:201–217

Ross NLJ (1999) T-cell transformation by Marek's disease virus. Trends in Microbiol 7:22–29

Ross N, Binns MM, Sanderson M, Schat KA (1993) Alterations in DNA sequence and RNA transcription of the BamHI-H fragment accompany attenuation of oncogenic Marek's disease herpesvirus. Virus Genes 7:33–51

Ross LJ, Milne B, Biggs PM (1983) Restriction endonuclease analysis of Marek's disease virus DNA and homology between strains. J Gen Virol 64:2785–2790

Ross N, O'Sullivan G, Rothwell C, Smith G, Burgess SC, Rennie M, Lee LF, Davison TF (1997) Marek's disease virus EcoRI-Q gene (meq) and a small RNA antisense to ICP4 are abundantly expressed in CD4+ cells and cells carrying a novel lymphoid marker, AV37, in Marek's disease lymphomas. J Gen Virol 78:2191–2198

Schat KA, Calnek BW (1978) Characterization of an apparently nononcogenic Marek's disease virus. J Natl Cancer Inst 60:1075–1082

Schwyzer M, Ackermann M (1996) Molecular virology of ruminant herpesviruses. Vet Micro 53:17–29

Secchiero P, Nicholas J, Deng H, Xiaopeng T, van Loon N, Ruvolo VR, Berneman ZN, Reitz Jr MS, Dewhurst S (1995) Identification of human telomeric repeat motifs at the genome termini of human herpesvirus 7: structural analysis and heterogeneity. J Virol 69:8041–8045

Shimojima Y, Jang HK, Ono M, Kai C, Mikami T (1997a) Identification and DNA sequence analysis of the Marek's disease virus serotype 2 gene homologous to the herpes simplex virus type 1 glycoprotein H. J Vet Med Sci 59:629–634

Shimojima Y, Jang HK, Ono M, Maeda K, Tohya Y, Mikami T (1997b) Identification and DNA sequence analysis of the Marek's disease virus serotype 2 genes homologous to the thymidine kinase and UL24 genes of herpes simplex virus type 1. Virus Genes 14:81–87

Silva RF, Barnett JC (1991) Restriction endonuclease analysis of Marek's disease virus DNA: differentiation of viral strains and determination of passage history. Avian Dis 35:487–495

Smith GD, Zelnik V, Ross LJN (1995) Gene organization in herpesvirus of turkeys: identification of a novel open reading frame in the long unique region and a truncated homologue of pp38 in the internal repeat. Virology 207:205–216

Solomon JJ, Witter RL, Nazerian K, Burmester BR (1968) Studies on the etiology of Marek's disease. I. Propagation of the agent in cell culture. Proc Soc Exp Biol Med 127:173–177

Stow ND, Stow EC (1986) Isolation and characterization of a herpes simplex virus type 1 mutant containing a deletion within the gene encoding the immediate early polypeptide Vmw 110. J Gen Virol 67:2571–2585

Tamashiro JC, Filpula D, Friedmann T, Spector DH (1984) Structure of the heterogeneous L-S junction region of human cytomegalovirus strain AD169 DNA. J Virol 52:541–548

Terford EAR, Watson MS, Aird HC, Perry J, Davison AJ (1995) The DNA sequence of equine herpesvirus 2. J Mol Biol 249:520–528

Telford EAR, Watson MS, Mcbride K, Davison AJ (1992) The DNA sequence of equine herpesvirus-1. Virology 189:304–316

Thomson BJ, Dewhurst S, Gray D (1994) Structure and heterogeneity of the *a* sequences of human herpesvirus 6 strain variants U1102 and Z29 and identification of human telomeric repeat sequences at the genomic termini. J Virol 68:3007–3014

Torelli G, Barozzi P, Marasca R, Cocconcelli P, Merelli E, Ceccherini-Nelli L, Ferrari S, Luppi M (1995) Targeted integration of human herpesvirus 6 in the p arm of chromosome 17 of human peripheral blood mononuclear cells in vivo. J Med Virol 46:178–188

Tsushima Y, Jang HK, Izumiya Y, Cai JS, Kato K, Miyazawa T, Kai C, Takahashi E, Mikami T (1999) Gene arrangement and RNA transcription of the *Bam*HI fragments K and M2 within the non-oncogenic Marek's disease virus serotype 2 unique long genome region. Virus Res 60:101–110

Van Zaane D, Brinkhof JM, Gielkens AL (1982) Molecular-biological characterization of Marek's disease virus. II. Differentiation of various MDV and HVT strains. Virology 121:133–146

Visalli RJ, Brandt CR (1993) The HSV-1 UL45 18kDa gene product is a true late protein and a component of the virion. Virus Res 29:167–178

von Bulow V, Biggs PM (1975a) Differentiation between strains of Marek's disease viruses and turkey herpesvirus by immunofluorescence assays. Avian Pathol 4:133–146

von Bulow V, Biggs PM (1975b) Precipitating antigens associated with Marek's disease viruses and a herpesvirus of turkeys. Avian Pathol 4:147–162

von Bulow V, Biggs PM, Frazier JA (1975) Characterization of a new serotype of Marek's disease herpesvirus. IARC Sci Publ 11:329–336

Wild MA, Cook S, Cochran M (1996) A genomic map of infectious laryngotracheitis virus and the sequence and organization of genes present in the unique short and flanking regions. Virus Genes 12:107–116

Witter RL, Burgoyne GH, Solomon JJ (1969) Evidence for a herpesvirus as an etiologic agent of Marek's disease. Avian Dis 13:171–184

Witter RL, Nazerian K, Purchase HG, Burgoyne GH (1970) Isolation from turkeys of a cell-associated herpesvirus antigenically related to Marek's disease virus. Am J Vet Res 31:525–538

Wu P, Lee LF, Reed WM (1997) Serological characteristics of a membrane glycoprotein gp82 of Marek's disease virus. Avian Dis 41:824–831

Wu P, Sui D, Lee LF (1996) Nucleotide sequence analysis of a 9-kb region of Marek's disease virus genome exhibits a collinear gene arrangement with the UL29 to UL36 of herpes simplex virus. In: Silva RF, Cheng HH, Coussens PM, Lee LF, Velicer LF (eds) Current Research on Marek's Disease. American Association of Avian Pathologists, pp 219–224

Xia K, DeLuca NA, Knipe DM (1996) Analysis of phosphorylation sites of herpes simplex virus type 1 ICP4. J Virol 70:1061–1071

Xie Q, Anderson AS, Morgan RW (1996) Marek's disease virus (MDV) ICP4, pp38, and *meq* genes are involved in the maintenance of transformation of MDCC-MSB1 MDV-transformed lymphoblastoid cells. J Virol 70:1125–1131

Yanagida N, Yoshida S, Nazerian K, Lee LF (1993) Nucleotide and predicted amino acid sequences of Marek's disease virus homologues of herpes simplex virus major tegument proteins. J Gen Virol 74:1837–1845

Yoshida S, Lee LF, Yanagida N, Nazerian K (1994) Identification and characterization of a Marek's disease virus gene homologous to glycoprotein L of herpes simplex virus. Virology 204:414–419

Zelnik V, Darteil R, Audonnet JC, Smith GD, Riviere M, Pastorek J, Ross LJ (1993) The complete sequence and gene organization of the short unique region of herpesvirus of turkeys. J Gen Virol 74:2151–2162

Zelnik V, Tyers P, Smith GD, Jiang CL, Ross NLJ (1995) Structure and properties of a herpesvirus of turkeys recombinant in which US1, US10 and SORF3 gene have been replaced by a *lacZ* expression cassette. J Gen Virol 76:2903–2907

Zhu GS, Iwata A, Gong M, Ueda S, Hirai K (1994) Marek's disease virus type 1-specific phosphorylated proteins pp38 and pp24 with common amino acid termini are encoded from the opposite junction regions between the long unique and inverted repeat sequences of viral genome. Virology 200:816–820

Ziemann K, Mettenleiter TC, Fuchs W (1998) Infectious laryngotracheitis herpesvirus expresses a related pair of unique nuclear proteins which are encoded by split genes located at the right end of the U_L genome region. J Virol 72:6867–6874

Zimmermann J, Hammerschmidt W (1995) Structure and role of the terminal repeats of Epstein-Barr virus in processing and packaging of virion DNA. J Virol 69:3147–3155

Marek's Disease Virus Latency

R.W. Morgan[1], Q. Xie[2], J.L. Cantello[3], A.M. Miles[1], E.L. Bernberg[1], J. Kent[1], and A. Anderson[1]

1 Introduction

Marek's disease (MD) is one of the most fascinating herpetic disorders because infection of susceptible chickens results in the rapid formation of T-cell lymphomas that are preventable by vaccination (Calnek and Witter 1997). For almost half a century, the commercial poultry industry has been dependent on effective MD vaccines. Because Marek's disease virus (MDV) is a relatively difficult herpesvirus to study, information regarding it has lagged behind that for some other herpes-

[1] Delaware Agricultural Experiment Station, Department of Animal and Food Sciences, College of Agriculture and Natural Resources, University of Delaware, Newark, DE 19717-1303, USA
[2] Institute for Human Gene Therapy, The Wister Institute, University of Pennsylvania, Philadelphia, PA 19104, USA
[3] Regeneron Pharmaceuticals, 777 Old Saw Mill River Road, Tarrytown, NY 10591, USA

viruses. Latency is no exception to this general statement. MDV latency appears somehow connected to transformation, although the relationship between these two states is by no means clear. For example, it is not known whether latency is a precursor to transformation or whether it is a parallel manifestation of disease. In this review, I will attempt to define latency for MDV, describe systems in which it can be studied, review what is known about transcription during latency, and point out gaps in our existing understanding of this mode of infection.

1.1 Definition of Latency

Herpesvirus latency is defined as the presence of the viral genome without production of infectious virus except during episodes of reactivation (Feldman 1991; Fraser et al. 1992; Garcia-Blanco and Cullen 1991; Stevens 1989; Wagner 1991). Latency is distinguished from both abortive and nonpermissive infections in that it is reversible. Maintenance of a reservoir of latently infected cells within an immunocompetent host is critical to viral persistence. The molecular details of latency vary among herpesviruses and are intimately related to the nature of the host cell that is latently infected. Herpesviruses that latently infect neurons, such as herpes simplex virus type 1 (HSV-1), exhibit very little viral gene expression during latency since the neuron is a nondividing cell. On the other hand, herpesviruses that latently infect lymphocytes, such as Epstein-Barr virus, exhibit more viral gene expression during latency. These lymphotropic herpesviruses must replicate to the extent that they ensure their continued presence in dividing lymphocytes.

MDV assumes a latent posture in T cells (Calnek and Witter 1997; Calnek et al. 1981, 1984) and can be reactivated by cell culture propagation of T cells isolated from infected chickens (Calnek et al. 1981a, 1984). In transformed T cells, whether tumors or lymphoblastoid cells, MDV is in some form of latent state. However, it is not clear that latency in transformed cells precisely mirrors latency in non-transformed T cells. In addition, it seems reasonable that various states of latency exist certainly among lymphoblastoid cell lines, perhaps among tumors, and even perhaps among non-transformed latently infected cells.

1.2 Systems for Studying MDV Latency

Systems for studying MDV latency can be divided into those that involve non-transformed cells and those that use transformed cells. One system that falls in the former class is simply the use of non-transformed T cells isolated from MDV-infected chickens (Calnek et al. 1981a). Latency is difficult to study in non-transformed T cells because of problems with achieving long-term culture of these cells. Maintaining latency in T-cell cultures derived from MDV-infected chickens is even more difficult since the virus readily reactivates once the cells are placed in culture. This system has been most useful for defining factors that maintain latency of T cells (discussed in Sect. 5) (Buscaglia and Calnek 1988). In addition, herpesvirus of turkeys (HVT) has been used to study latency of chicken herpesviruses outside the context of transforming infections (see Sect. 2) (Holland et al. 1996, 1998, 1999).

Most studies regarding latency have been done on lymphoblastoid cell lines. These cell lines are derived from MDV-induced tumors and many have been characterized to varying degrees (Nazerian 1987). In lymphoblastoid cell lines, MDV gene expression appears to be carefully regulated with the majority of transcription being derived from the repeat regions of the viral genome. Most lytic genes are not expressed in these cell lines, but treatment with iododeoxyuridine and other nucleoside analogs induces expression of lytic genes and production of viral antigens (Dunn and Nazerian 1977; Calnek et al. 1981).

Recent work involving lymphoblastoid cell lines derived from tumors induced by recombinant MDV strains indicates that marker genes expressed from a lytic promoter (SV40 early) inserted into the MDV genome are not efficiently expressed in lymphoblastoid cells, indicating that the genome in these cells is maintained in a latent state. However, marker gene expression is efficient in MDV that has been chemically reactivated from these cells, regardless of whether the marker gene is *lacZ* (Parcells et al. 1999) or *gfp* (Dienglewicz and Parcells 1999). Regulation of gene expression in these recombinant virus-derived cell lines does not appear to be mediated by methylation. Since they contain a marked virus, these cell lines provide a very useful system for studying MDV latency and reactivation.

Another model system for studying MDV latency is the fibroblastic OCL cell line system. These cell lines are derived from OU2 cells, chemically transformed chicken embryo fibroblasts (CEF) (Ogura and Fujiwara 1987). Two of the cell lines, OU2.1 and OU2.2, harbor MDV in a latent state when the cells are sub-confluent (Abujoub and Coussens 1995, 1996, 1997). When these MDV-infected cell lines reach confluency, MDV is reactivated. Related cell lines have been described that harbor serotype 2 MDV (SB1-OCL) and serotype 3 HVT (HVT-OCL) also in the latent state until the cells reach confluency (Abujoub and Coussens 1999). Although these cell lines have not been widely accepted as CEF substitutes for MDV vaccine production, they do offer a model system that has yet to be fully exploited for studying latency and reactivation of all three serotypes.

Avian leukosis virus (ALV)-induced B cell tumors can harbor latent serotype 2 MDV. In one report, latent SB-1 was discovered in B-cell lines established from ALV-induced tumors present in MDV-vaccinated chickens (Fynan et al. 1993). Fewer than 3% of these cells express MDV antigens unless 5-azacytidine is present during culture to prevent methylation. A similar finding has been reported for LSCC-BK3 clone A (BK3A), an ALV-induced B-cell line that harbors a serotype 2 MDV designated 2H (Hihara et al. 1998). About 40% of BK3A cells express a serotype 2 MDV antigen, but another cell line established from the same ALV-infected chicken (LSCC-BK3 clone 2C) does not express serotype 2 MDV antigens.

2 The MDV Genome During Latency

Ross pointed out that latently infected, non-transformed T cells contain very few copies of the MDV genome. In particular, MDV DNA is not detectable in these cells by in situ hybridization, with the limit of detection being five genome copies/cell. These cells do appear to be latently infected, however, for when cultured, viral antigens are produced and viral DNA is detectable by in situ hybridization (Ross 1985).

Latently infected, transformed cells from both lymphoblastoid cell lines and tumors harbor greater numbers of the MDV genome (Ross et al. 1981). For lymphoblastoid cell lines, the number of MDV genome copies present reflects the nature of the cell line. In particular, MDCC-HPRS1 and MKT1 have been reported to harbor 8 to 15 genome copies (Kaschka-Dierich et al. 1979; Tanaka et al. 1978), but MSB1 cells can harbor from 60 to more than 100 genome copies (Kaschka-Dierich et al. 1979; Nazerian and Lee 1976).

Viral DNA in lymphoblastoid cells and in tumor cells appears to be generally intact (Ross 1985). In lymphoblastoid cells, the viral genome exists in both episomal and integrated forms depending on the cell line (Kaschka-Dierich et al. 1979; Tanaka et al. 1978). Using Southern hybridization, Gardella gel electrophoresis, and in situ hybridization of metaphase and interphase chromosomes, it has been demonstrated that integration of viral DNA into the host chromosome occurs readily (6/6 cell lines examined) (Delecluse and Hammerschmidt 1993). The preferred integration sites vary among integration events but generally lie near the telomeres of large- and mid-sized chromosomes or on the minichromosomes. Covalently closed circular DNA is occasionally seen (1/6 cell lines examined), and most of the cell lines (4/6) have a small population of cells undergoing lytic infection and producing linear MDV genomes. When similar techniques were applied to the analysis of the MDV genome in lymphomas, episomal forms of the viral DNA were not found, viral DNA was randomly integrated at multiple chromosomal sites, and clonality of tumors was obvious from the integration patterns (Delecluse et al. 1993).

3 Transcription During Latency

3.1 General Patterns of Transcription in Latently Infected Cells

3.1.1 MDV

An early report indicated that 12–14% of the MDV genome is transcribed in lymphoblastoid cells (Silver et al. 1979). Northern analysis on polyA+ RNA purified from three lymphoblastoid cell lines (MDCC-MSB1, MDCC-RP1, and MDCC-LS1) indicated that the majority of transcripts hybridize to the repeats flanking the unique long (UL) region of the genome (Hirai et al. 1988). Another report indicated that only a few transcripts (< 10) are detectable in lymphoblastoid cell lines [MDCC-HP1 (non-producer) and MDCC-CU41 (non-expression)], and these are derived from immediate-early genes that map in the unique short (US) region and in the repeat sequences flanking the UL (IRL) and US (IRS) regions (Schat et al. 1989). Others have reported that 29–32 transcripts are derived from many sites on the genome (Maray et al. 1988). Using polyA + RNA and probing Northern blots with sequences representing about 95% of the MDV genome, Sugaya et al. (1990) reported that 29 viral transcripts are present in kidney lymphomas and 32 in MKT-1 cells. These transcripts hybridize to only 20% of the genome, the relevant regions being the IRL, IRS, and the adjacent unique sequences (Sugaya et al. 1990).

Given the difficulties of the MDV system, and the lack of genomic sequence information at the time of these studies, they are remarkably consistent. Minor differences could reflect the nature and passage history of the lymphoblastoid cell lines or the state of latency/reactivation of the tumors at the time of harvest for RNA purification. Taken together, these results indicate that transcription of the MDV genome in MDV-transformed lymphoblastoid cells and tumors is more extensive than that of the latent HSV-1 genome, but less extensive than that of the MDV genome in productively infected cells. The repeat regions of the genome are particularly active transcriptionally during latency and these RNAs are, to some extent, also present in cytolytically infected cells.

3.1.2 HVT

Transcription of the HVT genome during latent infections has been studied by in situ hybridization (Holland et al. 1996). Latent infections in these studies are defined as infections that are positive for in situ hybridization with probes from the repeat regions of the genome but negative for in situ hybridization with a glycoprotein B (gB)-specific probe, gB transcription being indicative of productive infection. Latent infections have been reported for lymphoid, nerve, and feather tissues of HVT-infected chickens (Holland et al. 1998). Moreover, the patterns of latent gene expression were found to differ among spleen, thymus, and nerves, suggesting that even in latency, there is tissue specificity to gene expression patterns (Holland and Silva 1999).

3.2 Analysis of cDNAs from Lymphoblastoid Cells and Tumors

Analysis of cDNAs derived from the *Bam*HI-H region of the genome led to the identification of the pp38/pp24 gene (Chen et al. 1992). See Sect. 4.2 for a more extensive discussion of pp38/pp24.

Two cDNAs derived from the *Bam*HI-I2 region of the MDV genome present in MKT-1 cells have been analyzed in detail (Peng et al. 1995). One of these cDNAs extends into the adjacent *Bam*HI-Q2 and -L fragments, represents an abundantly expressed transcript in lymphoblastoid cells, and corresponds to the *meq* gene. The other cDNA is also related to the *meq* transcript but contains only the DNA-binding domain sequence spliced to sequences present in *Bam*HI-L. Both transcripts are present in both latently infected lymphoblastoid cells and lytically infected cells. A more detailed discussion of Meq is given in Sect. 4.1 and also in another chapter of this volume.

Analysis of cDNAs prepared from polyA + RNA purified from MDCC-CU41 lymphoblastoid cells led to the identification of two cDNA clones that correspond to transcripts derived from the *Bam*HI-Q2 and -L regions (Ohashi et al. 1994b). These transcripts are 2.5, 0.8, and 0.6kb in size. The 0.6-kb transcriptional unit contains an ORF, termed ORF L1, that could encode a 107-amino-acid protein. ORF L1 corresponds to the *meq* gene and is related to the cDNAs described by Peng et al. (1995) (see above). Analysis of two MDV strain CVI988-based mutants in which ORF L1 has been disrupted indicated that ORF L1 is not required for virus replication in cell culture or in chickens, nor is it required for the establishment or reactivation from latency in chickens (Schat et al. 1998).

This same MDCC-CU41 cDNA library yielded three groups of MDV-specific clones that map to the IRS located within *Bam*HI-A. It is not entirely clear that these cDNAs represent latent transcripts since the amounts of the corresponding transcripts present in lymphoblastoid cells are much less than in lytically infected cells. One of the clones (A41) contained a small open reading frame (ORF). Antibody prepared against a bacterial fusion protein consisting of glutathione-*S*-transferase and the A41 ORF binds to a cytoplasmic antigen in lytically infected cells, but the antigen is not detectable in lymphoblastoid cells (Ohashi et al. 1994a).

3.3 Latency-Associated Transcripts and the RB1BLAT*lac* Mutant

We (Cantello et al. 1994, 1997), and others (Li et al. 1994, 1998; McKie et al. 1995), identified and characterized a group of latency-related RNAs all of which map antisense to the MDV ICP4 gene. Figure 1 shows a summary of our current understanding of the structures of the latency-associated transcripts (LATs) of MDV. The LATs reported so far include two small, spliced RNAs (0.9 and 0.75kb), named MSRs (Marek's disease virus small RNAs) (Cantello et al. 1994, 1997). The difference between the MSRs is unknown. The MSRs have also been referred to as SARs (Ross et al. 1996) and S RNA (Li et al. 1998). The splice junctions of the MSRs during latent and late productive infection were determined by sequencing RNA-PCR products generated with primers that flank the 3′ splice region (Cantello et al. 1997). The MSR contains four introns, the largest of which is 4852nt long and overlaps the ICP4 putative translational start site. In addition, a large 10-kb RNA first discovered by Northern analysis (Cantello et al. 1994) and also referred to as L RNA (Li et al. 1998) has been characterized. Other latency-related RNAs include three spliced, 3′ coterminal RNAs discovered by analyzing cDNAs (2.2, 1.8, and 0.5kb in size) that were grouped into classes based on three different splicing patterns (McKie et al. 1995). An additional 2.7-kb cDNA having

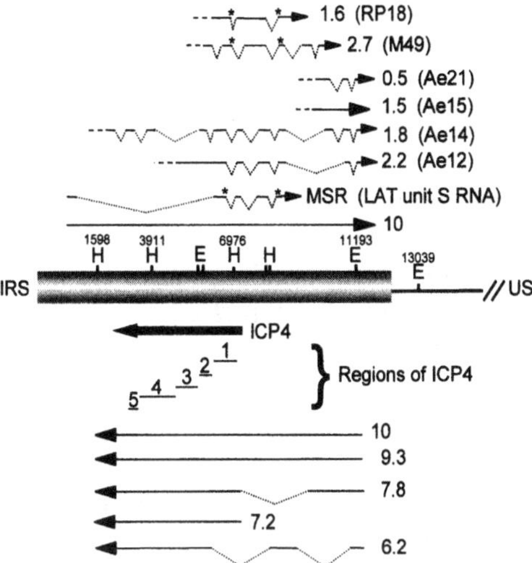

Fig. 1. Diagram of the known latency associated transcripts (*LATs*) and ICP4 transcripts of serotype 1 MDV. The *center portion* of the figure shows a portion of the MDV IRS and the US region. The positions of *Eco*RI and *Hind*III recognition sites are indicated and, in some cases, numbered using the IRS sequence published by McKie et al. (1995). The regions of the predicted ICP4 coding region are diagrammed (Anderson et al. 1992). The *upper portion* of the figure shows the LATs of MDV and their splicing patterns. *Asterisks* denote common splice sites. The *dashes* to the left of spliced LATs indicate that the precise 5′ ends of these species are unknown since they were identified by characterization of cDNAs. The sizes of the LATs are indicated to the *right* of the figure and alternative names for these RNAs used by Li et al. (1998) are indicated in parentheses. The 2.2-, 1.8-, and 0.5-kb cDNAs were identified by McKie et al. (1995), the 1.6- and 2.7-kb cDNAs were identified by Li et al. (1994 and 1998, respectively), the 10-kb RNA was identified by Cantello et al. (1994) and Li et al. (1994), and the MSR was identified by Cantello et al. (1994). The MSR is the same as SAR reported by Li et al. (Ross et al. 1996, 1997). In the *lower portion* of the figure, transcripts corresponding to the ICP4 coding region and its upstream region (which is in frame with the coding region) and their known splice sites are diagrammed (Morgan et al. 1996)

a different splicing pattern defines yet another LAT derived from this region of the genome (Li et al. 1994), as does a 1.6-kb cDNA (Li et al. 1998). Each of the small RNAs has a unique overall splicing pattern, although some splice donor and acceptor sites are shared. The MSRs are nonpolyadenylated, but the 10-kb RNA contains polyA (Cantello et al. 1997). The other small RNAs are presumed to be polyadenylated, since they were discovered by the analysis of oligo-dT-primed cDNAs (Li et al. 1994; McKie et al. 1995).

Based on RNase protection studies and Northern hybridizations, the 5′ end of the MSR has been placed at position 949 [numbering is based on the published sequence of the MDV IRS (McKie et al. 1995)] approximately 5kb upstream from the main body of the transcript. The nearest TATA box (TATAA) is located 26nt upstream at position 919–923, and another TATA box (TTATAT) is located 49nt upstream at positions 895–900. There is a potential αTIF-binding site 73nt upstream at position 866–876. αTIF is a component of the HSV-1 tegument that recognizes the core consensus sequence TAATGARAT in the promoter regions of immediate-early genes (Kristie and Roizman 1987), and a potential αTIF recognition site lies within the promoter regions of HSV-1 LATs (Batchelor and O'Hare 1990; Lagunoff and Roizman 1995; Zwaagstra et al. 1990, 1991). It is possible that the other small LATs initiate at this same promoter since they were defined by cDNAs that might not have been full-length.

Although the precise 5′ end of the MDV 10-kb LAT is not known, based on Northern hybridizations and RNase protection data, it appears that the end is identical to or lies very near to the 5′ end of the MSR (Cantello et al. 1997; Li et al. 1998). The 10-kb RNA lies entirely within the repeats flanking the US region of the genome. Furthermore, the large LAT overlaps almost perfectly a large 10-kb ICP4 transcript (Morgan et al. 1996; Li et al. 1998). It should be noted that remarkably similar results regarding the nature of this transcript have been obtained by two groups working independently with different cell lines and strains (Cantello et al. 1997; Li et al. 1998).

MDV LATs have been detected in at least three lymphoblastoid cell lines, namely MSB1 cells (Cantello et al. 1994; Li et al. 1994; McKie et al. 1995), MKT-1 cells (McKie et al. 1995), and RPL1 cells (Li et al. 1994), and in MDV-induced lymphoma tissue (Cantello et al. 1997; Li et al. 1994). Although these latency-related RNAs are abundant in MDV-induced lymphoblastoid cells, they are also detectable in productively infected CEF (Li et al. 1994; McKie et al. 1995), particularly late during infection (Cantello et al. 1997). At least for the MSRs and the 10-kb RNA, iododeoxyuridine treatment of MSB1 cells, which results in MDV reactivation, causes a decrease in the steady-state levels of the latency-related RNAs but an increase in the steady-state level of ICP4 RNA (Cantello et al. 1994). The pattern of expression of these LATs suggests that they are inversely related to MDV productive infection (Cantello et al. 1994; Li et al. 1994). Synthesis of the large RNA (and possibly all of the LATs) is sensitive to cycloheximide treatment and therefore requires protein synthesis (Li et al. 1998).

We have constructed a mutant, designated RB1BLAT*lac*, in which we disrupted the 5′ end of the MSR at the LAT promoter region. The transfer vector used to make the LAT mutant consisted of pMD206 (Cantello et al. 1997) in which the *Msc*I site at position 1055 was disrupted by addition of a *lacZ* cassette. The *Msc*I site lies just downstream from the TATA boxes, and disruption of this site should knock out transcripts that initiate at this particular promoter. The mutant was constructed in the genetic background of the oncogenic RB1B strain of MDV using procedures that we have used previously to construct RB1B-based mutants.

Southern hybridization indicated that the mutant had the expected genomic structure and did not contain parent virus (Fig. 2). Total DNAs from CEF infected with the parent virus (RB1Bp19), a derivative of the parent virus that was passed alongside the mutant during its construction (RB1Bp28), or three of the ten putative LAT mutants (RB1BLAT*lac*1-10) were digested with *Spe*I and *Nco*I, and Southern blots were prepared using standard methods. The MDV-specific probe detected a 2.2-kb fragment in DNA from the parent viruses. This fragment was not present in DNA from CEF infected with the putative RB1BLAT*lac* mutants and instead the MDV-specific probe detected two fragments of 5.5kb and 0.6kb, as expected. The *lacZ*-specific probe detected nothing in DNA from CEF infected

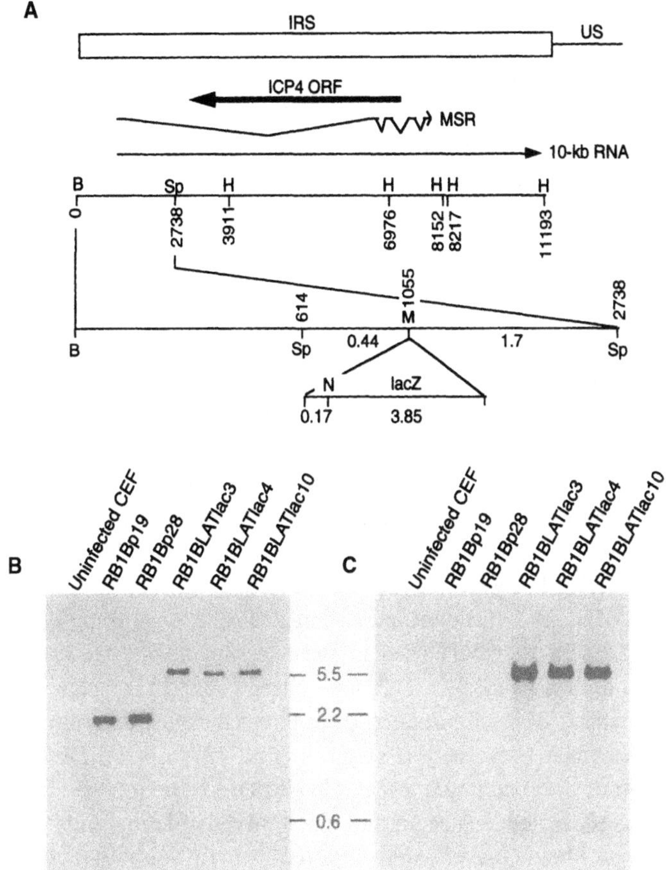

Fig. 2A–C. Southern hybridization analysis of the RB1BLAT*lac* mutant and the corresponding parent MDV strain. **A** Diagram of the relevant region of the MDV genome. Nucleotide positions correspond to the published IRS sequence (McKie et al. 1995). The positions of the ICP4 ORF, the MSR, and the 10-kb LAT are indicated. *B*, *Bam*HI; *H*, *Hind*III, *M*, *Msc*I; *N*, *Nco*I; *Sp*, *Spe*I. **B** The MDV-specific probe was a 1-kb *SpeI*/*Hind*III fragment purified from pMD206 (Cantello et al. 1997), which spans the *Msc*I insertion site. **C** The *lacZ*-specific probe was a 4.0-kb *Bam*HI fragment purified from p*lacZ* (Sondermeijer et al. 1993). For **B** and **C**, each lane contains 10μg of digested DNA and sizes are indicated in kb pairs. The MDV-specific probe detected a 2.2-kb fragment in DNA from the parent viruses. This fragment was not present in DNA from CEF infected with the putative RB1BLAT*lac* mutants and instead two fragments of 5.5kb and 0.6kb were present. The *lacZ*-specific probe detected nothing in DNA from CEF infected with the parent viruses but specifically hybridized to the 5.5-kb and 0.6-kb fragments in DNA from CEF infected with the putative RB1BLAT*lac* mutants

with the parent viruses but specifically hybridized to the 5.5-kb and 0.6-kb fragments in DNA from CEF infected with the putative RB1BLAT*lac* mutants. Thus, the Southern hybridization pattern indicated that insertion of the *lacZ* mutagenesis cassette occurred as expected and that the mutant virus stocks did not contain parent virus. In addition, the mutant stocks all appeared to be diploid; i.e., the *lacZ* cassette was present in both the IRS and TRS copies of the LAT promoter. We used PCR to confirm diploid insertion of the *lacZ* cassette and the absence of parental virus in our stock (data not shown). Finally, standard PCR reactions

(SILVA 1992; ZHU et al. 1992) were used to verify that the RB1BLAT*lac* mutant stocks had not sustained expansion of the 132-bp repeat region but had a pattern of repeats that was indistinguishable from that of the parent viruses. The parent viruses and all of the mutants had 1–3 copies of the 132-bp repeat (data not shown).

Northern hybridizations indicated that the defect in the RB1BLAT*lac* mutant had some effect on, but did not totally ablate, LAT transcription (data not shown). It is likely that the observed transcription stems from numerous promoter-like elements downstream from the cassette insertion site. Indeed, a TATA-less HSV-1 LAT promoter, designated LAP2, is located between LAP1 (the major LAT promoter) and the 5′ end of the 2.0-kb LAT (GOINS et al. 1994). LAP2 shows similarity to promoters of housekeeping genes. Secondary promoters for the MDV LATs remain to be described.

The RB1BLAT*lac* mutant and its parent virus have been tested in chickens and the results are shown in Table 1. One-day-old specific-pathogen-free single comb white leghorn chickens (SPAFAS, Norwich, Conn., USA) were wingbanded, randomly assigned to experimental groups, housed in isolation units, and maintained under negative pressure. At 1 day of age, chickens were inoculated intra-abdominally with the RB1Bp19 original parent virus, the equal passage level RB1B virus (RB1Bp28), the RB1BLAT*lac* mutant, mock-infected cells, or nothing at all. Titers were done on the actual inocula used in the experiment and are shown in Table 1. At 6, 12, and 18 days post-inoculation, spleen cells, thymocytes, and peripheral blood leukocytes (PBL) were isolated from three chickens per group for virus reisolations as described previously (PARCELLS et al. 1994a,b, 1995). Two weeks after the initiation of the experiment, 1-day-old chickens were obtained from the same breeder flock as the original chickens and placed in the isolation units with the 14-day-old inoculates. These chickens were exposed to MDV only by contact with the previously inoculated chickens, which at this time should have been shedding MDV. Two weeks after placement of the contact-exposed chickens (28 days after the initiation of the experiment), virus reisolations from spleen,

Table 1. Incidence of tumors in chickens exposed to RB1BLAT*lac*

Virus[a]	Inoculates		Contacts	
	No. of chickens[b]	No. of tumors (%)	No. of chickens[b]	No. of tumors (%)
None	4	0 (0)	4	0 (0)
RB1Bp19				
232 PFU	15	9 (60)	8	6 (75)
RB1Bp28				
151 PFU	12	0 (0)	10	0 (0)
1560 PFU	14	4 (29)	10	3 (30)
RB1BLAT*lac*p28				
169 PFU	15	0 (0)	9	0 (0)
2010 PFU	15	0 (0)	11	0 (0)

[a] For inoculates, the dose in PFU/chicken is indicated. For contacts, the indicated dose is that inoculated into cage-mates. RB1Bp19 is the original parent strain, RB1Bp28 is a derivative of the parent strain that was passaged in cell culture alongside the mutant during its construction, and RB1BLAT*lac* is the mutant.

[b] Number of chickens with gross Marek's disease lesions.

thymus, and PBL were done from three contact-exposed birds and three inoculates from each group. Both inoculates and contact-exposed chickens were observed daily for mortality until 8 weeks after the initiation of the experiment. At the end of the 8-week period, all remaining chickens were euthanized. Chickens were examined for gross MDV lesions such as tumors in the liver, spleen, gonads, kidney, heart, and intestines.

In vivo results indicated that the RB1BLAT*lac* mutant failed to cause tumors in either inoculates or contact-exposed chickens (Table 1). We do not know whether the lack of oncogenicity is due to the specific mutation at the putative LAT promoter or whether there have been other changes in this virus that result in an attenuated phenotype. We saw reduced tumorigenicity of the equally passaged parent virus, but tumors were clearly evident in both inoculates and contacts exposed to this parent virus derivative. Firm conclusions on the effect of impairing LAT expression in vivo await the construction of rescued derivatives of this mutant or of additional mutants.

4 Viral Proteins Expressed During Latency

The viral proteins discussed in this section are expressed in lymphoblastoid cells and/or tumors. It has not been possible to examine their expression in non-transformed latently infected cells. Nevertheless, they represent a group of MDV gene products whose expression parallels latency in some way and they have been considered to be potentially important in oncogenesis. The literature on these proteins is, in some cases, fairly extensive. Therefore, they are only briefly reviewed here.

4.1 Meq

An extensive discussion of Meq appears elsewhere in this volume. The *meq* gene was first identified because it is expressed in MDV-transformed lymphoblastoid cells (Jones et al. 1992; Tillotson et al. 1988). The gene encodes a 339-amino-acid bZIP protein that resembles the Fos/Jun family of oncoproteins. In addition to lymphoblastoid cells (Liu et al. 1998), the *meq* gene is expressed in lymphomas (Ross et al. 1997) and in lytically infected cells (Jones et al. 1992).

The Meq protein has domains for DNA binding, dimerization, transactivation, and intracellular localization (Qian et al. 1995, 1996). The amino terminal bZIP region of the protein is comprised of a basic region and a leucine zipper. The basic region has been further divided into two subregions; namely, BR1, which provides an auxiliary signal for nuclear translocation, and BR2, which has both the primary nuclear localization signal and the nucleolar localization signal (Liu et al. 1997). The leucine zipper can promote dimerization with other Meq molecules as well as association with c-Jun, p53, and probably other proteins (Brunovskis et al. 1996; Qian et al. 1996). The carboxyl-terminal region of Meq contains a proline-rich activation domain, which, when fused to the yeast Gal4 DNA-binding domain, transactivates the chloramphenicol acetyltransferase reporter. The last 33 amino acids of Meq are essential for transactivation and contain a consensus sequence for an RNA-binding motif.

A variant of Meq has been described, termed Meq-sp, which includes the DNA binding/dimerization domain of Meq but lacks the transactivation domain (Peng and Shirazi 1996a). This variant can bind DNA and complex with Meq or with Jun, but cannot transactivate. It has been proposed that Meq-sp regulates Meq activity by competing for heterodimer formation and DNA binding. The C-terminus of Meq-sp shares significant homology with the CXC chemokine IL-8 (Liu et al. 1999a). Further analysis facilitated the identification of vIL-8, an MDV gene that shares its second and third exons with Meq-sp, but which has a signal peptide encoded within its unique first exon (Liu et al. 1999).

The *meq* promoter lies on a 268-kb *Eco*RI-*Xmn*I fragment that contains two tail-to-tail copies of the consensus heat-shock regulatory element (Jones and Kung 1992) and an AP-1-like sequence (Qian et al. 1995). The *meq* promoter can be transactivated by Meq-cJun heterodimers but transactivation by Meq-Meq homodimers is inefficient (Qian et al. 1995). Two sequences have been described, MERE1 and MERE2, to which Meq can bind (Qing et al. 1996). MERE1 contains a TRE or CRE core flanked by specific sequences and has significant homology to the Maf oncoprotein recognition site (Katoaka et al. 1994; Kerppola and Curran 1994). MERE1 is located in the *meq* promoter region and, when bound by Meq and c-Jun heterodimers, transcription of *meq* can be activated (Liu et al. 1999). MERE2 contains an ACACACA core flanked by important but varying sequences. A MERE2 site lies in the putative MDV origin of replication suggesting a possible role for Meq in MDV replication.

To further complicate matters, two cDNAs which map to the right end of the *Bam*HI-I2 region and which are derived from mRNAs that are antisense to *meq* transcripts have been described. These cDNAs encode 135- and 195-amino-acid polypeptides. Antibodies raised against the 195-amino-acid polypeptide detected a 23-kDa nuclear protein in lymphoblastoid cells (Peng and Shirazi 1996b). The function of this protein or of these transcripts remains unknown.

Meq is probably the strongest candidate oncoprotein or co-oncoprotein known for MDV. Its expression is important for the maintenance of transformation of MDCC-MSB1 lymphoblastoid cells (Xie et al. 1996). Overexpression of Meq results in transformation of Rat-2 cells (Liu et al. 1998). In transformed Rat-2 cells, apoptosis is inhibited (Liu et al. 1997). This inhibition of apoptosis has been attributed, at least in part, to induction of *bcl* expression and suppression of *bax* expression. Meq requires a complementing oncoprotein such as vRas to transform primary cells such as fibroblasts, suggesting that Meq requires a cooperating oncoprotein(s) to realize its full oncogenic potential (Liu et al. 1998).

The intracellular location of Meq further suggests a role in transformation. Meq has been shown to localize in the coiled bodies of the nucleolus of infected cells (Liu et al. 1997). In transformed Rat-2 cells, Meq can co-localize with cyclin-dependent kinase 2 (CDK2) in the coiled bodies and nucleolar periphery, suggesting that Meq may affect the subcellular localization of CDK2 in these cells (Liu et al. 1999). The interaction between Meq and CDK2 is particularly evident during the G_1/S boundary and early S phase of the cell cycle. In addition, CDK2 can phosphorylate Meq at serine 42 and phosphorylated Meq has decreased DNA-binding activity. Thus, it appears that Meq and CDK2 may modify each other.

Meq exhibits a number of activities common to the DNA tumor virus oncoproteins such as SV40 T antigen, adenovirus E1A and E1B, and human papillomavirus E6 and E7 (McCance 1998). These activities include sequestering p53 and simultaneously inhibiting apoptosis. These activities are important for DNA tumor viruses since these viruses depend on the cell for their propagation. They must poise cells for DNA replication and simultaneously short-circuit apoptotic pathways that become stimulated as a result of inappropriate cell cycle progression.

4.2 pp38/pp24

The pp38 gene was first identified by λgt11 cloning using a monoclonal antibody that detected a 38-kDa viral antigen expressed in lymphoblastoid cell lines and in tumor cells (Cui et al. 1990). The pp38 gene maps to the *Bam*HI-H fragment, spans the IRL/UL junction, and encodes a 290-amino-acid ORF (Cui et al. 1991; Chen et al. 1992). The promoter for the pp38 gene overlaps that of the major 1.8-kb family of transcripts (see Sect. 4.3) and the gene is transcribed into a 1.8-kb unspliced mRNA leftward from this promoter in the direction opposite that of transcription of the 1.8-kb family. Antisera from chickens with MD will immunoprecipitate pp38. Homologs to the pp38 gene have been found in the genomes of serotype 2 (Ono et al. 1994; Cui et al. 1992) and 3 (HVT) (Cui et al. 1992) strains. A direct role of pp38 in MDV transformation is complicated by the fact that the gene is also expressed by nononcogenic attenuated serotype 1 strains and by nononcogenic serotype 2 and 3 strains, and there is no evidence showing that pp38 is related to known oncoproteins. Nevertheless, an indirect role of pp38 in pathways leading to transformation is quite possible, and pp38 has been called an MDV tumor antigen (Chen et al. 1992).

A related ORF, which maps to the *Bam*HI-D region of the genome, has been described (Becker et al. 1994; Makimura et al. 1994; Zhu et al. 1994). This 465-bp ORF encodes a 155-amino-acid protein named pp24. Along with pp38, pp24 is a member of a group of MDV phosphorylated proteins that can be immunoprecipitated using monoclonal antibodies against this phosphorylated protein complex (Zhu et al. 1994). The amino-terminal 65 amino acids of pp24 are shared with the amino-terminus of pp38 since the coding regions for these segments of the proteins are derived from sequences lying within TRL and IRL. In particular, residues 45–50 of the proteins comprise a shared antigenic domain. A function for pp24 is unknown at this time.

4.3 1.8-kb Gene Family

Comparison of oncogenic strains and their attenuated derivatives, which lose the ability to spread horizontally and to cause tumors in chickens, indicated that attenuated strains have sustained expansions in the *Bam*HI-D and *Bam*HI-H fragments, both of which map to the IRL (Fukuchi et al. 1985; Silva and Witter 1985). These expansions consist of the accumulation of tandem 132-bp direct repeats (Maotani et al. 1986). The region in the immediate vicinity of the 132-bp expansion is transcriptionally active. Bradley et al. (Bradley et al. 1989a,b) have reported that a major 1.8-kb family of transcripts produced by oncogenic strains is replaced by a group of truncated transcripts in attenuated strains. These transcripts are derived from a bidirectional promoter shared with the pp38 gene. The 1.8-kb family and pp38 genes are transcribed in opposite directions from this bidirectional promoter.

Several groups have reported analysis of cDNAs derived from this region. Two to three copies of the 132-bp repeats have been found in cDNAs from RPL1 and MSB1 cells (Kopacek et al. 1993). Two copies have also been reported in cDNAs derived from MD5-infected CEFs (Iwata et al. 1992). Four cDNAs 1.69, 1.5, 1.9, and 2.2kbp in length corresponding to transcripts of the 1.8-kb family have been sequenced (Peng et al. 1992). Two of these (1.69kbp and 2.2kbp) are not spliced; whereas two (1.5kbp and 1.9kbp) are singly spliced species. The cDNAs contain a number of potential ORFs. Others have reported that four groups of transcripts are produced in the region and that they can initiate or terminate in both rightward and leftward directions and at multiple sites relative to the 132-bp repeat region (Chen and Velicer 1991). Hong and Coussens reported two cDNAs, designated C1 and C2, derived from mRNAs spanning the *Bam*HI-H and *Bam*HI-I_2 regions (Hong and Coussens 1994). These cDNAs appear to be derived from a major 1.6-kb immediate-early transcript, and each of these cDNAs contains two small ORFs.

Evidence that this region is related to oncogenicity includes the following. First, as stated previously, the 1.8-family of transcripts is present only in oncogenic strains and is replaced by truncated transcripts in attenuated derivatives (Bradley et al. 1989b). Second, some of the ORFs associated with the 1.8-kb family have been reported to have some homology to cellular oncoproteins, particularly the mouse T-cell lymphoma oncoprotein and the Fes/Fps family of protein kinase-related oncoproteins (Peng et al. 1992). However, these homologies have not been noted by all investigators (Hong and Coussens 1994). Third, oligonucleotides complementary to the predicted splice donor site in the 1.8-kb family can inhibit proliferation of MDV-induced lymphoblastoid cells (Kawamura et al. 1991). Fourth, when introduced into primary CEF by transfection, two of the transcripts from the 1.8-kb family can reduce the serum dependence and prolong proliferation of these cells (Peng et al. 1993).

Although several potential ORFs have been observed in cDNAs corresponding to the 1.8-kb family, only two proteins have been reported to date. Antibodies raised against a fusion protein consisting of an ORF (ORF A) from the 1.7-kb unspliced transcript fused to glutathione-*S*-transferase immunoprecipitated a 7-kDa protein, called BHa (PENG et al. 1994). BHa was detectable from lymphoblastoid cells or from lysates of CEF infected with oncogenic MDV strains but not nononcogenic strains (PENG et al. 1994). In another study, antisera were raised against glutathione-*S*-transferase fusion proteins containing two ORFs derived from C1 and C2 cDNAs (HONG and COUSSENS 1994). Both antisera detected a 14-kDa species, designated pp14, in Western blots. The 14-kDa species was detected in lysates of cells infected with oncogenic and attenuated serotype 1 MDV strains as well as in MSB1 lymphoblastoid cells. Furthermore, this protein is phosphorylated and localized to the cytoplasm (HONG et al. 1995). The functions of these proteins, the relationship between them, and the role of the region in general in MDV-induced transformation remain obscure.

The 1.8-kb gene family and the pp38/pp24 genes share a common bidirectional promoter element. Transient expression studies using this promoter indicate that it functions in infected CEF but that promoter activities vary according to the nature of the MDV strain being examined (SHIGEKANE et al. 1999). Furthermore, the short enhancer region shared by the two genes can be bound by a nuclear factor present in infected cells.

4.4 Others

Although expression of the MDV ICP4 gene has not been demonstrated during latency, this protein will be included in this discussion since it is so central to MDV gene expression in general. The MDV ICP4 homolog gene has been identified and completely sequenced (ANDERSON et al. 1992). The gene is 4245 nucleotides long, maps to the *Bam*HI-A fragment of the genome, and shows similarity to ICP4 homolog genes of alphaherpesviruses. The MDV ICP4 protein is predicted to have a structure similar to that of ICP4-like proteins of other herpesviruses. Several potential transcriptional regulatory sites are present in the MDV ICP4 gene sequence, one of which is a putative autoregulatory site (FABER and WILCOX 1986) located 525bp into the predicted coding region. Transcription of the MDV ICP4 gene is complex. At least five RNAs with sizes of 10, 9.3, 7.8, 7.2, and 6.2kb have been reported (MORGAN et al. 1996). At least two of these RNAs (7.8kb and 6.2kb) are spliced. The nature of the 10-kb RNA is not well understood but it overlaps almost exactly the large LAT transcript. The MDV ICP4 protein can transactivate the pp38 gene (PRATT et al. 1994).

A nuclear antigen (MDNA) present in MKT-1 and MSB-1 cells has been reported to bind to two sites on the MDV genome (WEN et al. 1988). One site was located within the US (present in both MKT-1 and MSB1 cells) and the other (present only in MSB-1 cells) was within the IRS. DNAse digestion indicated that binding occurred to 60-bp and 103-bp sequences, respectively, both of which spanned palindromes. The biochemical nature of this factor has not been pursued and probably warrants further investigation.

5 Latency and Immunocompetence

Immunocompetence is important for initiating and maintaining MDV latency (Buscaglia et al. 1988). Latency can be maintained in cultured lymphocytes (as determined by suppression of expression of virus internal antigen) by a factor present in conditioned medium (Buscaglia and Calnek 1988). Conditioned media from ConA-stimulated spleen cells or from mixed-lymphocyte reactions is active. Some of this activity has been attributed to interferon (Volpini et al. 1992, 1995).

6 Summary and Conclusions

MDV latency is defined as the persistence of the viral genome in the absence of production of infectious virus except during reactivation. A number of systems for studying MDV latency exist, and most involve the use of lymphoblastoid cells or tumors. It has been difficult to divorce latency and transformation. Understanding the relationship between these two states remains a major challenge for the MDV system.

Based on their patterns of expression, the MDV LATs are apt to be important in the balance between latent and lytic infections. The LATs are a complex group of transcripts. The profile of gene expression that characterizes latency differs among all herpesviruses, and MDV is no exception. MDV LATs bear little resemblance to LATs of other alphaherpesviruses or to the LATs of other lymphotropic herpesviruses. LAT splicing patterns are complex and the relationships among various spliced species or between these species and the large 10-kb transcript are unknown. In addition, the existence of any protein gene products of significance is unknown at this time. More work is needed to further investigate the significance and function of these RNAs. Better technology to construct mutants in the MDV system is badly needed, since the analysis of mutants in the chicken is a powerful and unique advantage of the MDV system.

References

Abujoub A, Coussens PM (1995) Development of a sustainable chick cell line system infected with Marek's disease virus. Virology 214:541–549

Abujoub AA, Coussens PM (1996) Marek's disease virus (MDV) genomes are stabilized and in a latent state within MDV OU2 cells. In: Silva RF, Cheng HH, Coussens PM, Lee LF, Velicer LF (eds) Current Research on Marek's Disease. American Association of Avian Pathologists, Kennett Square, PA, pp 296–301

Abujoub A, Coussens PM (1997) Evidence that Marek's disease virus exists in a latent state in a sustainable fibroblast cell line. Virology 229:309–321

Abujoub AA, Williams DL, Reilly JD (1999) Development of a cell line system susceptible to infection with vaccine strains of MDV. Acta Virol 43:186–191

Anderson AS, Francesconi A, Morgan RW (1992) Complete nucleotide sequence of the Marek's disease virus IPC4 gene. Virology 189:657–667

Batchelor AH, O'Hare P (1990) Regulation and cell-type-specific activity of a promoter located upstream of the latency-associated transcript of herpes simplex virus type 1. J Virol 64:3269–3279

Becker Y, Asher Y, Tabor E, Davidson I, Malkinson M (1994) Open reading frames in a 4556 nucleotide sequence within MDV-1 BamHI-D DNA fragment: evidence for splicing of mRNA from a new viral glycoprotein gene. Virus Genes 8:55–69

Bradley G, Hayashi M, Lancz G, Tanaka A, Nonoyama M (1989a) Structure of the Marek's disease virus BamHI-H gene family: genes of putative importance for tumor induction. J Virol 63:2534–2542

Bradley G, Lancz G, Tanaka A, Nonoyama M (1989b) Loss of Marek's disease virus tumorigenicity is associated with truncation of RNAs transcribed within BamHI-H. J Virol 63:4129–4135

Brunovskis P, Qian Z, Li D, Lee LF, Kung H-J (1996) Functional analysis of the MDV basic-leucine zipper product, Meq. In: Silva R, Cheng HH, Coussens PM, Lee LF, Velicer LF (eds) Proceedings of the 5th International Symposium on Marek's Disease. American Association of Avian Pathologists, Kennett Square, PA, pp 265–270

Buscaglia C, Calnek BW (1988) Maintenance of Marek's disease herpesvirus latency in vitro by a factor found in conditioned medium. J Gen Virol 69:2809–2818

Buscaglia C, Calnek BW, Schat KA (1988) Effect of immunocompetence on the establishment and maintenance of latency with Marek's disease herpesvirus. J Gen Virol 69:1067–1077

Calnek BW, Witter RL (1997) Marek's disease. In: Calnek BW (eds) Diseases of poultry. Iowa State University, Ames, Iowa, pp 369–413

Calnek BW, Shek WR, Schat KA (1981) Latent infections with Marek's disease virus and turkey herpesvirus. J Natl Cancer Inst 66:585–590

Calnek BW, Shek WR, Schat KA (1981) Spontaneous and induced herpesvirus genome expression in Marek's disease tumor cell lines. Infect Immun 34:483–492

Calnek BW, Schat KA, Ross LJN, Shek WR, Chen CH (1984) Further characterization of Marek's disease virus-infected lymphocytes. I. In vivo infection. Int J Cancer 33:389–398

Cantello JL, Anderson AS, Morgan RW (1994) Identification of latency-associated transcripts that map antisense to the ICP4 homolog gene of Marek's disease virus. J Virol 68:6280–6290

Cantello JL, Parcells MS, Anderson AS, Morgan RW (1997) Marek's disease virus latency-associated transcripts belong to a family of spliced RNAs that are antisense to the ICP4 homolog gene. J Virol 71:1353–1361

Chen X, Velicer LF (1991) Multiple bi-directional initiations and terminations of transcription in the Marek's disease virus long repeat regions. J Virol 65:2445–2451

Chen XB, Sondermeijer PJ, Velicer LF (1992) Identification of a unique Marek's disease virus gene which encodes a 38-kilodalton phosphoprotein and is expressed in both lytically infected cells and latently infected lymphoblastoid tumor cells. J Virol 66:85–94

Cui Z, Yan D, Lee LF (1990) Marek's disease virus gene clones encoding virus-specific phosphorylated polypeptides and serological characterization of fusion proteins. Virus Genes 3:309–322

Cui Z, Lee LF, Liu J-L, Kung H-J (1991) Structural analysis and transcriptional mapping of the Marek's disease virus gene encoding pp38, an antigen associated with transformed cells. J Virol 65:6509–6515

Cui A, Qin A, Lee LF (1992) Expression and processing of Marek's disease virus pp38 gene in insect cells and immunological characterization of the gene product. In: De Boer G, Jeurissen SHM (eds) Proceedings of the Fourth International Symposium of Marek's Disease. Ponsen & Looijen, Wageningen, The Netherlands, pp 123–125

Delecluse HJ, Hammerschmidt W (1993) Status of Marek's disease virus in established lymphoma cell lines: herpesvirus integration is common. J Virol 67:82–92

Delecluse HJ, Schuller S, Hammeschmidt W (1993) Latent Marek's disease virus can be activated from its chromosomally integrated state in herpesvirus-transformed lymphoma cells. EMBO J 12:3277–3286

Dienglewicz RL, Parcells MS (1999) Establishment of a lymphoblastoid cell line using a mutant MDV containing a green fluorescent protein expression cassette. Acta Virol 43:106–112

Dunn K, Nazerian K (1977) Induction of Marek's disease virus antigens by IdUrd in a chicken lymphoblastoid cell line. J Gen Virol 34:413–419

Faber WW, Wilcox KW (1986) Association of the herpes simplex virus regulatory protein ICP4 with specific nucleotide sequences. Nucl Acids Res 14:6067–6083

Feldman LT (1991) The molecular biology of herpes simplex virus latency. In: Wagner EK (ed) Herpesvirus transcription and its regulation. CRC, Boca Raton, pp 233–243

Fraser NN, Block TM, Spivack JG (1992) The latency-associated transcripts of herpes simplex virus: RNA in search of function. Virology 191:1–18

Fukuchi K, Tanaka A, Schierman LW, Witter RL, Nonoyama M (1985) The structure of Marek disease virus DNA: the presence of unique expansion in nonpathogenic viral DNA. Proc Natl Acad Sci USA 82:751–754

Fynan EF, Ewert DL, Block TM (1993) Latency and reactivation of Marek's disease virus in B lymphocytes transformed by avian leucosis virus. J Gen Virol 74:2163–2170

Garcia-Blanco MA, Cullen BR (1991) The molecular basis of latency in pathogenic human viruses. Science 254:815–820

Goins WF, Sternberg LR, Croen KD, Krause PR, Hendricks RL, Fink DJ, Straus SE, Levine M, Glorioso JC (1994) A novel latency-active promoter is contained within the herpes simplex virus type 1 UL flanking repeats. J Virol 68:2239–2252

Hihara H, Imai K, Tsukamoto K, Nakamura K (1998) Isolation of serotype 2 Marek's disease virus from a cell line of avian lymphoid leucosis. J Vet Med Sci 60:143–148

Hirai K, Kanamori A, Niikura M, Ikuta K, Kato S (1988) RNA transcribed from Marek's disease virus genomes in productively and latently infected cells. In: Kato S, Horiuchi T, Mikami T, Hirai K (eds) Advances in Marek's disease research. Japanese Association on Marek's Disease, Osaka, pp 140–147

Holland MS, Silva RF (1999) Tissue-specific restriction of latent turkey HVT transcription. Acta Virol 43:148–151

Holland MS, Mackenzie CD, Bull RW, Silva RF (1996) A comparative study of histological conditions suitable for both immunofluorescence and in situ hybridization in the detection of herpesvirus and its antigens in tissues of chickens. J Histochem Cytochem 44:259–265

Holland MS, Mackenzie CD, Bull RW, Silva RF (1998) Latent turkey herpesvirus infection in lymphoid, nervous, and feather tissues of chickens. Avian Dis 42:292–299

Hong Y, Coussens PM (1994) Identification of an immediate-early gene in the Marek's disease virus long internal repeat region which encodes a unique 14-kilodalton polypeptide. J Virol 68:3593–3603

Hong Y, Frame M, Coussens PM (1995) A 14-kDa immediate-early phosphoprotein is specifically expressed in cells infected with oncogenic Marek's disease virus strains and their attenuated derivatives. Virology 206:695–700

Iwata A, Ueda S, Ishihama A, Hirai K (1992) Sequence determination of cDNA clones of transcripts from the tumor-associated region of the Marek's disease virus genome. Virology 187:805–808

Jones D, Kung H-J (1992) A heat-shock-like sequence in the meq gene promoter binds a factor in MDV lymphoblastoid cells. In: De Boer G, Jeurissen SHM (eds) Proceedings of the Fourth International Symposium on Marek's Disease. Ponsen & Looijen, Wageningen, The Netherlands, pp 58–61

Jones DL, Lee LF, Liu J-L, Kung H-J, Tillotson JK (1992) Marek's disease virus encodes a leucine zipper gene resembling the fos/jun family of oncogenes that is highly expressed in lymphoblastoid tumors. Proc Natl Acad Sci USA 89:4042–4046

Kaschka-Dierich C, Nazerian K, Thomssen R (1979) Intracellular state of Marek's disease virus DNA in two tumour-derived chicken cell lines. J Gen Virol 44:271–280

Kataoka K, Noda M, Nishizawa M (1994) Maf nuclear oncoprotein recognizes sequences related to an AP-1 site and forms heterodimers with both Fos and Jun. Mol Cell Biol 14:700–712

Kawamura M, Hayashi M, Furuichi T, Nonoyama M, Isogai E, Namioka S (1991) The inhibitory effect of oligonucleotides complementary to Marek's disease virus mRNA transcribed from the BamHI-H region, on the proliferation of transformed lymphoblastoid cells, MDCC-MSB1. J Gen Virol 72:1105–1111

Kerppola TK, Curran T (1994) A conserved region adjacent to the basic domain is required for recognition of an extended DNA binding site by Maf/Nrl family proteins. Oncogene 9:3149–3158

Kopacek J, Ross LJ, Zelnik V, Pastorek J (1993) The 132bp repeats are present in RNA transcripts from 1.8kb gene family of Marek disease virus-transformed cells. Acta Virol 37:191–195

Kristi TM, Roizman B (1987) Host cell proteins bind to the cis-acting site required for virion-mediated induction of herpes simplex virus type 1 α genes. Proc Natl Acad Sci USA 84:71–75

Lagunoff M, Roizman B (1995) The regulation of synthesis and properties of the protein product of open reading frame P of the herpes simplex virus 1 genome. J Virol 69:3615–3623

Li D-S, Pastorek J, Zelnik V, Smith GD, Ross LJN (1994) Identification of novel transcripts complementary to the Marek's disease virus homologue of the ICP4 gene of herpes simplex virus. J Gen Virol 75:1713–1722

Li D, O'Sullivan G, Greenall L, Smith G, Jiang C, Ross N (1998) Further characterization of the latency-associated transcription unit of Marek's disease virus. Arch Virol 143:295–311

Liu J-L, Lee LF, Ye Y, Qian Z, Kung H-J (1997) Nucleolar and nuclear localization properties of a herpesvirus bZIP oncoprotein MEQ. J Virol 71:3188–3196

Liu J-L, Ye Y, Lee LF, Kung H-J (1998) Transforming potential of herpesvirus oncoprotein MEQ: morphological transformation, serum-independent growth, and inhibition of apoptosis. J Virol 72:388–395

Liu J-L, Lin S-F, Xia S, Brunovskis P, Li D, Davison I, Lee LF, Kung H-J (1999a) Meq and v-IL8: cellular genes in disguise? Acta Virol 43:94–101

Liu J-L, Ye Y, Qian Z, Templeton DJ, Lee LF, Kung H-J (1999) Functional interactions between herpesvirus oncoprotein Meq and cell cycle regulator CDK2. J Virol 73:4208–4219

Makimura K, Peng F-Y, Tsuji M, Hasegawa S, Kawai Y, Nonoyama M, Tanaka A (1994) Mapping of Marek's disease virus genome: identification of junction sequences between unique and inverted repeat regions. Virus Genes 8:15–24

Maotani K, Kanamori A, Ikuta K, Ueda S, Kato S, Hirai K (1986) Amplification of a tandem direct repeat within inverted repeats of Marek's disease virus DNA during serial in vitro passage. J Virol 58:657–660

Maray T, Malkinson M, Becker Y (1988) RNA transcripts of Marek's disease virus (MDV) serotype-1 in infected and transformed cells. Virus Genes 2:49–68

McCance DJ (1998) Human tumor viruses. American Society for Microbiology, Washington DC

McKie EA, Ubukata E, Hasegawa S, Zhang S, Nonoyama M, Tanaka A (1995) The transcripts from the sequences flanking the short component of Marek's disease virus during latent infection form a unique family of 3′ coterminal RNAs. J Virol 69:1310–1314

Morgan R, Xie Q, Cantello J (1996) The Marek's disease virus ICP4 gene: update on sense and antisense RNAs and characterization of the gene product. In: Silva RF, Cheng HH, Coussens PM, Lee LF, Velicer LF (eds) Current Research on Marek's Disease. American Association of Avian Pathologists, Kennett Square, PA, pp 160–163

Nazerian K (1987) An updated list of avian cell lines and transplantable tumours. Avian Pathol 16: 527–544

Nazerian K, Lee LF (1974) Deoxyribonucleic acid of Marek's disease virus in a lymphoblastoid cell line from Marek's disease tumours. J Gen Virol 25:317–321

Ogura H, Fujiwara T (1987) Establishment and characterization of a virus-free chick cell line. Acta Med Okayama 41:141–143

Ohashi K, O'Connell PH, Schat KA (1994a) Characterization of Marek's disease virus BamHI-A-specific cDNA clones obtained from a Marek's disease lymphoblastoid cell line. Virology 199:275–283

Ohashi K, Zhou W, O-Connell, Schat KA (1994b) Characterization of a Marek's disease virus BamHI-L-specific cDNA clone obtained from a Marek's disease lymphoblastoid cell line. J Virol 68:1191–1195

Ono M, Kawaguchi Y, Maeda K, Kamiya N, Tohya Y, Kai C, Niikura M, Mikami T (1994) Nucleotide sequence analysis of Marek's disease virus (MDV) serotype 2 homolog of MDV serotype 2 pp38, an antigen associated with transformed cells. Virology 201:142–146

Parcells MS, Anderson AS, Morgan RW (1994a) Characterization of Marek's disease virus insertion and deletion mutants that lack US1 (ICP22 homolog), US10, and/or US2 and neighboring short-component open reading frames. J Virol 68:8239–8253

Parcells MS, Anderson AS, Morgan RW (1994b) Characterization of a Marek's disease virus mutant containing a lacZ insertion in the US6 (gD) homologue gene. Virus Genes 9:5–13

Parcells MS, Anderson AS, Morgan RW (1995) Retention of oncogenicity by a Marek's disease virus (MDV) mutant lacking six unique-short region genes. J Virol 69:7888–7898

Parcells MS, Dienglewicz RL, Anderson AS, Morgan RW (1999) Recombinant Marek's disease virus (MDV)-derived lymphoblastoid cell lines: regulation of a marker gene within the context of the MDV genome. J Virol 73:1362–1373

Peng G, Bradley G, Tanaka A, Lancz G, Nonoyama M (1992) Isolation and characterization of cDNAs from BamHI-H gene family RNAs associated with the tumorigenicity of Marek's disease virus. J Virol 66:7389–7396

Peng F, Donovan J, Specter S, Tanaka A, Nonoyama M (1993) Prolonged proliferation of primary chicken embryo fibroblasts transfected with cDNAs from the BamHI-H gene family of Marek's disease virus. Intern J Oncol 3:587–591

Peng F, Specter S, Tanaka A, Nonoyama M (1994) A 7 kDa protein encoded by the BamHI-H gene family of Marek's disease virus is produced in lytically and latently infected cells. Int J Oncol 4: 799–802

Peng Q, Zeng M, Bhuiyan ZA, Ubukata E, Tanaka A, Nonoyama M, Shirazi Y (1995) Isolation and characterization of Marek's disease virus (MDV) cDNAs mapping to the BamHI-I2, BamHI-Q2, and

BamHI-L fragments of the MDV genome from lymphoblastoid cells transformed and persistently infected with MDV. Virology 213:590–599

Peng Q, Shirazi Y (1996a) Characterization of the protein product encoded by a splicing variant of the Marek's disease virus Eco-Q gene (Meq). Virology 226:77–82

Peng Q, Shirazi Y (1996b) Isolation and characterization of Marek's disease virus (MDV) cDNAs from a MDV-transformed lymphoblastoid cell line: identification of an open reading frame antisense to the MDV Eco-Q protein (Meq). Virology 221:368–374

Pratt WD, Cantello J, Morgan RW, Schat KA (1994) Enhanced expression of the Marek's disease virus-specific phosphoproteins after stable transfection of MSB-1 cells with the Marek's disease virus homolog of ICP4. Virology 201:132–136

Qian Z, Brunovskis P, Fauscher F, Lee Ll, Kung H-J (1995) Transactivation activity of meq, a Marek's disease herpesvirus bZIP protein persistently expressed in latently infected transformed T cells. J Virol 69:4037–4044

Qian Z, Brunovskis P, Lee LF, Vogt PK, Kung H-J (1996) Novel DNA binding specificities of a putative herpesvirus bZIP oncoprotein. J Virol 70:7161–7170

Ross LJN (1985) Molecular biology of the virus. In: Payne LN (ed) Marek's disease. Martinus Nijhoff, Boston, pp 113–150

Ross NLJ, DeLorbe W, Varmus HE, Bishop MJ Brahie M, Haase A (1981) Persistence and expression of Marek's disease virus DNA in tumour cells and peripheral nerves studied by in situ hybridization. J Gen Virol 57:285–296

Ross N, O'Sullivan G, Rothwell C, Smith G, Rennie M, Lee LF, Davison TF (1996) Expression of MDV genes in lymphomas and their role in oncogenesis. In: Silva RF, Cheng HH, Coussens PM, Lee LF, Velicer LF (eds) Current Research on Marek's Disease. American Association of Avian Pathologists, Kennett Square, PA, pp 40–46

Ross N, O'Sullivan G, Rothwell C, Smith G, Burgess SC, Rennie M, Lee LF, Davison TF (1997) Marek's disease virus EcoRI-Q gene (meq) and a small RNA antisense to ICP4 are abundantly expressed in CD4+ cells and cells carrying a novel lymphoid marker, AV37, in Marek's disease lymphomas. J Gen Virol 78:2191–2198

Schat KA, Buckmaster A, Ross LJN (1989) Partial transcription map of Marek's disease herpesvirus in lytically infected cells and lymphoblastoid cell lines. Int J Cancer 44:101–109

Schat KA, Hooft van Iddekinge BJ, Boerrigter H, O'Connell P, Koch G (1998) Open reading frame L1 of Marek's disease herpesvirus is not essential for in vitro and in vivo virus replication and establishment of latency. J Gen Virol 79:841–849

Shigekane H, Kawaguchi Y, Shirakata M, Sakaguchi M, Hirai K (1999) The bi-directional transcriptional promoters for the latency-relating transcripts of the pp38/pp24 mRNAs and the 1l8 kb-mRNA in the long inverted repeats of Marek's disease virus serotype 1 DNA are regulated by common promoter-specific enhancers. Arch Virol 144:1893–1907

Silva RF (1992) Differentiation of pathogenic and non-pathogenic serotype 1 Marek's disease viruses (MDVs) by the polymerase chain reaction amplification of the tandem direct repeats within the MDV genome. Avian Dis 36:521–528

Silva RF, Witter RL (1985) Genomic expansion of Marek's disease virus DNA is associated with serial in vitro passage. J Virol 54:690–696

Silver S, Tanaka A, Nonoyama M (1979) Transcription of the Marek's disease virus genome in a nonproductive chicken lymphoblastoid cell line. Virology 93:127–133

Stevens JG (1989) Human herpesviruses: A consideration of the latent state. Microbiol Rev 53: 318–332

Sugaya K, Bradley G, Nonoyama M, Tanaka A (1990) Latent transcripts of Marek's disease virus are clustered in the short and long repeat regions. J Virol 64:5773–5782

Tanaka A, Silver S, Nonoyama M (1978) Biochemical evidence of the non-integrated status of Marek's disease virus DNA in virus-transformed lymphoblastoid cells of chickens. Virology 88:19–24

Tillotson JK, Lee LF, Kung H-J (1988) Accumulation of viral transcripts coding for a DNA binding protein in Marek's disease tumor cells. In: Kato S, Horiuchi T, Mikami T, Hirai K (eds) Advances in Marek's disease research. Japanese Association on Marek's Disease, Osaka, pp 128–134

Volpini L, Calnek BW, Sneath B (1992) Cytokine modulation of latency in Marek's disease virus-infected cells. In: Proceedings of the 19th World's Poultry Congress. Ponsen & Looijen, Wageningen, pp 254–257

Volpini L, Calnek BW, Sekellick MJ, Marcus PI (1995) Stages of Marek's disease virus latency defined by variable sensitivity to interferon modulation of viral antigen expression. Vet Microbiol 47:99–109

Wagner EK (1991) Herpesvirus transcription – general aspects. In: Wagner EK (ed) Herpesvirus transcription and its regulation. CRC, Boca Raton, pp 233–243

Wen LT, Tanaka A, Nonoyama M (1988) Identification of Marek's disease virus nuclear antigen in latently infected lymphoblastoid cells. J Virol 62:3764–3771

Xie Q, Anderson AS, Morgan RW (1996) Marek's disease virus (MDV) ICP4, pp38, and meq genes are involved in the maintenance of transformation of MDCC-MSB1 MDV-transformed lymphoblastoid cells. J Virol 70:1125–1131

Zhu G-S, Ojima T, Hironaka T, Ihara T, Mizukoshi N, Kato A, Ueda S, Hirai K (1992) Differentiation of oncogenic and nononcogenic strains of Marek's disease virus type 1 by using polymerase chain reaction DNA amplification. Avian Dis 36:637–645

Zhu G-S, Iwata A, Gong M, Ueda S, Hirai K (1994) Marek's disease virus type 1-specific phosphorylated proteins pp38 and pp24 with common amino acid termini are encoded from the opposite junction regions between the long unique and inverted repeat sequences of viral genomes. Virology 200: 816–820

Zwaagstra JC, Ghiasi JH, Slanina SM, Nesburn AB, Wheatley SC, Lillycrop K, Wood J, Latchman DS, Patel K, Wechsler SL (1990) Activity of herpes simplex virus type 1 latency associated transcript (LAT) promoter in neuron derived cells: evidence for neuron specificity and for a large LAT transcript. J Virol 64:5019–5028

Zwaagstra JC, Ghiasi JH, Nesburn AB, Wechsler SL (1991) Identification of a major regulatory sequence in the latency associated transcript (LAT) promoter of herpes simplex virus type 1 (HSV-1). Virology 182:287–297

Meq: An MDV-Specific bZIP Transactivator with Transforming Properties

H.-J. KUNG[1,2,3], L. XIA[1,2,3], P. BRUNOVSKIS[3], D. LI[3], J.-L. LIU[3], and L.F. LEE[4]

1 Introduction

The principal cause of chicken T-lymphomas and their accompanying demyelinating disease is infection with Marek's disease virus (MDV) (CALNEK et al. 1997), a herpesvirus. Marek's disease can be prevented by vaccination with an antigenically related nonpathogenic herpesvirus, HVT (turkey herpesvirus). MDV is among the most potent oncogenic herpesviruses, and induces tumors as early as 4 weeks post-inoculation. As such, the virus is likely to encode a direct-acting oncogene or transforming gene. To search for a possible candidate(s), early studies focused on genes expressed in tumor cells or transformed cells. In general, in these cells where

[1] Department of Biological Chemistry, School of Medicine, University of California, Davis, CA 95616, USA
[2] University of California Davis Cancer Center, Sacramento, CA 95817, USA
[3] Department of Molecular Biology and Microbiology, Case Western Reserve University, Cleveland, OH 44106, USA
[4] Avian Disease and Oncology Laboratory, Agricultural Research Station, US Department of Agriculture, East Lansing, MI 48823, USA

there is no virus production, the transcriptional activities are confined to the repeat regions only, which, based on *Bam*HI digestion map, span the *Bam*H, -I2, -Q2, -L, and -A fragments (Ross 1999). Within this region, most of the open reading frames are short and their existence as proteins not conclusively resolved. The exceptions are pp38, ICP4, and Meq, for which the protein products have been clearly identified. It should be noted that most of these studies were carried out using the entire population of tumor cells and cell lines; it is not clear whether all these proteins are expressed in the same or different cell populations. This is an important issue, as in any given latent state, there is usually a fraction of cells, spontaneously releasing viruses, which may contribute to the detection of some lytic gene products. In surveying through the literature, Meq, the focus of this chapter, emerges as one that is most consistently expressed in all tumors and transformed cell lines, both at the transcript and at the protein level. Meq has a structure resembling nuclear oncogenes and, as will be described in detail below, shared properties, characteristic of oncogenes.

1.1 Discovery and Identification of Meq

As discussed above, Meq was uncovered in the course of searching for MDV genes consistently expressed in MDV latently infected or tumor cells (Jones et al. 1992b). Meq protein is 339 amino acids (aa) long and carries a bZIP (basic-leucine zipper) domain at the N-terminal part, which is closely related to the Jun/Fos oncoproteins, and a C-terminal proline-rich domain structurally resembling the WT-1 tumor suppressor gene (Fig. 1). It is encoded by the genome of the oncogenic strain, serotype 1 MDV or MDV-1, but not by two other related strains, HVT and SB-1, serotypes 2 and 3. In this review, we will use the term MDV and MDV-1 interchangeably. As such, this is a gene that has evolved late and may account for some of the unique properties of MDV, including its potent oncogenic potential and lymphotropic properties. In the past 8 years since the discovery of Meq, various laboratories have contributed to the characterizations of this gene as a transcriptional factor, an oncogene, and an immunogen. Some of these studies will be summarized below.

1.2 The Genomic Organization of Meq

The major product of Meq is a molecule of 339 amino acids and is encoded by a single exon, completely contained in the MDV Eco RI Q fragment, and hence the name Meq. In the *Bam*HI digestion map of MDV, Meq open reading frame spans both *Bam*I2 and *Bam*Q fragments, which is upstream from the *Bam*L fragment (see the chapter by Silva et al., this volume). All these fragments are contained in the repeat regions flanking the unique long segment. Meq is thus in the diploid form (i.e., there are two copies of *meq* genes in the MDV genome). The recent sequence information of serotype 1, 2, and 3 reveals a high degree of conservation of genes

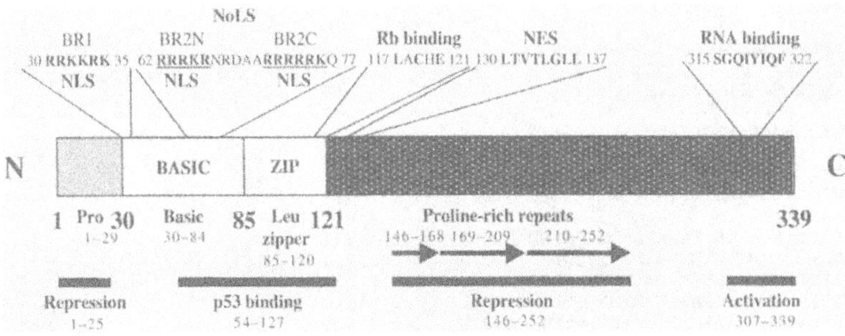

Fig. 1. Structure of the MDV Eco Q (Meq) protein. *Numbers* represent the positions of the amino acid residuals at the boundaries of each functional domain or motif. Meq protein consists of an N-terminal proline (*Pro*)-rich domain, a basic region (*BR*), a leucine (*Leu*) zipper, and a transactivation domain with two and a half proline (Pro)-rich repeats. The N-terminal 25 amino acids have transrepression activity, whereas the C-terminal 33 amino acids are essential but not sufficient for transactivation. Amino acids 54–127 are required for Meq–p53 interaction. Amino acid sequences for nuclear localization signal (*NLS*), nucleolar localization signal (*NoLS*), Rb binding, nuclear exit signal (*NES*), and RNA binding are depicted

and sequences in the unique-long (UL) and unique-short (US) regions, which encode primarily genes involved in replication and virus assembly. By contrast, the repeat regions, mostly encoding regulatory proteins, are almost completely different among these three serotypes. Meq is likely to play an important regulatory role, catered to the unique lifestyle of MDV. The promoter of *meq* gene contains several enhancer motifs including a heat-shock protein binding motif ((-18) GTTCGCGAAC (-9)), a TRE (TPA response element)/CRE (cAMP response element) hybrid motif ((-167) AGTCATGCATGACGT (-152)), a potential binding site for AP-1 transcriptional factors, and an SP1 binding site (GGCGGG) at 20 nucleotides preceding the initiation AUG (Jones et al. 1992a; Qian et al. 1995). The presence of these enhancer motifs suggests that Meq promoter may respond to different extracellular stimuli.

2 Biochemical and Biological Properties of Meq

2.1 The Dimerization Potential of Meq

The ZIP or leucine zipper domains of bZIP proteins are known to be engaged in transcriptional factor dimerizations, as the basic region preceding the ZIP domain in DNA binding. The ZIP domain of Meq contains five leucines and one histidine that are equally spaced (seven residues apart), and serves as the nucleation site for dimerization. The presence of histidine at the C-terminal end of the ZIP domain is a characteristic of the Jun/Fos family of transcriptional factors. Although the rules governing dimerization specificity are not completely understood, the association affinity is largely affected by the nature of the amino acids in the e and g positions of the leucine zipper-coiled-coil structure (Fig. 2) (Schuermann et al. 1991; Baxevanis et al. 1993; Glover et al. 1995). These are positions known to mediate

	---- *----* *----* *----* 1 2 3 4 5 defgabcdefgabcdefgabcdefgabcdef	homo- +, -	Meq +, -	c-Jun +, -	c-Fos +, -
Meq	**L**HEACEE**L**QRANEH**L**RKEIRD**L**RTECTS**L**RV	4,0	4,0	2,0	3,1
c-Jun	**L**EEKVKT**L**KAQNSE**L**ASTANM**L**REQVAQ**L**KQ	0,2	2,0	0,2	4,0
JunB	**L**EDKVKT**L**KAENAG**L**SSAAGL**L**REQVAQ**L**KQ	0,2	3,0	0,2	4,0
c-Fos	**L**QAETDQ**L**EDEKSA**L**QTEIAN**L**LKEKEK**L**EF	0,4	3,1	4,0	0,4
FosB	**L**QAETDQ**L**EEEKAE**L**ESEIAE**L**QKEKER**L**EF	0,6	3,1	4,0	0,6
CREB	**L**ENRVAV**L**ENQNKT**L**IEELKA**L**KDLYCH**K**SD	4,0	2,0	2,1	2,1
ATF1	**L**ENRVAV**L**ENQNKT**L**IEELKT**L**KDLYSN**K**SV	4,0	2,0	2,1	2,1
ATF2	**L**EKKAED**L**SSLNGQ**L**QSEVTL**L**RNEVAQ**L**KQ	4,0	4,0	2,1	3,1
ATF3	**L**QKESEK**L**ESVNAE**L**KAQIEE**L**KNEKQH**L**IY	0,2	2,0	3,0	2,3
B-ATF	**L**HLESED**L**EKQNAA**L**RKEIKQ**L**TEELKY**F**TS	0,2	2,0	4,0	2,3
v-Maf	**L**ESEKNQ**L**LQQVEH**L**KQEISR**L**VRERDA**Y**KE	2,0	3,0	3,0	2,2
MafB	**L**ENEKTQ**L**IQQVEQ**L**KQEVTR**L**ARERDA**Y**KL	2,0	3,0	3,0	2,2
MafF	**L**QKQKME**L**EWEVDK**L**ARENAA**M**RLELDT**L**RG	4,0	5,0	3,0	2,2
MafK	**L**ERQRVE**L**QQEVEK**L**ARENSS**M**KLELDA**L**RS	4,0	5,0	2,0	2,1
DBP	**I**SVRAAF**L**EKENAL**L**RQEVVA**V**RQELSH**Y**RA	8,0	5,0	3,1	4,2
C/EBP	**T**QQKVLE**L**TSDNDR**L**RKRVEQ**L**SRELDT**L**TG	2,0	2,1	1,2	3,1
NF/IL6	**T**QHKVLE**L**TAENER**L**QKKVEQ**L**SRELST**L**RN	2,0	3,1	1,2	2,1
CHOP	**L**AEENER**L**KQEIER**L**TREVET**T**RRALID**R**MV	4,0	3,0	2,1	2,1
ZEBRA	**Y**REVAAA**K**SSENDR**L**RLLLKQ**M**CPSLDV**D**SI	2,0	1,0	0,0	1,0
GCN4	**L**EDKVEE**L**LSKNYN**L**ENEVAR**L**KKLVGE**R**	4,0	2,1	1,1	2,1

Fig. 2. Dimerization index of bZIP proteins, based on charge–charge interactions. The numbers *1–5* denote the five heptad repeats, which correspond to five helical turns, of the zipper domain. The N-terminal leucine is set at number *1*. All the leucines or their substitutes are in *boldface*. Positions e and g of each helical turn are indicated. These two positions interact with each other between the two dimerization partners, such that opposite-charged amino acids favor dimerization, whereas like-charged amino acids repel each other. The number of opposite-charged pairs is given under the "+" column and the number of like-charged pairs is under the "–" column. Pairs with the highest number of + and lowest number of – are predicted to form stable dimer

the formation of salt bridges between adjacent bZIP helices. These salt bridges always occur between an amino acid at a g position of one molecule and that at an e position residing on the other molecule. Opposite-charged amino acids promote dimerization through electrostatic attraction; like-charged amino acids repel each other. Careful inspection of the Meq leucine zipper region revealed the presence of multiple complementary charged amino acid residues in the e and g positions. Numbers in the "+" column (Fig. 2) represent the number of opposite-charged pairs, whereas those in the "–" column indicate the number of like-charged pairs. A high number in the "+" and a low number in the "–" column denotes favorable association. Thus, Fos-Fos homodimer has four unfavorable pairs where as Fos-Jun heterodimer carries four favorable pairs, consistent with the ability of Fos to form heterodimer only. Biochemical tests using gel shift, GST fusion protein pull-down and co-precipitation of in vitro translated products demonstrate that Meq forms homodimer with itself and heterodimers with c-Jun, JunB, CRE-B, ATF1, ATF2, ATF3, Fos, FosB, SNF, and c/EBP, consistent with the generally favorable numbers predicted in Fig. 2 (Brunovskis et al. 1996). Thus, Meq seems to be a promiscuous bZIP transcriptional factor and by its propensity to bind multiple partners, its potential to reprogram viral and host gene expressions is enormous. The results described above are based primarily on in vitro binding assays. A key issue in the future is to identify natural dimerization partners of Meq inside the transformed cells, as well as during latent and lytic MDV infections.

2.2 The Association of Meq with Other Cellular Factors

In addition to the formation of homo- and heterodimers with other bZIP proteins, Meq is able to interact with a number of other cellular factors, several of which are strongly implicated in oncogenesis. The proline-rich region of Meq contains several PXXP motifs, which are binding modules for SH3 containing proteins. It was found that the SH3 domain of Src selectively binds to Meq via this motif and phosphorylates Meq (Liu, unpublished data). In addition, Meq and c-Src are found to be co-localized in the nucleolus in Meq transformed cells. While the significance of these findings is unclear, they are tantalizing and suggest a new role of Meq in mislocating cellular proteins, a function shared by other herpesvirus gene products. Like other tumor virus transforming proteins, Meq binds p53 and RB, the two major tumor suppressors controlling cell cycle progression. The binding of Meq to p53 is specific and requires amino acids 54–127, the ZIP domain of Meq, and the C-terminal tetramerization domain of p53 (Brunovskis et al. 1996). The binding region of Meq toward Rb has not been precisely mapped, but there is an LXCXE motif, the RB-binding consensus, at the end of the ZIP domain. The capability of Meq to associate with p53 and RB is likely to be responsible for its anti-apoptosis function and the growth-stimulation potential. Viral proteins appear to evolve means to interact with multiple cellular factors and intercept multiple cellular pathways. If EBV BZLF1 (also called Zebra and Zta) is a pertinent precedent, a viral protein can complex with as many as seven cellular factors, including transcriptional factors, tumor suppressors, and steroid hormone receptor, to effect multiple cellular responses (Gutsch et al. 1994; Zhang et al. 1994; Pfitzner et al. 1995; Adamson et al. 1999). Meq therefore may have more association partners than currently identified. A cautionary note to these observations is that in vitro association may not be directly translated into in vivo functions; genetic studies are required to determine the significance of the association of Meq to Src, p53, RB as well as other proteins.

2.3 The Transactivation Potential of Meq

The major transactivation domain of Meq has been mapped to amino acids 129–339, the portion, C-terminal to the ZIP domain. This portion, referred to as the C-terminal domain, when fused to Gal4 DNA-binding domain, acts as a transactivator on a reporter driven by a promoter containing Gal4 recognition motifs. Based on this assay, the last 33 amino acids of Meq are essential for the transactivation function. Other than some homology to RNA-binding motifs, the last 33 amino acids do not have any remarkable structural feature. Within the C-terminal domain, there are two and a half proline-rich repeats located at amino acids 146–252. Meq variants that carry additional copies of the repeat have recently been identified in the latent MDV genomes of the MSB-1 cell line (Kamil, personal communication). These repeats each contain 37% of proline residues and are rich in PXXP and PPPP motifs. This unusually long proline-rich domain rivals that of

WT-1, a strong transcriptional repressor encoded by Wilm's tumor suppressor gene (CALL et al. 1990). While based on deletion analysis, at least one repeat is required for the full transactivation activity of Meq (QIAN et al. 1995), these proline-rich repeats in isolated form have strong transrepression function, similar to that of WT-1 tumor suppressor. The N-terminal 29 amino acids of Meq before the basic region also contain a domain somewhat rich in proline and is recently shown to have transrepression activity (Xia, unpublished data). It thus seems that Meq contains both activation and repression modules. These modules may contribute to the transactivation or repression functions of Meq, depending on the conformation of Meq, i.e., whether these modules are exposed or not. Interaction with other molecules or phosphorylation by kinases are some of the possible means that can modulate the confirmation and hence the activities of Meq. Meq has a number of potential phosphorylation sites for serine/threonine kinases and is shown to be a good substrate for PKA, PKC, MAPK, CDK1, and CDK2 (LIU et al. 1999b). These authors showed that PKC phosphorylation seems to increase its DNA-binding activity (on MERE I), whereas CDK phosphorylation has the opposite effect. The transactivation potentials of the phosphorylated Meq, however, were not tested in that study. In another interesting study, Askovic and Baumann have compared the transactivation potential of various viral transactivation domains (those of Meq, HSV VP-16, EBV EBNA3C, and EBV Rta) by substituting the transactivation domain of EBV ZEBRA protein individually with the heterologous domains (ASKOVIC et al. 1997). It was found that the Meq C-terminal domain (aa 149–339) functions very well in this context, and enables the fusion ZEBRA to transactivate, to disrupt EBV latency, and to stimulate replication at the EBV lytic origin. It also confirms that the C-terminal amino acids are essential for these functions, as a construct that deletes the last 56 aa is completely non-functional in all three assays. Since the transactivation function of EBV ZEBRA is mediated by interacting with the basic transcriptional machinery (i.e., TFIID and TFIIA) (CHI et al. 1995), it is likely that the C-terminal domain of Meq also targets these molecules.

2.4 The DNA-Binding Motifs of Meq

A number of approaches have been applied to identify the DNA sequences recognized by Meq. These approaches include CASTing (cyclic amplification of selected targets), footprint (or methylation interference) analysis, and identifications of motifs in promoters responding to Meq (QIAN et al. 1996). Gratifyingly, all three approaches yield converging results. The optimal binding motifs for Meq/Meq homodimer selected by the PCR-based CAST approach contain two types of consensus sequences referred to as MERE (Meq response element) I and II. MERE I, GAGTGATGAC(G)TCATC, contains a TRE/CRE core (underlined) but with additional flanking sequence critical for binding, confirmed by the methylation interference and mutational analyses. This sequence coincides with the response element of Maf oncoprotein. MERE II is entirely different with a CACAC core sequence, but the binding critically depends on the DNA bent, modulated by flanking sequences. Perhaps this is not too surprising as CACAC motif has an intrinsic tendency to bend, but the degree depends on the environment. Thus, in the homodimer form, Meq recognizes a unique binding motif not shared by any other bZIP proteins. Using a similar approach, the binding motifs of Meq/Jun heterodimer were also determined to be TRE and CRE, confirming the notion that Meq is an immediate member of the Jun/Fos family. With a more quantitative assay, interestingly, preference of Meq binding to TRE vs CRE, depending on the heterodimer partner, has been observed. For instance, Meq/ATF2 heterodimer recognizes TRE with a much higher affinity than CRE, whereas ATF2 homodimer does not discriminate between TRE and CRE. Thus, Meq can potentially modulate the transcriptional regulation of host genes by modifying the target selectivity of host bZIP proteins. Knowing the target sequences, the MDV promoters that carry MERE I, II, TRE, and CRE were scanned, in an effort to identify potential genes modulated by Meq. Many genes including Meq itself contains TRE, CRE, or TRE/CRE hybrid motifs. In transactivation assay, the Meq promoter is activated by Meq/Meq and Meq/Jun dimers, confirming the functional role of these binding motifs. Interestingly, the MERE II motif, CACAC, is found located near the replication origin of MDV (Qian et al. 1996). This region coincides with the divergent promoter of viral genes pp38 and pp14. Direct Meq/Meq binding to an oligonucleotide corresponding to the replication origin was demonstrated. Thus, Meq/Meq may serve a dual role of transcriptional regulation and replication modulation for the MDV genome.

It is noteworthy that Meq also has a strong affinity toward cellular RNA. Most of the RNA-binding molecules carry long stretches of arginines, called arginine fork (Lazinski et al. 1989; Sacher et al. 1989; Delling et al. 1991; Zapp et al. 1991; Hatton et al. 1992; Hammes et al. 1993; Lee et al. 1993). As will be discussed later, Meq has a basic region with an unusually high content of arginines. In addition, there is a motif at the C-terminus, 315SGQIYIQF322, which shares a consensus with other RNA-binding proteins. Thus, Meq is likely to be a bona fide RNA-binding molecule. While it is not clear whether there is specificity or selectivity toward certain RNA species, the RNA-binding property of Meq may be partially responsible for its tight association with the nucleolus and coiled body, two subnuclear structures involved in RNA storage.

2.5 The Subcellular Localizations of Meq

Perhaps as a reflection of its multifunctional roles, Meq has an interesting subcellular localization pattern. In cycling cells, Meq is seen in nucleoplasm, nucleolus, coiled bodies as well as in cytoplasm (Liu et al. 1997). The cytoplasmic location (mostly in the nuclear periphery) of Meq is cell-cycle-dependent and detected only in S phase. The signal sequences that transport transcriptional factors to the nucleus are usually a cluster of basic residues. In Meq, there are two stretches of basic amino acids, BR1 (basic region 1) and BR2 (basic region 2). BR1 (30RRKKRR35) contains six consecutive basic amino acids, typical of NLS (nuclear localization signal) and functions as such (Liu et al. 1997). BR2 (62RRRKRNRDA ARRRRRK77) is much longer with 16 amino acids and 12 of them are basic, of which 10 are arginines. BR2 is required for the nucleolar translocation of Meq. A fusion of BR2 to v-raf, a molecule localized in the cytoplasm, allows the latter to be translocated to the nucleolus, providing strong evidence that BR2 is an autonomous nucleolar localization signal (NoLS). BR2 has a bipartite structure with two consecutive basic amino acids, BR2N and BR2C, interrupted by five amino acids. When tested individually, either BR2N or BR2C alone is sufficient to mediate nuclear, but not nucleolar, transport. This is consistent with the accumulated evidence that the NoLS requires a much longer stretch of basic amino acids than NLS. A key question then is: what is Meq doing in the nucleolus? Meq mutants, which fail to localize in the nucleolus, would be most valuable in this regard. Unfortunately, this is difficult to construct, as BR2 completely overlaps with the basic region of the bZIP domain, which is responsible for DNA binding. Mutants that delete this region will necessarily affect the transcription potential of Meq as well. Further delineation of the required amino acids for nucleolar localization vs DNA binding is required to make more subtle mutations which may affect one but not the other function. At present, we can only speculate on what Meq might be doing in the nucleolus, based on analogy to other nucleolar proteins. Among proteins reported to be localized in the nucleolus are viral proteins HIV rev, HIV1 rex, tax, HSV Us11, and EBV EBNA5; cellular factors YY1, TBP, PCNA, LYAR, and HSP70; tumor suppressors Rb, and p19-ARF; and oncogene Mdm2, just to name a few (see Table 1 in Liu et al. 1997). Meq may be involved in viral RNA transport, ribosomal RNA genesis, or Pol-I-mediated transcription. There is also an interesting similarity between Meq and Mdm2 oncogene: both bind RNA and p53, are localized in the nucleolus, and can shuttle between nucleus and cytoplasm (Brunovskis et al. 1996; Elenbaas et al. 1996; Liu et al. 1997, 1998; Roth et al. 1998; Tao et al. 1999a). Mdm2 transports p53 to cytoplasm and targets p53 for degradation (Freedman et al. 1998; Roth et al. 1998; Tao et al. 1999a). Meq could perform a similar function to facilitate oncogenesis.

Meq is also localized in coiled body, a subnuclear structure defined by the presence of p80 coilin (Liu et al. 1999b). The presence of coiled body, like nucleolus, is correlated with transformation and increased metabolic activity. It has been suggested that coiled bodies are involved in some aspects of ribonucleoprotein assembly, transport, and recycling. What Meq regulates in coiled body is still a mystery. An intriguing lead is its co-localization with CDK2, a cyclin-dependent kinase usually localized in the cytoplasm (Liu et al. 1999b). Meq or the transformation process apparently mislocates this kinase to the coiled bodies. The consequence of this co-localization is the phosphorylation of Meq by CDK2 at serine 42, a residue in the proximity of BR1 and BR2. This phosphorylation results in reducing the DNA-binding capacity (presumably by the addition of a negative charge) and the exit of Meq to the cytoplasm. It is noteworthy that there is a putative nuclear exit signal (130LTVTLGLL137) located in the Meq. This may account for the cell-cycle-dependent translocation of Meq.

2.6 Transforming, Mitogenic, and Anti-Apoptotic Potentials of Meq

For the lack of a proper chicken T-cell transformation system, the transforming properties of Meq have been studied on rodent fibroblast cell lines such as Rat-2 and NIH 3T3. Overexpression of Meq by infection of cells with a retrovirus vector carrying Meq leads to serum-independent and anchorage-independent growth, which is accompanied by striking morphological changes (Liu et al. 1997). On the other hand, if Meq is introduced by DNA-mediated transfection, the expression level is lower and a complementary oncogene such as *ras* is required to achieve full transformation phenotype. This behavior is identical to c-Jun, which also complements *ras* in this type of transformation assay. Chimeric molecule, Meq (bZIP)-Jun (TA), which fuses the bZIP domain of Meq and the transactivation domain (TA) domain of Jun is as potent as wild-type Meq or wild-type Jun in this assay. Likewise, the reciprocal construct, Meq (TA)-Jun (bZIP), also exhibits transforming properties. These results, coupled with the finding that Meq and c-Jun co-localize in the transformed cells, provide strong evidence that Meq and c-Jun participate in a similar signal transduction pathway, and that the different domains of these molecules are interchangeable with respect to this particular function (Liu et al. 1999a).

In addition to transforming and mitogenic properties empowered by Meq, Meq-infected cells are highly resistant to apoptosis induced by serum-withdrawl, TNF-α, UV-irradiation, and C2-ceramide. Thus, Meq seems to induce a general protective pathway, and given its transactivation ability, a plausible mechanism is that it mediates its transcriptional regulation of apoptosis-related molecules such as bcl2, Bcl-xL, and Bax. In MDV-transformed chicken T-cells, Bcl-xL, an anti-apoptotic factor, is often up-regulated (Ohashi et al. 1999), whereas in Meq-transformed Rat-2 cells, Bcl-2 (anti-apoptotic factor) is transcriptionally up-regulated and Bax (pro-apoptotic factor) is down-regulated (Liu et al. 1999a). Whether Meq is directly involved in the transcriptional regulation of the apoptosis-related genes is not clear and awaits the isolation and sequencing of the chicken and rat promoters for these genes. We note that MERE I and II sequences are both found in the human Bcl2 promoter and the p53-binding site in human Bax promoter. In the latter regard, Meq can potentially transmit apoptosis signal by its ability to bind and sequester p53, a molecule strongly implicated in the regulation of apoptotic pathway. Many DNA tumor virus-transforming proteins have targeted p53 for mis-location or degradation as their means to prevent apoptosis (Szekely et al. 1993; Miyashita et al. 1994; Zhang et al. 1994). As described above, there are some similarities between Meq and oncogene Mdm2. While the primary sequences of these two molecules are quite different, both are associated with p53 and localized in the nucleolus (Momand et al. 1992; Brunovskis et al. 1996; Liu et al. 1997; Roth et al. 1998). Mdm2 is localized in the nucleolus via its binding to p19ARF tumor suppressor gene (Tao et al. 1999b; Weber et al. 1999, 2000; Sherr et al. 2000). In the absence of p19ARF (due to deletion in human cancers) or when Mdm2 is overexpressed (due to amplification in human cancers), Mdm2 binds p53 and shuttles p53 to cytoplasm for degradation. It is tantalizing to postulate that Meq may participate in a similar function.

3 Replication and Transforming Functions of Meq

3.1 Meq and MDV Latency

As Meq is one of the few proteins expressed in latent cells, it is tempting to suggest that Meq plays a role in maintaining latency. If so, this would be the opposite of what other ZIP proteins do in EBV (the Zta product, also called ZEBRA and BZLF1 product) or in KHSV (the K-bZIP product) (Lin et al. 1999), where the bZIP proteins are expressed predominantly in lytic phase and, at least in EBV, forced introduction of Zta triggers latency. The paradigm set by Zta suggests that reactivation from latency is orchestrated by Zta through reprogramming of gene expression, i.e., Zta activates Rta, a transcriptional factor which together with Zta activates Mta, another transcriptional factor, and the cascade goes on and on. If this is the case, we can suggest that Meq suppresses the transcription of a factor similar to Zta, thereby maintaining the latent state (although sequence analysis of the entire MDV genome does not reveal any obvious candidate, other than Meq itself). Upon reactivation, Meq transcription is down-modulated and the suppression is removed. Consistent with this notion is the finding that after butyrate treatment of MSB-1, Meq expression is down-modulated. TPA treatment however does not affect the Meq expression that much during the same time period. Alternatively, Meq can be the Zta-like factor itself. This seemingly paradoxical notion can be rationalized by the fact that the activity as an activator or a repressor of transcriptional factor can often be modulated (or switched) by the phosphorylation status and/or interacting partners. Thus, in latent state, Meq assumes a conformation, unable to activate the transcription of lytic genes, but upon treatment by inducing agents such as TPA, its activity changes due to post-translational modification or association with different partners, and becomes an activator for lytic genes. This change can be mediated by the activation of a kinase/phosphatase, which phosphorylates/de-phosphorylates Meq, or of a Meq partner such as the Jun/Fos/CREB family of proteins that dimerizes with Meq. The finding that Meq is phosphorylated by PKC (the receptor for TPA) and the well-known fact that Jun and Fos are activated by PKC lend support to this notion. If EBV Zta is any indication, phosphorylation of EBV Zta at a specific serine residue is critical to the reactivation process (Baumann et al. 1998; Francis et al. 1999). The demonstration that Meq activity toward MDV lytic gene promoter increases upon phosphorylation by PKC or by the formation of heterodimers with the Jun/Fos/CREB family of molecules has yet to come. Indeed, what the downstream MDV target genes of Meq are, is still unknown, which hampers the test of the hypothesis. Based on the Meq/Meq homodimer-binding motifs (MERE I and II), the promoters for Meq, pp38, pp24, and pp14 are all found to be potential targets (Qian et al. 1996). Among them, only the Meq promoter has been tested and shown to be activated by transfected Meq (in the absence of TPA treatment). There are AP-1 (TRE) and CRE sites in numerous promoters of the MDV genome, which are potential targets for Meq/Jun and Meq/ATF2 heterodimer. A challenge in the near future is to identify Meq target genes either in the homodimer or the heterodimer configuration. The advent of DNA chip/microarray technology and the availability of the complete MDV DNA sequences, which permit genome-wide examination of potential targets, should greatly facilitate this process.

3.2 Meq and MDV Replication

Does Meq play a role in replication? There is no concrete evidence either way. Although Meq knockout mutant appears to replicate well in vitro, its replication in vivo is extremely poor compared to wild-type viruses (Morgan, personal communication). In addition, while Meq is predominantly a latent protein, there were data suggesting the expression of Meq in the early phase of MDV infection (Tillotson et al. 1988). As described in the previous section, viral protein can serve a dual role in replication and latency. An intriguing observation is that Meq/Meq homodimer has a functional binding site (MERE I) near the lytic replication origin of MDV (Qian et al. 1996). This site is close to the binding site of UL9, a protein, and by analogy to HSV it recognizes specific sequences near DNA replication origin and directs DNA polymerase to the origin. It is conceivable that Meq binding either inhibits (as a latent protein) or facilitates (as a replication protein) the binding of UL9/DNA polymerase to the origin of replication. Alternatively, Meq serving as a transcriptional factor with its possible association with chromatin-remodeling proteins may regulate the conformation of the DNA replication origin and thus the replication initiation process. In the case of EBV Zta protein, it is found that Zta binds to the lytic replication origin of EBV and that such a binding is critical in the initiation of viral DNA replication (Sarisky et al. 1996; Gao et al. 1998). At present the direct involvement of Meq in MDV DNA replication is pure conjecture; more likely, given the transactivation function of Meq and the numerous AP-1 sites present in the MDV genome, Meq is needed at some phase to transcribe these genes, especially in cell types lacking the proper AP-1 components.

3.3 Meq and MDV Oncogenesis

As described before, Meq, when overexpressed alone, is capable of inducing morphological transformation and anchorage-independent growth of Rat cells, and confer anti-apoptotic activities to these cells. When expressed at a lower level, it complements transformation mediated by ras, in a manner similar to c-Jun. Furthermore, retroviruses carrying Meq are able to induce lymphocytic tumors, although the tumors are not as aggressive as those induced by retroviral oncogenes (Ewert, personal communications). The above properties are hallmarks of an oncogene. Yet, compared to oncogenes derived from acute retroviruses, which are mutated copies of cellular proto-oncogenes, Meq is only weakly oncogenic. This however is not a total surprise, as the transforming genes of most DNA tumor viruses are weakly oncogenic and need to be complemented by each other (e.g., adenovirus E1A and E1B, SV 40 Large T and small T etc.) to induce full transforming phenotypes. The lack of oncogenicity of MDV mutant with Meq deleted is consistent with the notion that Meq plays a role in transformation. This interpretation however is somewhat complicated by the possibility that Meq may also be involved in the in vivo replication and infection. The low infection of Meq mutant virus precludes the analysis of lymphoma induction. If and when the various functional domains of Meq can be dissected, it may be possible to engineer a mutant that affects only transformation without impeding replication. A second piece of evidence supporting a role of Meq in transformation is the interesting report by Xie et al. (Xie et al. 1996), who showed that anti-sense of Meq inhibits the growth of MDV-transformed T cells, suggesting Meq activity is required for the maintenance of the transformed phenotype. A prerequisite for this conclusion is that Meq inhibition did not activate the latent genome leading to cell lysis, a phenomenon the authors would have detected during their studies. On balance, the current evidence suggests Meq play a significant role in transformation and oncogenesis.

How does Meq mediate transformation? Several possibilities come to mind. First, given its similarity to Jun/Fos-like oncogene and that it can bind the AP-1 site after dimerization with Jun, it is possible that Meq activates the same set of genes v-Jun or v-Fos induces, leading to the oncogenic pathway. Second, the Meq/Meq homodimer binds the extended AP-1 site, which interestingly resembles the DNA-binding motif of Maf, another bZIP oncogene of avian sarcoma virus. C-Maf, the cellular form, has recently been implicated in the oncogenesis of multiple myeolomas due to translocation of this gene (Chesi et al. 1998). Meq can thus mediate its transformation by turning on the genes activated by Maf. The third possibility is that Meq forms complex with p53 and RB, and in a manner similar to other DNA tumor virus oncogenes, it overcomes the G1/S restriction point and induces unscheduled proliferation as well as genomic stability. In addition, since p53 mediates apoptosis, complexing and inactivating p53 can account for the reduced expression of Bax (a target gene for p53) and the anti-apoptotic functions. An intriguing possibility of how Meq inactivates p53 is that Meq, by virtue of its localization in the nucleolus, can "trap" p53 and RB in the nucleolus, preventing them from reaching the chromatin targets. The above possibilities are not mutually exclusive and may account for the pleiotropic properties of Meq in transformation.

4 The Origin and Evolution of Meq

Earlier hybridization studies as well as recent genomic sequences indicate that Meq is uniquely present in MDV-1, the virulent serotype, but not in serotype 2 and 3, the vaccine strains. The gene is located in the repeat region flanking the UL sequences. Genes present in the repeat regions show greater diversity than those in the unique region. The former encodes mostly regulatory genes unique to MDV-1 whereas the latter, structural and replication genes, common to other MDV serotypes as well as herpes-simplex virus (see the chapter by Silva et al., this volume). It is indeed remarkable that MDV-1 has a gene organization (the number of genes, the relative positions of the genes, and the polarities of the transcripts) in the UL and US region nearly identical to that of HSV1, yet, the genes in the repeat regions are almost completely different. This suggests that the unique region encodes genes essential for the basic functions in viral replication and assembly, whereas the repeat regions are the products of continual evolution and adaptation to the host environment. Changes and mutations of the genes in the repeat region, because of the diploid nature, are more readily tolerated than those in the unique regions, accounting for their faster rate of evolution. In this regard, studying the genes in the repeat region is likely to reveal the special virus–host interactions and relations. Meq is one of such proteins that seem to confer MDV with outstanding features such as oncogenicity, immunogenicity, and perhaps latency in target cells. The evolution of herpes virus genome involves transduction or recombination with the host genes or with other viral genes. Based on the structural similarity to the Jun/Fos family of proteins, especially the SNF gene (involved in HTLV-1 mediated transformation of T-cells; Waschman, personal communication), Meq is likely to be derived from the host bZIP family of genes through recombination. The proline-rich part has no counterpart in chicken genome, but its structural feature is similar to the WT-1 gene. It is possible that Meq is a result of two recombination events, fusing two proteins together. Another interesting feature is that the major form of Meq is encoded by one long open reading frame without intervening sequences, which appears to be at odds with the notion that it was derived from host genomes via recombination, as most host genes contain introns. It may be argued that herpesvirus, due to its size limit in packaging, can only afford transducing genes without introns. This is not exactly true, as vIL-8 (Liu et al. 1999a), a chemokine gene encoded by MDV, retains the intron-exon structure of the host gene. It is possible that Meq is derived from reverse-transcribed host pseudogene. Alternatively, it is transduced by retrovirus, which has an uncanny ability to capture and transduce host genes, and because of its life cycle involving RNA intermediate, the genes being transduced will be in the spliced or intronless form. The bZIP-transcriptional factors such as v-Jun, v-Fos, and v-Maf are all oncogenes transduced by retroviruses in the spliced form. In the case of Meq, two transduction events need to be contemplated as the bZIP domain and the C-terminal proline-rich domain may be derived from two different genes. Retroviruses that transduce two genes, such as AEV (erbA and erbB) have been time and again isolated. The recent discovery that MDV-1 genome harbors the LTR remnants of REV (reticuloendotheliosis virus), an avian retrovirus, and the demonstration that avian retroviruses integrate efficiently into the genome of MDV make this hypothesis more tenable (Isfort et al. 1992, 1994a,b; Jones et al. 1993, 1996).

5 Concluding Remarks

In this review, we described the properties and functions of Meq. It is clear that Meq is a versatile protein and likely to participate in both transforming and replication functions of MDV. Yet, despite all these studies, the crucial evidence that Meq is involved in oncogenesis or latency is still lacking, largely due to the difficulty in conducting genetic analyses of MDV. The cell-associated nature of MDV makes the cloning of genetic mutants more difficult. The elegant technique developed by Morgan and Parcells (Parcells et al. 1999) has overcome some of the difficulties and provides means to alter individual genes of MDV. The recent completion of the sequence determination of the MDV genome should also greatly facilitate the genetic analysis (Lee et al. 2000). In addition, the development of high throughput analyses such as DNA chip and microarray will allow rapid identification of Meq targets and the study of the genes involved in MDV replication and latency. Thus, although Meq was identified almost a decade ago, this is an exciting juncture when we can put all the information together to understand its in vivo functions.

Acknowledgements. We thank many of the colleagues who have contributed to Meq research. Dr. Joanne Kivela and Dr. Dan Jones were founding investigators of this work. Dr. Mark Parcells and Dr. Robin Morgan's outstanding work in generating MDV recombinants add greatly to our understanding of MDV genes including Meq. We also thank Dr. Su-Fang Lin for contributing to the information of the MDV vIL-8 studies and Jeremy Kamil for insightful discussion. The work was supported by NIH and USDA grants to H.J.K.

References

Adamson AL, Kenney S (1999) The Epstein-Barr virus BZLF1 protein interacts physically and functionally with the histone acetylase CREB-binding protein. J Virol 73(8):6551–6558

Askovic S, Baumann R (1997) Activation domain requirements for disruption of Epstein-Barr virus latency by ZEBRA. J Virol 71(9):6547–6554

Baumann M, Mischak H, Dammeier S, Kolch W, Gires O, Pich D, Zeidler R, Delecluse HJ, Hammerschmidt W (1998) Activation of the Epstein-Barr virus transcription factor BZLF1 by 12-O-tetradecanoylphorbol-13-acetate-induced phosphorylation. J Virol 72(10):8105–8114

Baxevanis A, Vinson C (1993) Interactions of coiled coils in transcription factors: where is the specificity? Curr Opin Gen Dev 3:278–285

Brunovskis P, Qian Z, Li D, Lee LF, Kung HJ (1996) Functional analysis of the MDV basic leucine zipper product, Meq. The 5th International Symposium on Marek's Disease. American Association of Avian Pathologists, Kellogg Center, Michigan State University, East Lansing, Michigan

Call KM, Glaser T, Ito CY, Buckler AJ, Pelletier J, Haber DA, Rose EA, Kral A, Yeger H, Lewis WH, et al. (1990) Isolation and characterization of a zinc finger polypeptide gene at the human chromosome 11 Wilms' tumor locus. Cell 60(3):509–520

Calnek BW, Witter RL (1997) Marek's disease. Iowa State University Press, Ames, Iowa, USA

Chesi M, Bergasagel PL, Shonokun O, Martelli ML, Brents LA, Chen T, Schrock E, Ried T, Kuehl WM (1998) Frequent dysregulation of the c-maf proto-oncogene at 16q23 by translocation to an Ig locus in multiple myeloma. Blood 91(19):4457–4463

Chi T, Lieberman P, Ellwood K, Carey M (1995) A general mechanism for transcriptional synergy by eukaryotic activators. Nature 377(6546):254–257

Delling U, Roy S, Sumner-Smith M, Barnett R, Reid L, Rosen CA, Sonenberg N (1991) The number of positively charged amino acids in the basic domain of Tat is critical for trans-activation and complex formation with TAR RNA. Proc Natl Acad Sci USA 88(14):6234–6238

Elenbaas B, Dobbelstein M, Roth J, Shenk T, Levine AJ (1996) The MDM 2 oncoprotein binds specifically to RNA through its RING finger domain. Mol Med 2(4):439–451

Francis A, Ragoczy T, Gradoville L, Heston L, El-Guindy A, Endo Y, Miller G (1999) Amino acid substitutions reveal distinct functions of serine 186 of the ZEBRA protein in activation of early lytic cycle genes and synergy with the Epstein-Barr virus R transactivator. J Virol 73(6):4543–4551

Freedman DA, Levine AJ (1998) Nuclear export is required for degradation of endogenous p53 by MDM 2 and human papillomavirus E6. Mol Cell Biol 18(12):7288–7293

Gao Z, Krithivas A, Finan JE, Semmes OJ, Zhou S, Wang Y, Hayward SD (1998) The Epstein-Barr virus lytic transactivator Zta interacts with the helicase-primase replication proteins. J Virol 72(11):8559–8567

Glover J, Harrison S (1995) Crystal structure of the heterodimeric bZIP transcription factor c-Fos-c-Jun bound to DNA. Nature 373:257–261

Gutsch DE, Holley-Guthrie EA, Zhang Q, Stein B, Blanar MA, Baldwin AS, Kenney SC (1994) The bZIP transactivator of Epstein-Barr virus, BZLF1, functionally and physically interacts with the p65 subunit of NF-kappa B. Mol Cell Biol 14(3):1939–1948

Hammes SR, Greene WC (1993) Multiple arginine residues within the basic domain of HTLV-I Rex are required for specific RNA binding and function. Virology 193(1):41–49

Hatton T, Zhou S, Standring DN (1992) RNA- and DNA-binding activities in hepatitis B virus capsid protein: a model for their roles in viral replication. J Virol 66(9):5232–5241

Isfort R, Jones D, Kost R, Witter R, Kung HJ (1992) Retrovirus insertion into herpesvirus in vitro and in vivo. Proc Natl Acad Sci USA 89(3):991–995

Isfort RJ, Qian Z, Jones D, Silva RF, Witter R, Kung HJ (1994a) Integration of multiple chicken retroviruses into multiple chicken herpesviruses: herpesviral gD as a common target of integration. Virology 203(1):125–133

Isfort RJ, Witter R, Kung HJ (1994b) Retrovirus insertion into herpesviruses. Trends Microbiol 2(5): 174–177

Jones D, Brunovskis P, Witter R, Kung HJ (1996) Retroviral insertional activation in a herpesvirus: transcriptional activation of US genes by an integrated long terminal repeat in a Marek's disease virus clone. J Virol 70(4):2460–2467

Jones D, Isfort R, Witter R, Kost R, Kung HJ (1993) Retroviral insertions into a herpesvirus are clustered at the junctions of the short repeat and short unique sequences. Proc Natl Acad Sci USA 90(9):3855–3859

Jones D, Kung HJ (1992a) A heat-shock-like sequence in the meq gene promoter binds a factor in MDV lymphoblastoid cells. The 19th World's Poultry Congress. Ponsen and Looijen, Wageningen, Amsterdam, The Netherlands

Jones D, Lee L, Liu JL, Kung HJ, Tillotson JK (1992b) Marek disease virus encodes a basic-leucine zipper gene resembling the fos/jun oncogenes that is highly expressed in lymphoblastoid tumors [published erratum appears in Proc Natl Acad Sci USA 1993 Mar 15; 90(6):2556]. Proc Natl Acad Sci USA 89(9):4042–4046

Lai Z, Freedman DA, Levine AJ, McLendon GL (1998) Metal and RNA binding properties of the hdm2 RING finger domain. Biochemistry 37(48):17005–17015

Lazinski D, Grzadzielska E, Das A (1989) Sequence-specific recognition of RNA hairpins by bacteriophage antiterminators requires a conserved arginine-rich motif. Cell 59(1):207–218

Lee CZ, Lin JH, Chao M, McKnight K, Lai MM (1993) RNA-binding activity of hepatitis delta antigen involves two arginine-rich motifs and is required for hepatitis delta virus RNA replication. J Virol 67(4):2221–2227

Lee LF, Wu P, Sui D, Ren D, Kamil J, Kung HJ, Witter RL (2000) The complete unique long sequence and the overall genomic organization of the GA strain of Marek's disease virus. Proc Natl Acad Sci USA 97(11):6091–6096

Lin SF, Robinson DR, Miller G, Kung HJ (1999) Kaposi's sarcoma-associated herpesvirus encodes a bZIP protein with homology to BZLF1 of Epstein-Barr virus. J Virol 73(3):1909–1917

Liu JL, Lee LF, Ye Y, Qian Z, Kung HJ (1997) Nucleolar and nuclear localization properties of a herpesvirus bZIP oncoprotein, MEQ. J Virol 71(4):3188–3196

Liu JL, Lin SF, Xia L, Brunovskis P, Li D, Davidson I, Lee LF, Kung HJ (1999a) MEQ and V-IL8: cellular genes in disguise? Acta Virol 43(2–3):94–101

Liu JL, Ye Y, Qian Z, Qian Y, Templeton DJ, Lee LF, Kung HJ (1999b) Functional interactions between herpesvirus oncoprotein MEQ and cell cycle regulator CDK2. J Virol 73(5):4208–4219

Miyashita T, Krajewski S, Krajewska M, Wang HG, Lin HK, Liebermann DA, Hoffman B, Reed JC (1994) Tumor suppressor p53 is a regulator of bcl-2 and bax gene expression in vitro and in vivo. Oncogene 9(6):1799–1805

Momand J, Zambetti GP, Olson DC, George D, Levine AJ (1992) The mdm-2 oncogene product forms a complex with the p53 protein and inhibits p53-mediated transactivation. Cell 69(7):1237–1245

Ohashi K, Morimura T, Takagi M, Lee SI, Cho KO, Takahashi H, Maeka Y, Sugimoto C, Onuma M (1999) Expression of bcl-2 and bcl-x genes in lymphocytes and tumor cell lines derived from MDV-infected chickens. Acta Virol 43(2–3):128–132

Parcells MS, Dienglewicz RL, Anderson AS, Morgan RW (1999) Recombinant Marek's disease virus (MDV)-derived lymphoblastoid cell lines: regulation of a marker gene within the context of the MDV genome. J Virol 73(2):1362–1373

Pfitzner E, Becker P, Rolke A, Schule R (1995) Functional antagonism between the retinoic acid receptor and the viral transactivator BZLF1 is mediated by protein-protein interactions. Proc Natl Acad Sci USA 92(26):12265–12269

Qian Z, Brunovskis P, Lee L, Vogt PK, Kung HJ (1996) Novel DNA binding specificities of a putative herpesvirus bZIP oncoprotein. J Virol 70(10):7161–7170

Qian Z, Brunovskis P, Rauscher III F, Lee L, Kung HJ (1995) Transactivation activity of Meq, a Marek's disease herpesvirus bZIP protein persistently expressed in latently infected transformed T cells. J Virol 69(7):4037–4044

Ross NL (1999) T-cell transformation by Marek's disease virus. Trends Microbiol 7(1):22–29

Roth J, Dobbelstein M, Freedman DA, Shenk T, Levine AJ (1998) Nucleo-cytoplasmic shuttling of the hdm2 oncoprotein regulates the levels of the p53 protein via a pathway used by the human immunodeficiency virus rev protein. Embo J 17(2):554–564

Sacher R, Ahlquist P (1989) Effects of deletions in the N-terminal basic arm of brome mosaic virus coat protein on RNA packaging and systemic infection. J Virol 63(11):4545–4552

Sarisky RT, Gao Z, Lieberman PM, Fixman ED, Hayward GS, Hayward SD (1996) A replication function associated with the activation domain of the Epstein-Barr virus Zta transactivator. J Virol 70(12):8340–8347

Schuermann M, Hunter J, Hennig G, Muller R (1991) Non-leucine residues in the leucine repeats of Fos and Jun contribute to the stability and determine the specificity of dimerization. Nucl Acids Res 19:739–746

Sherr CJ, Weber JD (2000) The ARF/p53 pathway. Curr Opin Genet Dev 10(1):94–99

Szekely L, Selivanova G, Magnusson KP, Klein G, Wiman KG (1993) EBNA-5, an Epstein-Barr virus-encoded nuclear antigen, binds to the retinoblastoma and p53 proteins. Proc Natl Acad Sci USA 90(12):5455–5459

Tao W, Levine AJ (1999a) Nucleocytoplasmic shuttling of oncoprotein Hdm2 is required for Hdm2- mediated degradation of p53. Proc Natl Acad Sci USA 96(6):3077–3080

Tao W, Levine AJ (1999b) P19(ARF) stabilizes p53 by blocking nucleo-cytoplasmic shuttling of Mdm2. Proc Natl Acad Sci USA 96(12):6937–6941

Tillotson JK, Lee LF, Kung HJ (1988) Accumulation of viral transcripts coding for a DNA binding protein in Marek's disease tumor cells. In: Kato S, Horiuchi T, Mikami T, Hirai K (eds) Advance in Marek's Disease Research Proc. 3rd Intl. Marek's Disease Symposium. pp 128–134

Weber JD, Kuo ML, Bothner B, DiGiammarino EL, Kriwacki RW, Roussel MF, Sherr CJ (2000) Cooperative signals governing ARF-mdm2 interaction and nucleolar localization of the complex. Mol Cell Biol 20(7):2517–2528

Weber JD, Taylor LJ, Roussel MF, Sherr CJ, Bar-Sagi D (1999) Nucleolar Arf sequesters Mdm2 and activates p53. Nat Cell Biol 1(1):20–26

Xie Q, Anderson AS, Morgan RW (1996) Marek's disease virus (MDV) ICP4, pp38, and meq genes are involved in the maintenance of transformation of MDCC-MSB1 MDV-transformed lymphoblastoid cells. J Virol 70(2):1125–1131

Zapp ML, Hope TJ, Parslow TG, Green MR (1991) Oligomerization and RNA binding domains of the type 1 human immunodeficiency virus Rev protein: a dual function for an arginine-rich binding motif. Proc Natl Acad Sci USA 88(17):7734–7738

Zhang Q, Gutsch D, Kenney S (1994) Functional and physical interaction between p53 and BZLF1: implications for Epstein-Barr virus latency. Mol Cell Biol 14(3):1929–1938

Polyvalent Recombinant Marek's Disease Virus Vaccine Against Poultry Diseases

K. HIRAI[1] and M. SAKAGUCHI[1,2]

1 Introduction

In 1982, MACKETT et al. (1982) and PANICALI and PAOLETTI (1982) constructed a recombinant vaccinia virus expressing a foreign gene, the thymidine kinase (TK) gene of herpes simplex virus (HSV), which was inserted into the nonessential regions of the virus genome by homologous recombination. Since then, a variety of foreign genes were inserted into large DNA viruses such as poxviruses and herpesviruses, and many attempts have been made to use recombinant viruses as polyvalent vaccines against human and animal diseases. Why are recombinant viruses expected to produce better vaccines against infectious disease? Firstly,

[1] Department of Tumor Virology, Division of Virology and Immunology, Medical Research Institute, Tokyo Medical and Dental University, Bunkyo-ku, Tokyo 113-8510, Japan

[2] The Chemo-Sero Therapeutic Research Institute, Kikuchi Research Center, Kyokushi Kikuchi, Kumamoto 869-1298, Japan

polyvalent recombinant viruses may save labor for vaccination and cost for production. Secondly, a live vaccine is required for effective vaccination. It is especially important for pathogens which cannot grow in culture. In this case, the subunit vaccine and recombinant proteins have been developed against hepatitis B and malaria for human use. Insertion of the gene for protection against these diseases into virus vectors may produce better protective efficacy. In addition, care is taken to use the attenuated strain of pathogens for safety, especially for retroviruses, because of the increased virulency due to spontaneous mutation or the induction of cellular transformation by retroviral transduction of cellular oncogenes. RNA viruses generally produce much higher rates of spontaneous mutation than DNA viruses. In that respect, DNA viruses are preferable to use as virus vectors for recombinant vaccines. The rabies-recombinant vaccine using vaccinia virus as a virus vector is the most successful case in the veterinary field which has been used in Europe and North America (Blancou et al. 1986). This is mainly because it can be used as an oral vaccination and is heat-stable.

Large DNA viruses have a greater capacity for insertion of foreign DNAs without destruction of infectivity, whereas smaller DNA viruses have a limited capacity for foreign DNA due to severe packaging constraints imposed by the icosahedral virus capsid. Thus, two large DNA viruses, Marek's disease virus (MDV), avian herpesvirus, and fowlpox virus (FPV), have mostly been used as virus vectors to express a foreign gene for the protection of chickens against poultry diseases.

We began developing recombinant MDV vaccines in 1988 with the ultimate intention of producing safe and effective vaccines against various poultry diseases. Recently, we succeeded in the construction of recombinant MDV which can sufficiently protect commercial chickens with maternal antibodies against Newcastle disease (ND) as well as Marek's disease (MD) by only one-time inoculation at 1 day of age (Sonoda et al. 2000). This chapter reviews the development of recombinant MDV against poultry diseases.

2 MDV as a Virus Vector

MD is a malignant T-lymphomatosis of chickens caused by Marek's disease virus serotype 1 (MDV1) and it has been largely controlled by use of live attenuated or naturally avirulent vaccines (Witter 1988). MD vaccine viruses are of three types: attenuated MDV1; naturally apathogenic MDV serotype 2; and serotype 3 (MDV3), the naturally apathogenic herpesvirus of turkeys (HVT). Most viral proteins in cells productively infected with these three virus serotypes possess cross-reactive epitopes (Ikuta et al. 1981, 1983; Van Zaane et al. 1982).

MD vaccine viruses have been considered to be one of the most potent vectors for polyvalent live vaccines expressing foreign antigens related to vaccine-induced immunity against poultry diseases for the following reasons: (a) the viruses induce lifetime protection against MD by just one time vaccination in 1-day-old chickens (Witter 1985); (b) the viruses have a natural host range limited to several avian species, and therefore the vectors would be safe for other domestic animals and people working in the poultry industry; and (c) techniques for generating recombinant MDVs have been well established (Cantelo et al. 1991; Hirai et al. 1992). Among the vaccine viruses, HVT has been used worldwide both as live vaccines and polyvalent vaccine vectors. However, attenuated MDV1 strains such as CVI988 c/R6 (de Boer et al. 1988) and R2/23 (Witter 1991) are clearly superior to HVT (Witter 1992). Since the MDV1 vaccine has the closest antigenic similarity to the very virulent (vv) MDV1, including field virulent strains, attenuated MDV1 appears to be the best for prevention from MD, among the three serotypes of MDV vaccine viruses, and is suitable for construction of recombinant vaccines against avian diseases.

MDV1 DNA of about 180kbp consists of unique long (U_L) and unique short (U_S) regions, each bound by a set of inverted repeats (TR_L, IR_L, IR_S, and TR_S) (Fig. 1a). Several sets of direct repeats consisting of over 100bp are present in MDV1 DNA. One is a tandem direct repeat of a 132-bp repeat unit located within the TR_L and IR_L of MDV1 DNA (Maotani et al. 1986). Serial passages of oncogenic MDV1 strains in culture result in a loss of oncogenicity concomitantly with expansion of the tandem direct repeat (Hirai et al. 1981, 1984a; Ross et al.

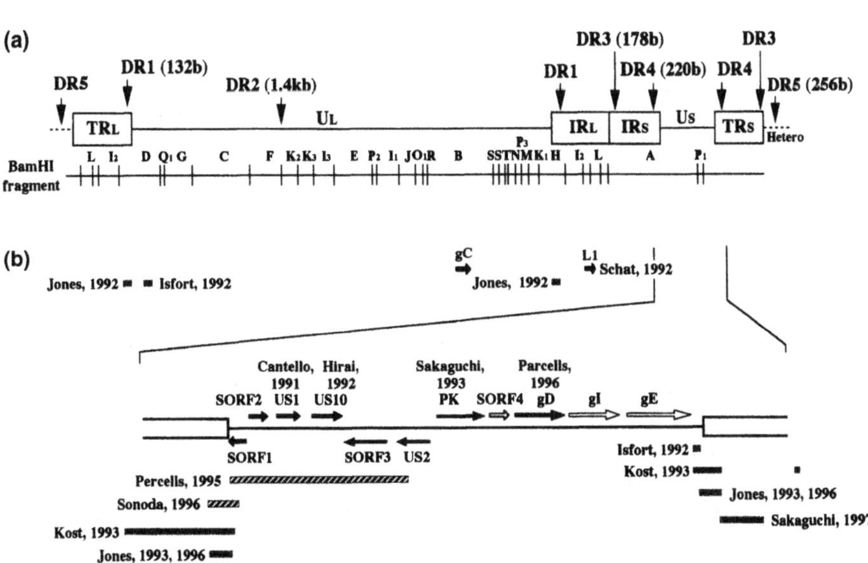

Fig. 1a,b. Location of nonessential regions for virus growth on the MDV1 genome. **a** Structure of the MDV1 genome. *DR*, direct repeat (size in base); TR_L, IR_L, IR_S, and TR_S, terminal and internal long inverted repeats and internal and terminal short inverted repeats, respectively; U_L and U_S, long and short unique sequences, respectively; *hetero*, heterogeneous in size. **b** Nonessential regions or viral genes for virus growth. *Hatched boxes* represent regions identified by the location of inserted retrovirus DNA; *shadow boxes* represent regions identified by insertion of the *lacZ* gene expression cassette. *Filled arrows*, nonessential genes identified as described in the text; *open arrows*, homologs of nonessential genes of HSV1. Name and year next to each box and arrow indicates the reference

1983; FUKUCHI et al. 1985; SILVA and WITTER 1985; MAOTANI et al. 1986). The second repeat is a direct repeat of about 1.4kbp, which is located within the *Bam*HI-F of the U_L (FUKUCHI et al. 1985). The third consists of a direct 178-bp repeat which is located within the IR_S and TR_S adjacent to the IR_L-IR_S junction and terminal repeats (HAYASHI et al. 1988). The fourth consists of several 220-bp tandem direct repeats which are also located at the IR_L-IR_S junction and terminal regions (SONODA et al. 1996). The fifth is located within the U_S-IR_S/TR_S junction and terminal regions (KISHI et al. 1991). The region contains multiple sets of a 256-bp repeat which is structurally similar to the a-sequence of herpes simplex virus (HSV) DNA. Although MDV2 and HVT genomes show similar DNA structure and size to the MDV1 genome, very few homologies among the three serotype DNAs were detected using Southern blot hybridization under stringent conditions. However, Southern blot hybridization under a less stringent condition revealed that weak homologies among these viral DNAs are dispersed throughout these viral genomes and that the DNA regions encoding each viral protein of these serotypes are mostly collinear (HIRAI et al. 1984b, 1986; FUKUCHI et al. 1984; IGARASHI et al. 1987; ONO et al. 1992).

3 Nonessential Sites for Insertion of Foreign Genes on the MDV Genome

In the first step for insertion and expression of foreign genes, the foreign gene linked to an appropriate promoter is inserted within the viral DNA sequence of a plasmid vector. Secondly, this gene linked to a promoter is inserted into the viral genome by homologous recombination between virus and plasmid DNAs in cultured cells (Fig. 2). Foreign genes of interest must be inserted into the nonessential region of the viral genome. The insertion sites should be nonessential for virus growth in chickens as well as in cultured cells. Since the size of the MDV genome is as large as about 180kbp, it is expected to contain many dispensable regions suitable for insertion of foreign genes. First, the TK gene of non-pathogenic HVT was proposed as an insertion site for expression of foreign genes (ROSS et al. 1993). Most viral genes in the U_S region of the HSV1 genome are dispensable for viral growth in cultured cells (LONGNECKER and ROIZMAN 1987; WEBER et al. 1987). Therefore, the U_S region of MDV1 genome may provide a suitable site for the insertion of foreign genes. The locations of the nonessential sites on the MDV genome were determined as described in Sects. 3.1–3.4 and indicated in Fig. 1b.

3.1 In Vitro Site-Specific Mutagenesis

There are several approaches to determine the nonessential sites for insertion of foreign genes on the MDV genome. The approach frequently used is to introduce

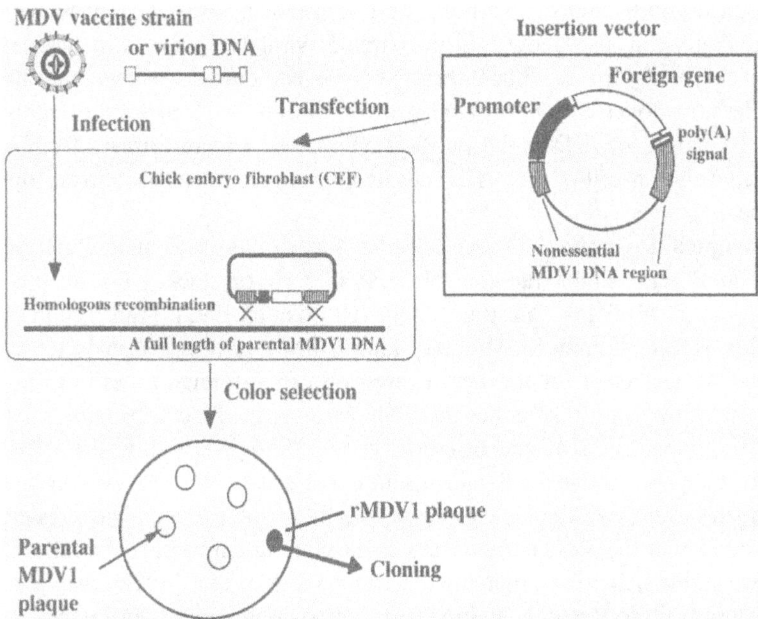

Fig. 2. Schematic representation of the generation of recombinant MDV1 expressing a foreign gene

mutations into the viral genome by marker transfer using homologous recombination and to test whether the mutated virus can grow in cultured cells (Fig. 2).

Insertion of the *Escherichia coli* (*E. coli*) *lacZ* gene encoding β-galactosidase has been frequently employed to locate the nonessential sites on the MDV1 genome for viral growth in culture. First, MDV1 DNA fragments containing a cutting site for a unique restriction enzyme are cloned within the plasmid vector. Then, the *lacZ* gene under the control of various promoters, such as human cytomegalovirus major immediate-early promoter and SV40 early and late promoters, is inserted at the unique restriction enzyme site within the MDV1 DNA. MDV1 DNA sequences on both sides of the inserted gene are preferred to be larger than 1kbp for homologous recombination with a full length of MDV1 DNA. Secondly, the insertion vectors containing the *lacZ* gene are cotransfected with a full length of viral DNA to chick embryo fibroblasts (CEF) or transfected directly to MDV1-infected cells for homologous recombination. To select the plaques expressing the *lacZ* gene, the transfected cells are overlaid with 1% agarose-culture medium containing 5-bromo-4-chloro-3-indolyl-β-D-galactoside (X-gal) or *o*-nitrophenyl-β-galactopyranoside (ONPG) as a β-galactosidase indicator. ONPG is less toxic to CEF than X-gal. Blue (X-gal) or red (ONPG) plaques are plated on fresh CEF for further passages. Since MDV is cell-associated, the cloning process is repeated until 100% of the plaques are positive. When the clones are stable for expression of *lacZ* gene, the recombinant viruses are purified from plaques produced by cell-free viruses obtained by sonication of infected cells. DNA from the recombinant MDV1 should be analyzed by Southern blot hybridization or PCR to test whether the *lacZ* gene is stably integrated at predicted sites into MDV1 DNA by double homologous recombination on both sides of the insert. Southern blot hybridization of recombinant viral DNA can also show that no parental viral DNA exists in plaque-purified recombinant viruses. When these are proved, one can show that the insertion sites are nonessential for viral growth in culture. Due to the highly cell-associated nature of MDV, the purification of recombinant viruses from a plaque infected with a cell-free virus is repeatedly required to separate from the parental virus.

We attempted to locate nonessential sites for viral growth in culture by insertion of *lacZ* genes under the control of SV40 early or chicken β-actin promoters to 22 sites in MDV1 (K554 strain) DNA (Hirai et al. 1992). Transfection of these insertion vector plasmids into monkey kidney BMT-10 cells revealed that all insertion sites were efficient for transient expression of the foreign genes in transfected cells under the control of either the SV40 early or the β-actin promoter. At 5–7 days after transfection of the insertion vectors into MDV1-infected CEF, expression of plaques positive for β-galactosidase was usually 0.5–3%. Among the 22 insertion sites examined, the *lacZ* genes at three insertion sites stably expressed β-galactosidase under the SV40 early promoter throughout ten passages in culture. Although five stable sites were originally determined (Hirai et al. 1992), two later became unsuccessful to grow in culture for unknown reasons. Several clones expressing the *lacZ* gene inserted at other sites were passaged serially in CEF and then suddenly became unstable after five to ten passages. All three stable sites were located within the S component of MDV1 DNA.

Most of the viral genes in the U_S region of the HSV1 genome were shown to be nonessential for viral growth in cultured cells (Longnecker and Roizman 1987; Weber et al. 1987). Nucleotide sequence determination of the U_S region in the MDV1 genome has revealed that there are 12 ORFs, seven of which are homologous to those of HSV1 (Ross et al. 1991; Sakaguchi et al. 1992; Brunovskis and Velicer 1995). Among them, the US1 (Parcells et al. 1994a), US10 (Hirai et al. 1992; Sakaguchi et al. 1994; Parcells et al. 1994a), US2 (Cantello et al. 1991; Parcells et al. 1994a), US3 (Sakaguchi et al. 1993), and US6 (gD) (Parcells et al. 1994b) homologs of HSV1 have been shown to be nonessential for viral growth in culture by insertion of the *lacZ* gene within these ORFs. Construction of recombinant MDV1 by deletion of the 4.8-kbp region within the U_S and insertion of the *lacZ* gene into the region revealed that the US1, US10, and US2 genes as well as three putative MDV-specific genes, SORF1, SORF2, and SORF3, were nonessential for viral growth in culture (Parcells et al. 1994a, 1995). However, the 4.8-kbp deletion mutant showed growth impairment in cultured cells similar to the US1 mutant, which lacks a US1 gene function by the *lacZ* gene insertion. Therefore, the US1 gene may be required for efficient replication of MDV1 in cultured cells, but is nonessential. Among these nonessential genes, the US3 gene was shown to encode 44- and 45-kDa proteins exhibiting protein kinase activity (Sakaguchi et al. 1993). The SORF3 encodes a 40-kDa protein in MDV1-infected cells (Urakawa et al. 1994). A 37-kDa protein, cross-reactive to the MDV1 protein, was detected in HVT-infected cells. The HVT gene homologous to the SORF3 of MDV1 is located within the U_S region of the HVT genome (Zelnik et al. 1993).

In addition to these recombinant MDV1s, stable recombinant MDV1s lacking the U_S-IR_S junction region were obtained by deletion of the region and insertion of the *lacZ* gene (Sonoda et al. 1996). The deleted region contains the SORF1 and DR4 consisting of several 220-bp tandem direct repeats within the IR_S. Therefore, the junction regions are not essential for viral growth in culture. Furthermore, Southern blot hybridization of the recombinant MDV1 indicated an asymmetrical deletion in the IR_S-U_S junction, suggesting that a set of the small inverted repeats of the same size are not necessary for viral growth in cultured cells. When the *lacZ* gene was inserted in the opposite left direction to the stable recombinant MDV1, it did not produce stable recombinant viruses. Therefore, the insertion of the foreign gene in the opposite direction may interfere with the surrounding viral gene such as the ICP4 homolog gene within the IR_S, possibly by readthrough transcription. The ICP4 gene of HSV1 is known to be essential for expression of viral genes and for viral replication (Roizman and Sears 1996). An ORF, ORF L1, located in the *Bam*HI Q2 and L fragments within the IR_L region was also shown to be non-essential for viral growth in culture by the *lacZ* gene insertion (Schat et al. 1998). The ORF L1 encodes one of the alternatively spliced mRNAs, while another one of the mRNAs encodes a viral oncoprotein, Meq (Jones et al. 1992; Liu et al. 1998).

The nonessential sites in the HVT genome for viral growth were found by insertion of foreign genes within the US10 gene (Morgan et al. 1992; Sondermeijer et al. 1993), the thymidine kinase gene (Ross et al. 1993), the small subunit of ribonucleotide reductase and gI (US7) genes (Darteil et al. 1995), two intergenic loci in the *Bam*HI I and pp38 gene-homolog of MDV1 (Bublot et al. 1999), and at the L-S junction (Marshall et al. 1993). The insertion site at the L-S junction was determined by a positive selection method to produce the recombinant HVT by homologous recombination using the insertion vector expressing the *E. coli* xanthine-guanine phosphoribosyltransferase (gpt) gene under the control of human cytomegalovirus (HCMV)-major immediate-early (MIE) promoter. Only cells expressing the gpt gene can grow in the selected media containing inhibitors for de novo purine synthesis such as mycophenolic acid and aminopterin. This positive selection method was also used for construction of stable recombinant MDV2 containing the gpt gene inserted within both long inverted repeats TR_L and IR_L (Marshall et al. 1993). Therefore, two copies of the foreign gene are inserted into one molecule of MDV2 DNA and expected to yield twice as many products.

Experiments using in vitro mutagenesis indicate that the stable insertion sites for foreign genes are mainly localized within the S component and long inverted repeats of the MDV DNA, where nonessential genes and latency, or tumor-associated genes, are clustered.

3.2 Inserted Retrovirus DNA Sequences

The presence of foreign genes in the MDV genome has been found in viruses from infected cultured cells. To find such a foreign gene in the MDV genome is another way to locate stable insertion sites and nonessential sites for viral growth. First, the presence of a full-size replicating MDV2 genome containing the MDV1 DNA sequence was observed in MD lymphoblastoid cell line (HIRAI et al. 1990). The inserted MDV1 DNA consisted of sequences mainly from the long inverted repeat including the *Bam*HI fragments H, I2, and L. The recombinant MDV2 could be derived by recombination between the latent MDV1 and infected replicating MDV2 DNAs. Although DNAs of oncogenic MDV1 and latent MDV1 in T-lymphoblastoid cell lines from MD tumors contain a few copies of the 132-bp tandem direct repeats within the long inverted repeats (MAOTANI et al. 1986; KANAMORI et al. 1986), the virus contained a multiple copy of the 132-bp tandem direct repeats, similar to the nononcogenic MDV1 strain DNA, and failed to induce MD tumors in chickens.

The first evidence for retrovirus DNA insertion into MDV1 DNA was obtained in attenuated MDV1 strains at a high passage level in culture (ISFORT et al. 1992). Two reported cases showed that a single copy of the LTR of the reticuloendotheliosis virus (REV) was inserted within the TR_L-U_L and the U_S-TR_S border regions of the MDV1 genome. Since the REV sequence was not detected in the same MDV strain of DNA at earlier passages, the LTR sequence found in higher passaged viruses was possibly due to a contamination by REV. When cultured cells were experimentally coinfected with MDV1 and REV, a similar insertion of REV-LTR sequence was also found within the U_S region close to the TR_S of the MDV1 genome during five passages after infection. It has been reported in several cases (COFFIN 1996) that insertional activation or inactivation of cellular genes under the transcriptional control of retroviral LTR occurs during development of tumors. However, insertion of the REV-LTR sequence within the U_S-TR_S did not affect the expression of MDV1 genes located downstream of the inserted LTR (KOST et al. 1993). One plaque-purified clone Cl.59 of the MDV1 with REV-LTR contains a large deletion of 5.5kbp in the IR_S-U_S border region including SORF1, and was shown to grow well in culture. Further, extensive studies on coinfection of MDV1 and REV indicated that retroviral insertion is common and most of the insertion sites are clustered within the small region of the U_S-IR_S/TR_S (JONES et al. 1993). The inserted REV sequences are mostly a single copy of the LTR. DNA arrangements such as small deletions and duplications occur at the insertion sites of REV-LTR and MDV1 DNAs, respectively. The rearrangement may occur by recombination between these viral DNAs. In addition, the REV proviral DNA sequence, except the LTR, may be deleted during serial passages, probably by homologous recombination between the LTRs at both ends of proviral DNA (JONES et al. 1993). All of these MDV1 isolates examined containing the inserted REV-LTR sequence have growth advantages over the parental strain in cultured cells. Using one of the MDV1 isolates, polycistronic mRNA including the coding sequences for US1, US10, and SORF2 were shown to be transcribed under the control of the LTR promoter (JONES et al. 1996). However, it is not known whether expression of these viral proteins is correlated with enhanced growth of MDV1 carrying the REV-LTR.

Furthermore, Southern blot hybridization of natural MDV isolates using an REV-LTR as a probe under low-stringency washing conditions revealed that only

MDV1 DNA but not MDV2 and HVT DNAs contained 70–81% homology of a length of approximately 20 nucleotides with the R and U3 regions of REV LTR within the *Bam*HI D, H, and Q1 of MDV1 DNA (Isfort et al. 1992). The homology in MDV1 DNA was not observed in the LTR sequences of avian leukosis virus (ALV) or murine leukemia virus, indicating that REV and MDV1, which share T-cell tropism, may exchange their genetic elements during their evolution.

However, a later report (Isfort et al. 1994) showed that coinfection of cultured cells with ALV and MDV1 also resulted in ALV-LTR insertion within the U_S-IR_S/TR_S border regions. One of these insertion sites was located within the gD gene, which had been shown to be nonessential for viral growth in culture (Parcells et al. 1994b). In contrast to REV-LTR, the inserted ALV sequence was only a portion of the U3 sequence of the LTR. We also reported that tandem direct repeats of the complete LTR of ALV-RAV0 strain was inserted within the short terminal repeats (Sakaguchi et al. 1997). The ALV-RAV0 strain is of endogenous origin, indicating that the retroviral insertion may occur via molecular interaction between the virus and host cells. The LTR sequence of avian erythroblastosis virus was also found in MDV1 DNA (Endoh et al. 1998).

The REV-LTR insertion is also found at several sites within the U_S-TR_S border region of the HVT genome in cells coinfected with REV and HVT (Isfort et al. 1994). Analysis of cloned DNAs from coinfected cells showed that the inserted REV sequence varied in cloned HVT DNAs containing the inserted REV sequence. One of the DNA clones contained a full length of the proviral DNA.

Infection with MDV1, REV, or ALV produces tumors in chickens while MDV, REV, and lymphoproliferative disease virus induces tumors in turkeys. In addition, the incidence of MD is increased when birds are dually infected with MDV and ALV (Frankel and Groupe 1971; Peters et al. 1973; Jakovleva and Mazurenko 1979). A 5-year survey of tumor-bearing chickens and turkeys in the field showed 25% of chickens and 24% of turkeys carried multiple infections of these oncogenic viruses (Davidson and Borenstein 1999). PCR with MDV1 and REV-LTR primers identified the presence of chimeric DNA products between the REV-LTR or ALV-LTR and the adjoining sequence of the 132-bp tandem direct repeats within the TR_L and IR_L of MDV1 DNA in coinfected chickens, indicating that the LTR insertion may occur in chickens and turkeys under natural conditions.

These findings indicate that the retroviral insertion into MDV genomes is a rather general phenomenon. The hot spots for retroviral insertion into the MDV genome is in and around the terminal and internal repeat regions. Since many cellular gene homologs in herpesviruses are also clustered at similar genomic locations, this suggests that the gene transfer between the host and viral genomes may occur by retrotransposon during the long evolutionary process of herpesviruses (Brunovskis and Kung 1996). The DNA element, known as a retrotransposon, is first copied into RNA. This RNA encodes a reverse transcriptase activity, which copies it back into DNA. The DNA copy is then integrated by site-specific cleavage into a new location by specific integrase activity. The full or near full-length of proviral REV DNA is also inserted in the field and vaccine strains of

FPV (Hertig et al. 1997). Therefore, the retroviral insertion into avian DNA viruses appears to be circulating widely in chickens.

Together with the findings obtained by in vitro mutagenesis, these observations support the idea that the evolutionary ancestor of herpesvirus genome consisted of only the L component containing viral genes essential for viral growth.

3.3 Nonessential Sites for In Vivo Growth

To characterize the recombinant MDV1 expressing the *lacZ* gene in vivo, we examined its immunogenicity and protective efficacy against virulent MDV1 challenge (Hirai et al. 1992; Sakaguchi et al. 1993, 1994; Sonoda et al. 1996). One-day-old SPF chickens were inoculated with recombinant MDV1 containing the *lacZ* gene inserted within the US3 (PK), US10, or the IR_S region. Antibodies against MDV1 antigens and β-galactosidase were present in all the sera of chickens vaccinated with these recombinant MDV1s even 16 weeks after immunization. Recombinant MDV1s expressing the *lacZ* gene were isolated from peripheral blood monocytes of the immunized chickens. These findings indicate that insertion of the *lacZ* gene into these insertion sites did not block immune responses and viral growth in chickens. When uninfected chickens were bred for 11 weeks in the same isolator with chickens inoculated with recombinant MDV1 containing the *lacZ* gene within the US10 gene, neither virus recovery nor induction of antibodies against MDV1 and β-galactosidase were observed (Sakaguchi et al. 1994). Therefore, recombinant MDV1 cannot be horizontally transmitted from inoculated to uninoculated chickens. Among the nonessential MDV1 genes examined, the US10 gene appears to be the most stable insertion site for foreign gene expression in cultured cells and birds (Sakaguchi et al. 1994). The oncogenic recombinant MDV1 strain RB1BΔ45*lac* lacking US1, US10, US2, and three putative MDV1-specific genes SORF1, SORF2, and SORF3 (Parcells et al. 1995), and nononcogenic MDV1 strains $CVIL_1$*LacZ*-A and -B lacking ORF L1 (Schat et al. 1998), were also shown to replicate in inoculated birds. Furthermore, the gD (US6) gene of MDV1 was also shown to be nonessential for replication of SPF chickens inoculated with recombinant MDV1 containing the inserted *lacZ* gene within the US6 ORF (Parcells et al. 1994b). However, the oncogenic RB1BΔ45*lac* showed a growth impairment in chickens as well as in cultured cells compared with the parental strain (Parcells et al. 1995). It is noteworthy that *lacZ* gene expression is downregulated in tumor cells. The promoter used to express the *lacZ* gene is the SV40 early promoter, which is expected to be constitutive in most of cell types. Therefore, the promoter or the insertion site may not be suitable to express foreign genes in latently infected cells. Among in vivo isolates from chickens inoculated with $CVIL_1$*LacZ*, there were some white plaques lacking the *lacZ* gene expression which appeared to be rearranged during in vivo replication (Schat et al. 1998).

The glycoprotein C (gC) gene (formally A antigen or gA) of MDV1 is nonessential for viral growth in cultured cells because attenuation of MDV1 in culture results in loss of gC gene expression (Churchill et al. 1969; Ikuta et al.

1983). The attenuated MDV1 strain grows as well as the unattenuated strain. In contrast, recombinant MDV1 lacking gC expression was not isolated from spleen and peripheral blood cells in inoculated chickens (MORGAN et al. 1996). However, further studies are required to clarify the role of the gC gene on in vivo growth and vaccinal immunity in inoculated chickens.

3.4 Nonessential Sites for Protection Against Marek's Disease

It is important to examine whether insertion of foreign genes into the MDV1 genome does not result in the decreased protective efficacy of MDV vaccine against virulent MDV1 challenge. To examine protective efficacy, SPF chickens vaccinated with recombinant MDV at 1 day of age are challenged with virulent or vvMDV1 strain about 1 week after vaccination. Ten weeks after challenge, vaccinated chickens are first examined for gross lesions and then histologically for the presence of MD lymphomas in visceral organs and lymphoproliferative infiltration in peripheral nerves. A sufficient level of protection (80–100%) can be obtained in SPF chickens vaccinated with recombinant MDV1 containing the following: the *lacZ* gene within the US3 gene, the US10 gene, and the IR_L region (SAKAGUCHI et al. 1993, 1994; SONODA et al. 1996); the fusion (F) gene of Newcastle disease virus (NDV) within the US10 gene (SAKAGUCHI et al. 1995, 1998); and the VP2 gene of infectious bursal disease virus (IBDV) within the US2 gene (TSUKAMOTO et al. 1999). The protective efficacy is comparable to the control MDV1 vaccine strains.

In contrast, recombinant HVT showed relatively low protection against virulent MDV1 challenge in SPF chickens. Very low (less than 10%) protection was obtained in SPF chickens compared with the parental HVT strain (84%) (DARTEIL et al. 1995). The recombinant HVTs containing the VP2 gene of infectious bursal disease virus within the ORFs encode either the small subunit of ribonucleotide reductase or glycoprotein I, respectively. When the NDV-F gene was inserted within the US10 gene of HVT, protection (64% and 76%) by the recombinant HVT was comparable to the control HVT vaccine strain (50%). The HVT FC126 strain contains a population of infectious viruses including the genome lacking 650bp in both the TR_L and IR_L (BUBLOT et al. 1999). The 650-bp deletion containing the pp38 homolog gene did not affect its protective efficacy.

Glycoprotein B (gB) of three MDV serotypes is responsible for virus neutralization and cross-reaction among MDV serotypes (IKUTA et al. 1984; HIRAI et al. 1986). Chickens immunized with the purified protein were protected partially against virulent MDV1 challenge (ONO et al. 1985). In addition, recombinant FPV expressing the MDV1-gB protected chickens without maternal antibodies against virulent MDV1 challenge (NAZERIAN et al. 1992, 1996). Therefore, recombinant HVT expressing MDV1-gB is expected to provide better protective efficacy than commercial HVT vaccine. Recombinant HVT expressing the MDV1-gB gene which was inserted within the thymidine kinase (TK) gene can grow well as a control HVT strain in culture (ROSS et al. 1993). However, the recombinant HVT induced lower HVT-antibody titers than control HVT. This indicates that the recombinant HVT grew less efficiently than control HVT in chickens. In addition, the spontaneous TK-less mutant also induced low antibody titers similar to the recombinant HVT. Therefore, the TK gene of HVT may be required for its efficient replication of HVT in chickens, although it is not required for viral replication in culture, as found in the HSV-TK gene (Field and Wildy 1978; Tenser and Dunstan 1979).

4 Protection of Poultry Diseases by Recombinant MDV Vaccine

Most vaccines for poultry diseases are used to control viral diseases such as MD, ND, infectious bronchitis, laryngotracheitis, fowlpox, and infectious bursal disease. The laying bird may receive as many as 7–10 vaccinations against 6–11 diseases from hatching till the beginning of egg-laying. These vaccination programs cause various problems such as side effects, labor costs, and stress that may lead to reduction in egg-laying or to induction of bacterial infections.

4.1 Newcastle Disease and NDV

Newcastle disease (ND) is a severe respiratory, neurological, and enteric disease caused by Newcastle disease virus (NDV) which belongs to the *Paramyxovirus* genus. ND shows high frequency of death among infected young chickens. The disease continues to have a substantial economic impact on the poultry industry throughout the world. Vaccination of chickens against ND has been used since the 1940s. The most commonly used vaccines are live virus vaccines, particularly nonvirulent (lentogenic) or selected intermediate virulent (mesogenic) strains. The main disadvantage of these live vaccines is that they cause mild respiratory disease, resulting in increased bacterial infections. Since the immunity produced by these live vaccines cannot last long, chickens have to be repeatedly vaccinated. In addition, just-hatched chicks have high levels of maternal antibodies against NDV from parent birds, which interfere with the replication of NDV vaccine strains. After the levels of the antibodies decrease at around 4–5 weeks old, chickens may then be at risk when exposed to virulent (velogenic) NDVs in the field. Therefore, vaccinations are required several times before 5 weeks of age and then at least once more before 12 weeks of age. Vaccinations by live vaccine are required every 3 months thereafter. Thus, construction of recombinant MDV vaccines expressing NDV antigens related to vaccinal immunity may solve these problems because a one-time vaccination of 1-day-old chicks using recombinant MDV vaccines is expected to induce lifetime protection against ND as well as MD.

NDV contains a nonsegmented negative-stranded RNA genome of 15,186 nucleotides encoding at least six major proteins. Of these proteins, two viral glycoproteins, fusion (F) protein and hemagglutinin-neuraminidase (HN), are expressed on the surfaces of viral particles and infected cells. The F protein is required for membrane fusion between the viral envelope and plasma membranes of host cells and is involved in cell-to-cell infection by membrane fusion between infected and uninfected cells. The HN protein mediates attachment of the viral particle to the host cell receptor and has receptor-destroying activity. Both viral glycoproteins are able to elicit neutralizing antibodies, but only antibodies to F protein prevent cell-to-cell spread of infection (AVERY and NIVEN 1979; MERZ et al. 1981). Antibodies against the NDV-F were shown to protect chickens from velogenic NDV challenge better than HN antibodies (MEULEMANS et al. 1986; UMINO et al. 1990). We showed that injection of a DNA vaccine consisting of a plasmid vector expressing the NDV-F gene into SPF chickens efficiently produced the anti-NDV-F antibody, and the chickens that had the antibody were protected from lethal NDV challenge (SAKAGUCHI et al. 1996). Therefore, the NDV genes encoding these glycoproteins are suitable for insertion into the MDV genome.

4.1.1 Construction of Recombinant MDV Expressing NDV-F or -HN

Field outbreaks of MD and ND have caused the most damage, among avian diseases, to the poultry industry. Therefore, construction of recombinant MDV expressing NDV antigens has been in increasing demand in vaccination programs.

To construct recombinant expressing NDV-F, cDNA to a monocistronic mRNA of 1.8kb is inserted at the nonessential site in the MDV DNA sequence of the insertion vector. The cDNA expressed under the control of various promoters encodes a precursor (F_0), which is activated upon proteolytic cleavage to produce disulfide-linked F_1 and F_2 by host cellular proteases (HOMMA and OUCHI 1973; NAGAI et al. 1976). The cleavage is required for the progeny virus to become infective, and the cleavage site is a major determinant for virulence. HN protein is translated from a monocistronic mRNA of 1.9kb. Table 1 summarizes the findings of the protective efficacy of recombinant MDVs against challenges of NDV, MDV1, or infectious bursal disease virus.

4.1.2 Protection of SPF Chickens Against ND

We constructed recombinant MDV1 expressing NDV-F and examined the protective efficacy against ND in SPF chickens (NAKAMURA et al. 1992; SAKAGUCHI et al. 1995, 1998). Although the *lacZ* gene under the control of SV40 early promoter induces effective immunity, the SV40 early promoter was less effective in expression of NDV-F than the late promoter. Therefore, the NDV-F gene under the control of SV40 late promoter was inserted at the stable insertion site within the US10 gene of the MDV1 (CVI988 strain) genome. For cloning the recombinant MDV1, plaques expressing the NDV-F were detected by staining with anti-NDV-F monoclonal antibody. Their plaque sizes were a similar size to those of the parental strain. The plaque-purified clone was named rMDV1-US10L(F).

SPF chickens vaccinated with rMDV1-US10L(F) at 1 day of age were sufficiently protected from velogenic NDV challenge through intramuscular routes at 3, 9, and 24 weeks after vaccination or through ocular, intranasal, intratracheal, or

Table 1. Protection of chickens with and without maternal antibodies from poultry diseases with recombinant MDV

MDV vector	Inserted DNA			Vaccination			Challenge[d]					Reference
	Foreign gene	Insertion site	Promoter	Maternal antibody	Route[c]	Age of chicken	A	B	Time	Route	Protection (%)	
HVT	NDV-HN	US10	LTR	–	IA	1 day	MDV		9 days	IA	64	Morgan et al. 1992
								NDV	28 days	IM	58	
	NDV-F	US10	LTR	–	IA	1 day	MDV		9 days	IA	76	
								NDV	28 days	IM	96	
	NDV-F	US10	LTR	+	IA	1 day	MDV		8 days	IA	50–76	Morgan et al. 1993
								NDV	29 days	IA	70–100	
	IBD-VP2	UL40 (RR)	UL40	–	IM	1 day	MDV		7 days	IP	0–6	Darteil et al. 1995
								IBDV	21 days	OC	0–50	
	IBD-VP2	US7 (gI)	HCMV-IE	–	IM	1 day	MDV		7 days	IP	10	
								IBDV	21 days	OC	100	
	MDV1-gC, gB, NDV-F, HN	ND[a]	See footnote[b]	–	SC	1 day		NDV	2 weeks	OC	94–100	Heckert et al. 1996
	MDV1-gC, gB, NDV-F, HN	ND	See footnote[b]	–	AF	18-day egg	MDV		1 week	IA	91	Reddy et al. 1996
				–	AF	18-day egg		NDV	4 weeks	OC	100	
MDV1	NDV-F	US10	SV40 late	–	SC	1 day	MDV1		1 week	IP	100	Sakaguchi et al. 1998
								NDV	3–24 weeks	IM	100	
				+	SC	1 day	MDV1		1 week	IP	86	
								NDV	6 weeks	IM	60–90	
			MDV1-gB	+		1 day	MDV1		1 week	IP	89	Sonoda et al. 2000
								NDV	6 weeks	IM	100	
	IBDV-VP2	US2	SV40 early	–	SC	1 day	MDV		1 week	IP	100	Tsukamoto et al. 1999
								IBDV	6 weeks	Oral	55	

[a] ND, no details available;
[b] Their own promoters for MDV1-gC and gB genes; the promoters of PRV-gX for NDV-HN and HCMV-IE for NDV-F;
[c] IM, intramuscular; IA, intra-abdominal; IP, intraperitoneal; OC, ocular, oculonasal, or eye drop; SC, subcutaneous; AF, amniotic fluid
[d] Challenged with two virulent viruses, either A or B.

subcutaneous routes at 4 weeks after vaccination. Thus, protective immunity persisted for at least 24 weeks. The recombinant MDV1 also prevented infection of velogenic NDV. Detection of antibodies against NDV-F using the ELISA test in sera of vaccinated chickens was useful for evaluating the protective efficacy, since vaccinated chickens that survive or do not survive from NDV challenge can be estimated by the ELISA values of the sera (SAKAGUCHI et al. 1998). The recombinant MDV1s were isolated even after ten blind passages in birds which showed expression of NDV-F in infected cells. These findings indicate the stability of rMDV1-US10L(F) expressing NDV-F in SPF chickens.

Protective efficacy of recombinant HVT expressing NDV-F or -HN has been extensively studied by MORGAN et al. (1992, 1993). The NDV-F or -HN gene was inserted within the US10 gene and expressed under the control of the long terminal repeat (LTR) promoter of Rous sarcoma virus. More than 90% of 1-day-old SPF chickens vaccinated with recombinant HVT expressing NDV-F were sufficiently protected from lethal intramuscular challenge by the neurotropic velogenic NDV strain at 28 days after vaccination. In contrast, recombinant HVT expressing NDV-HN provided partial protection (35%) against NDV challenge. However, antibody responses against NDV-F or -HN at the time of NDV challenge, 28 days after immunization, were very low in sera of SPF chickens vaccinated with recombinant HVT expressing NDV-F or -HN compared with the control NDV vaccine. Combination of these two recombinant HVTs did not increase the levels of antibody responses and protection against NDV challenge. Together with the findings of rMDV1-US10L(F), these observations indicate that expression of the NDV-F gene alone is sufficient for protection of chickens against velogenic NDV challenge.

NDV replication in the trachea of chickens vaccinated with these recombinant HVTs is not prevented by ocular NDV challenge (MORGAN et al. 1992). Therefore, the recombinant HVT induces host systemic immune responses but not local respiratory immune responses to protect chickens from NDV challenge. A study also reported that vaccination of recombinant HVT expressing NDV-F and -HN sufficiently protected SPF chickens from velogenic NDV challenge only when vaccinated chickens were challenged 14 days after vaccination, but not at earlier times (HECKERT et al. 1996). Since information on the genomic structure of recombinant HVT was not provided, it is difficult at present to evaluate their findings.

4.1.3 Effects of Maternal Antibodies on Protective Efficacy

The presence of maternal antibodies to NDV in commercial chickens interferes with the protective efficacy of NDV vaccine (GIAMBRONE and CLOSSER 1990a; SHARMA et al. 1989). If the NDV-F or -HN is expressed on the surface of a recombinant MDV particle, maternal antibodies may neutralize the virus. It is also possible that maternal antibodies against NDV-F or -HN kill cells infected with recombinant MDV by antibody-dependent cell-mediated cytotoxicity.

The effect of maternal antibodies on protection against ND and MD by recombinant MDV has been examined by two groups (MORGAN et al. 1993;

Sakaguchi et al. 1995, 1998; Sonoda et al. 2000). Firstly, protection by the recombinant HVT expressing NDV-F did not differ between SPF chickens and broiler chickens with maternal antibodies to MDV (Morgan et al. 1993). However the protective indexes (PIs) of the recombinant HVT against MD were 75% and 56%, respectively, in chickens with and without maternal antibodies, while those of the control HVT vaccine were 33% and 77%, respectively. The recombinant HVT and control NDV vaccine provided 73% and 80% protection, respectively, against ND in chickens with maternal antibodies, while both vaccines protected all SPF chickens from lethal NDV challenge. Similar protection was observed by vaccination of chickens with maternal antibodies to NDV and MDV using rMDV1-US10L(F) expressing NDV-F under the control of SV40 late promoter (Sakaguchi et al. 1995, 1998). However, the recombinant MDV1 provided 86% and 100% protection, respectively, against MD in chickens with and without maternal antibodies.

In these MDV-based recombinant vaccines against ND, the expression of the NDV-F or -HN gene was controlled by the heterologous promoters such as SV40 late promoter and RSV LTR. These promoters are known to show very strong activity in various types of cells. Therefore, high expression levels of NDV antigens in vaccinated chickens may be easily neutralized by maternal antibodies against NDV antigens and result in lower protective efficacy against ND. This suggests that use of an appropriate promoter that properly regulates the expression of foreign genes would improve the protective efficacy of the recombinant viruses in the presence of maternal antibodies. For this, we chose the gB promoter of MDV1. MDV1-gB is one of the viral antigens responsible for virus neutralization (Ikuta et al. 1984), and chickens immunized with the purified protein were partially protected against virulent MDV1 challenge (Ono et al. 1985).

We constructed recombinant MDV1, rMDV1-US10P(F), expressing the NDV-F gene controlled by MDV1-gB promoter which was inserted within the US10 gene of the MDV1 genome. The protective efficacy against MD and ND in chickens with maternal antibodies was compared with rMDV1-US10L(F) expressing NDV-F controlled by the SV40 late promoter (Sonoda et al. 2000). In cultured cells transfected with insertion vectors, the expression of the NDV-F gene controlled by the MDV1-gB promoter was much lower than that controlled by the SV40 late and chicken β-actin promoters. Similarly, rMDV1-US10P(F) expressed significantly less NDV-F than rMDV1-US10L(F) in infected cells. In commercial chickens used, maternal antibodies against NDV and MDV1 could be detected until 5 weeks of age. The antibody titers against NDV-F and MDV1 antigens in commercial chickens vaccinated with rMDV1-US10P(F) increased from 5 weeks after vaccination (equivalent to 5 weeks of age), much earlier and higher than rMDV1-US10L(F). The ELISA values of antibodies against NDV-F were also much higher than the minimum ELISA value (0.6) which provides 100% protection against NDV challenge. The protective efficacy of rMDV1-US10P(F) against velogenic NDV challenge was 100% in all four experiments using 20 birds each, whereas that of rMDV1-US10L(F) varied from 60 to 74% which was similar to that (40–100%) reported before (Sakaguchi et al. 1995, 1998). Protection of commercial chickens against vvMDV1 challenge by rMDV1-US10P(F) was as good as the parental MDV1 strain, CVI988, and the commercial vaccine Rispens strain. Thus, rMDV1-US10P(F) is an effective and stable polyvalent vaccine against both MD and ND even in the presence of maternal antibodies. Importantly, the ELISA values of antibody against NDV-F always gave sufficient levels of protection against ND for over 80 weeks post-vaccination. This suggests that rMDV1-US10P(F) can protect commercial chickens against ND for over 20 months after vaccination by a one-time inoculation at 1 day of age.

4.1.4 In Ovo Vaccination

Since Sharma and Burmester (1982) demonstrated that in ovo vaccination by HVT provided better protection than chickens vaccinated at 1 day of age, in ovo vaccination became popular in Western countries. More than 80% of the US broiler industry employs in ovo vaccination against MD (Ricks 1999). Inoculation of MDV1 and MDV2 vaccines to 18-day-old embryonated eggs also resulted in better protection against lethal MDV1 challenge at 3 days after hatching than with vaccines given to 1-day-old chickens (Sharma and Witter 1983). However, the in ovo inoculation of MDV1 resulted in a lower growth rate than MDV2 and HVT (Gagic et al. 1999).

Protective efficacy of an in ovo vaccine against ND and MD was also attempted using recombinant HVT expressing NDV-F and -HN, as well as MDV1 gC and gB (Reddy et al. 1996). In ovo vaccination of the recombinant HVT was performed by deposition in the amniotic fluid at 18 days of embryonation (Reddy et al. 1996). All SPF chickens vaccinated in ovo with the recombinant HVT were protected from challenges with velogenic NDV at 4 weeks of age as well as with virulent MDV1 at 7 days. However, further challenge tests at later times were required for evaluation of the protective efficacy as a vaccine. We have also attempted in ovo vaccination with rMDV1-US10P(F) (K. Yokogawa et al., manuscript in preparation). In ovo inoculation at the 18th day of embryonation with cell-free rMDV1-US10P(F) provided better virus recovery and induction of anti-NDV-F antibody than that with infected cells. In ovo inoculation with the cell-free viruses of less than 100 PFU provided good virus recovery and anti-NDV-F antibody induction to all inoculated chickens, while these responses were observed in only one quarter of chickens inoculated in ovo with cells infected at a similar level of PFU.

4.2 Infectious Bursal Disease and Others

Next to ND, infectious bursal disease (IBD) or Gumboro disease is an important target for protection of chickens from diseases by polyvalent recombinant virus vaccines. The etiological agent of IBD is infectious bursal disease virus (IBDV), a member of the *Birnaviridae* family, whose genome consists of two segments of double-stranded RNA. Virulent IBDV infection to young chickens causes destruction of lymphocytes in the bursa of Fabricius, resulting in severe immunosuppression and high mortality. Chickens can be protected from IBD by vaccination of attenuated live IBDV vaccine at 2–4 weeks of age. However, these commercial vaccines may induce some degree of virulence and immunosuppression in chickens (Mazariegos et al. 1990). In addition, some IBDV vaccine strains do not sufficiently protect commercial chickens with maternal antibodies to IBDV (Tsukamoto et al. 1995). Thus, construction of recombinant MDV expressing IBDV antigens was attempted to protect chickens from IBD (Darteil et al. 1995; Tsukamoto et al. 1999). The inserted IBDV gene for protection is the VP2 gene which encodes one of the major IBDV structural proteins. VP2 is responsible for induction of neutralizing antibodies and the major host protective antigen (Becht et al. 1988; Fahey et al. 1989).

The first attempt was recombinant HVT expressing IBDV-VP2 controlled by human cytomegalovirus (HCMV) major immediate-early (MIE) promoter (Darteil et al. 1995). The VP2 expression cassette was inserted within either the HVT genes of the ribonucleotide reductase (RR) small subunit or gI. However, these recombinant HVTs provided very low protection against both IBD and MD in SPF chickens. This may indicate that both deleted viral genes are responsible for viral replication in chickens. Similar in vivo limitations were reported in pseudorabies virus lacking the RR small subunit gene (De Wind et al. 1994) and gI gene (Kimman et al. 1992). Next, the VP2 gene was expressed under the control of SV40 early promoter and the expression cassette was inserted within the nonessential US2 gene of the MDV1 (CVI988 strain) genome (Tsukamoto et al. 1999). The recombinant MDV1 provided partial (55%) protection against vvIBDV challenge in SPF chickens, while the control commercial IBDV vaccine gave full protection. This partial protection could be due to lower serum antibody titers to IBDV-VP2 in SPF chickens. In contrast, the recombinant vaccine provided full protection against vvMDV1 challenge in SPF chickens and the serum antibody responses to MDV1 of chickens vaccinated by the recombinant MDV1 were similar to those of chickens vaccinated using control MDV1 vaccine. Therefore, we surmise that the recombinant MDV1 persists stably in SPF chickens after vaccination. Further improvement of the recombinant MDV1 against IBD is required to obtain more effective protection in chickens with and without maternal antibodies to IBDV and MDV.

Coccidiosis caused by protozoan parasites of the genus *Eimeria* in chickens is also known to cause economical losses in the poultry industry. Four or five species of the genus cause diseases in chickens. *Eimeria acervulina* is the main species and causes significant weight loss and growth retardation by infection of the chicken intestinal epithelium. One of the *Eimeria* antigens used for induction of protection against *Eimeria* challenge is the Ea1A antigen, a parasite refractile body transhydrogenase. A DNA fragment encoding a fusion protein of NDV-HN and Ea1A under the control of RSV LTR promoter was inserted within the US10 gene of HVT, and the broiler chickens vaccinated by recombinant HVT at 1 day of age were challenged at day 24 (Cronenberg et al. 1999). Three weeks after challenge, the individual body weight of chickens vaccinated with the recombinant HVT was about 150g higher than that of the controls. However, this attempt is preliminary and more tests are required for evaluation of protective efficacy using recombinant live vaccine.

5 Recombinant Fowlpox Virus Against Poultry Diseases

FPV, canarypox, and pigeonpox viruses used as virus vectors belong to the *Avipoxvirus* genus of the *Chordpoxvirinae* subfamily. Members of the genera are genetically and antigenically related and have a similar morphology and host range. Viruses recovered from various species of birds are given names related to their hosts. FPV has been also used as a vector for construction of polyvalent recombinant vaccines against poultry diseases such as ND (Edbauer et al. 1990; Ogawa et al. 1990; Taylor et al. 1990, 1996; Iritani et al. 1991; McMillen et al. 1994; Nagy et al. 1994; Parks et al. 1994; Heckert et al. 1996), MD (Nazerian et al. 1992, 1996), avian influenza (Taylor et al. 1988; Beard et al. 1991; Tripathy et al. 1991; Webster et al. 1991), infectious bursal disease (Bayliss et al. 1991, Heine and Boyle 1993), avian leukosis (Nazerian et al. 1995), and avian reticuloendotheliosis (Calvert et al. 1993).

The gnome of FPV is about 240–270kbp (Gafford et al. 1978). The recombinants of both FPV and canary poxvirus have also been attempted for use as vaccines in humans and other mammalian species, although both viruses cannot replicate completely in these cells. At least three types of avipox viruses – FPV, pigeonpox virus, and canarypox virus – are known to cause avian pox, which is characterized by wart-like nodules on the skin and death by asphyxiation in any age of birds. Avipox infection is usually mild and healing occurs within 3 weeks.

Chickens as young as 1-day-old can be vaccinated by attenuated FPV using the wing-web puncture method. One-time immunization by FPV vaccine provides sufficient protection in broiler birds. However, laying birds require at least twice the vaccination dose because the vaccinal immunity usually persists for 2 or 3 months. Recombinant FPV was constructed on the basis of the strategies used for construction of recombinant vaccinia virus by homologous recombination. The promoter derived from FPV or vaccinia virus is required for expression of foreign genes because the promoter sequences of poxviruses are different from the consensus ones of eukaryotic cells conserved in the poxvirus genera (Boyle 1992). In addition, the cDNA encoding a foreign gene should be inserted downstream of the promoter because the viral transcription in the cytoplasm of cells infected with poxviruses excludes the usage of cellular splicing apparatus. The promoters used for expression of foreign genes for protection of chickens against poultry diseases were the vaccinia virus early/late H6 promoter (Edbauer et al. 1990), early P7.5 promoter (Ogawa et al. 1990), and FPV-TK promoter (Nagy et al. 1993). However, the expression of the NDV-HN gene controlled by the FPV-TK promoter was probably not detectable in cells infected with recombinant FPV due to the weak promoter activity (Nagy et al. 1994).

Inoculation of the recombinant FPV expressing the NDV-F or -HN gene through intramuscular or wing-web routes provided sufficient protection against velogenic NDV challenge in SPF chickens (Edbauer et al. 1990; Tayler et al. 1990; McMillen et al. 1994). Recombinant FPV expressing the NDV-HN gene controlled by the vaccinia early P7.5 promoter did not induce immune responses to NDV in chickens with maternal antibodies to FPV because immunity to FPV blocked the vaccine taken at the inoculation site (Iritani et al. 1991). In contrast, recombinant FPV expressing NDV-F and -HN genes controlled by the vaccinia virus H6 promoter provided better protection against virulent FPV challenge in chickens with maternal antibodies to FPV (Taylor et al. 1996). However, only partial protective efficacy (40–70%) against velogenic NDV challenge in commercial chickens was observed. The difference in vaccinal immunity between these recombinant FPVs might not be due to coexpression of NDV-F and -HN genes because equivalent findings were obtained in commercial chickens vaccinated with recombinant FPV expressing the NDV-HN gene alone (Taylor et al. 1996). However, the longevity of vaccinal immune responses against FPV and NDV was determined only up to 28 days after vaccination. Despite partial protection against ND, the humoral responses to NDV were not observed in commercial chickens at 21 and 28 days after vaccination. At present it is difficult to evaluate the findings of these experiments because of the lack of control of the ND and MD vaccines, and the protection provided by recombinant FPVs may not be as good as that induced by conventional NDV vaccine (Boyle and Heine 1993).

Protection of chickens against MD by recombinant FPV was extensively analyzed in SPF chickens (Nazerian et al. 1992) and in chickens with maternal antibodies against all three serotypes of MDV (Nazerian et al.1996). Recombinant FPV expressing MDV1-gB provided perfect protection against virulent MDV1 challenge in SPF chickens, but only 18% protection in chickens with maternal antibodies. In the same trial, control MDV1 vaccine R2/23 provided 69% protection in chickens with maternal antibodies. No protection was observed in chickens with maternal antibodies by recombinant FPV expressing the MDV1 gene encoding one of the MDV1 glycoproteins gC and gD and two tegument proteins UL47 and UL48. Coexpression of MDV1-gB and these viral proteins did not increase protective efficacy against MD.

6 Conclusions

Recently, polyvalent vaccines have been extensively studied especially in the field of poultry diseases. This could be mainly due to the important problem of how more than 100 thousand chickens should be efficiently vaccinated for reduction of labor. In addition to MDV and FPV described here, avian adenovirus has also been used as a virus vector for construction of recombinant vaccine (Sheppard et al. 1998).

Table 2. Comparison of MDV with FPV as vectors

Characteristics	Vector virus		
	MDV1	HVT	FPV
Foreign gene insertion capacity	Large	Large	Large
Utilization of eukaryotic promoter	Yes	Yes	No[a]
Heat stability	Unstable	Unstable	Stable[b]
Oral or spray vaccination	No	No	ND
Cost of vaccine production	High	High	Low
In ovo vaccination	Possible	Proven suitable	Possible
Persistency of vaccinal immunity	Long	Long	Short
Protective efficacy in the presence of maternal antibodies	Excellent	Good	Fair

ND, no data available.
[a] Since poxviruses use their own transcriptional machinery, inserted foreign genes should be expressed under the control of the poxvirus promoter.
[b] However, commercialized recombinant FPV against ND (Taylor et al. 1996) has to be stored in liquid nitrogen (Jackwood 1999).

Among these avian recombinant vaccines, recombinant FPV has been put to practical use (Sheppard 1999; Jackwood 1999). Table 2 shows comparisons of MDV and FPV as virus vectors.

Efforts to improve the technology for construction of recombinant vaccines against MD, ND, and the other poultry diseases are in progress. However, most studies on protective efficacy of recombinant vaccines are based on the findings obtained from SPF chickens without maternal antibodies, although commercial chickens in the field are positive for maternal antibodies to these virus vaccines. As reviewed here, rMDV1-US10P(F) constructed by us appears to produce better protective efficacy in the presence of maternal antibodies and is ready for practical use. There are advantages to use recombinant MDV1s other than those described here. Since the recombinant MDV1 vaccine does not induce anti-NDV-HA antibody, it is easy to distinguish from natural infection of NDV. In addition, the interferon production induced by NDV vaccines can be minimized so that the protective efficacy of the other vaccines may be expected to increase. Despite these advantages, development of better designed recombinant vaccines which show sufficient protective efficacy against various poultry diseases in the presence of maternal antibodies and against very virulent viral strains, which may appear in the near future, is urgently required.

Acknowledgements. We sincerely appreciate the support of many people who have contributed to the construction of our effective recombinant MDV1 over the years. Thanks are due to colleagues in our laboratories, in particularly to G.S. Zhu and Y. Kawaguchi at Tokyo Medical and Dental University, K. Sonoda, H. Nakamura, H. Okamura, K. Yokogawa, T. Urakawa, Y. Hirayama, N. Miki, K. Naruse, H. Sakamoto, M. Fujimoto, H. Maeda, Y. Kino, E. Tokunaga, K. Matsuo and M. Yamamoto at The Chemo-Sero Therapeutic Research Institute, and G.F. de Boer at Serendip B.V. We also wish to thank our wives M. Hirai and S. Sakaguchi for their continuous support of our work. This study (K.H.) was supported by a Grant-in-Aid for Scientific Research and a Grant-in-Aid for Scientific Research in Priority Areas from the Ministry of Education, Science, Sports and Culture of Japan, and by The Japan Health Science Foundation.

References

Avery RJ, Niven J (1979) Use of antibodies to purified Newcastle disease virus glycoproteins for strain comparisons and characterizations. Infect Immun 26:795–801

Bayliss CD, Peters RW, Cook JK, Reece RL, Howes K, Binns MM, Boursnell ME (1991) A recombinant fowlpox virus that expresses the VP2 antigen of infectious bursal disease virus induces protection against mortality caused by the virus. Arch Virol 120:193–205

Beard CW, Schnitzlein WM, Tripathy DN (1991) Protection of chickens against highly pathogenic avian influenza virus (H5N2) by recombinant fowlpox viruses. Avian Dis 35:356–359

Becht H, Muller H, Muller HK (1988) Comparative studies on structural and antigenic properties of two serotypes of infectious bursal disease virus. J Gen Virol 69:631–640

Blancou J, Kieny MP, Lathe R, Lecocq JP, Pastoret PP, Soulebot JP, Desmettre P (1986) Oral vaccination of the fox against rabies using a live recombinant vaccinia virus. Nature 322:373–375

Boyle DB (1992) Quantitative assessment of poxvirus promoters in fowlpox and vaccinia virus recombinants. Virus Genes 6:281–290

Boyle DB, Heine HG (1993) Recombinant fowlpox virus vaccines for poultry. Immunol Cell Biol 71: 391–397

Brunovskis P, Velicer LF (1995) The Marek's disease virus (MDV) unique short region: alphaherpesvirus-homologous, fowlpox virus-homologous, and MDV-specific genes. Virology 206:324–338

Brunovskis P, Kung HJ (1995) Retrotransposition and herpesvirus evolution. Virus Genes 11:259–270

Bublot M, Laplace E, Audonnet J-C (1999) Nonessential loci in the BamHI-I and -F fragments of the HVT FC126 genome. Acta Virologica 43:181–185

Calvert JG, Nazerian K, Witter RL, Yanagida N (1993) Fowlpox virus recombinants expressing the envelope glycoprotein of an avian reticuloendotheliosis retrovirus induce neutralizing antibodies and reduce viremia in chickens. J Virol 67:3069–3076

Cantello JL, Anderson AS, Francesconi A, Morgan RW (1991) Isolation of a Marek's disease virus (MDV) recombinant containing the lacZ gene of *Escherichia coli* stably inserted within the MDV US2 gene. J Virol 65:1584–1588

Churchill AE, Chubb RC, Baxendale W (1969) The attenuation, with loss of oncogenicity, of the herpes-type virus of Marek's disease (strain HPRS-16) on passage in cell culture. J Gen Virol 4: 557–564

Coffin JM (1996) Retroviridae: The viruses and their replication. In: Fields BN, Knipe DM, Howley PM (eds) Fields Virology. Lippincott-Raven Publishers, Philadelphia, pp 1767–1847

Cronenberg AM, Van Geffen CEH, Dorrestein J, Vermulen AN, Sondermeijer PJA (1999) Vaccination of broilers with HVT expressing an Eimeria Acervulina antigen improves performance after challenge with Eimeria. Acta Virologica 43:192–197

Darteil R, Bublot M, Laplace E, Bouquet JF, Audonnet JC, Riviere M (1995) Herpesvirus of turkey recombinant viruses expressing infectious bursal disease virus (IBDV) VP2 immunogen induce protection against an IBDV virulent challenge in chickens. Virology 211:481–490

Davidson I, Borenstein R (1999) Multiple infection of chickens and turkeys with avian oncogenic viruses; Prevalence and molecular analysis. Acta Virologica 43:136–142

de Boer GF, Pol JMA, Jeurissen SHM (1988) In: Kato S, Horiuchi T, Mikami T, Hirai K (eds) Marek's disease vaccination strategies using vaccines made from three avian herpesvirus serotypes. 3rd International Symposium on Marek's disease, Advances in Marek's disease Research, Osaka, pp 21–42

De Wind N, Peeters BP, Zuderveld A, Gielkens AL, Berns AJ, Kimman TG (1994) Mutagenesis and characterization of a 41-kilobase-pair region of the pseudorabies virus genome: transcription map, search for virulence genes, and comparison with homologs of herpes simplex virus type 1. Virology 200:784–790

Edbauer C, Weinberg R, Taylor J, Rey-Senelonge A, Bouquet JF, Desmettre P, Paoletti E (1990) Protection of chickens with a recombinant fowlpox virus expressing the Newcastle disease virus hemagglutinin-neuraminidase gene. Virology 179:901–904

Endoh D, Ito M, Cho KO, Kon Y, Morimura T, Hayashi M, Kuwabara M (1998) Retroviral sequence located in border region of short unique region and short terminal repeat of Md5 strain of Marek's disease virus type 1. J Vet Med Sci 60:227–235

Fahey KJ, Erny K, Crooks J (1989) A conformational immunogen on VP-2 of infectious bursal disease virus that induces virus-neutralizing antibodies that passively protect chickens. J Gen Virol 70: 1473–1481

Field HJ, Wildy P (1978) The pathogenicity of thymidine kinase-deficient mutants of herpes simplex virus in mice. J Hyg (Lond) 81:267–277

Frankel JW, Groupe V (1971) Interactions between Marek's disease herpesvirus and avian leucosis virus in tissue culture. Nat New Biol 234:125–126

Fukuchi K, Sudo M, Lee YS, Tanaka A, Nonoyama M (1984) Structure of Marek's disease virus DNA: detailed restriction enzyme map. J Virol 51:102–109

Fukuchi K, Tanaka A, Schierman LW, Witter RL, Nonoyama M (1985) The structure of Marek disease virus DNA: the presence of unique expansion in nonpathogenic viral DNA. Proc Natl Acad Sci USA 82:751–754

Gafford LG, Mitchell EB Jr, Randall CC (1978) Sedimentation characteristics and molecular weights of three poxvirus DNAs. Virology 89:229–239

Gagic M, St Hill CA, Sharma JM (1999) In ovo vaccination of specific-pathogen-free chickens with vaccines containing multiple agents. Avian Dis 43:293–301

Giambrone JJ, Closser J (1990) Effect of breeder vaccination on immunization of progeny against Newcastle disease. Avian Dis 34:114–119

Hayashi M, Jessip J, Fukuchi K, Smith M, Tanaka A, Nonoyama M (1988) The structure of Marek's disease virus DNA: amplification of repeat sequence in IRs and TRs. Microbiol Immunol 32:265–274

Heckert RA, Riva J, Cook S, McMillen J, Schwartz RD (1996) Onset of protective immunity in chicks after vaccination with a recombinant herpesvirus of turkeys vaccine expressing Newcastle disease virus fusion and hemagglutinin-neuraminidase antigens. Avian Dis 40:770–777

Heine HG, Boyle DB (1993) Infectious bursal disease virus structural protein VP2 expressed by a fowlpox virus recombinant confers protection against disease in chickens. Arch Virol 131:277–292

Hertig C, Coupar BE, Gould AR, Boyle DB (1997) Field and vaccine strains of fowlpox virus carry integrated sequences from the avian retrovirus, reticuloendotheliosis virus. Virology 235:367–376

Hirai K, Ikuta K, Kato S (1981) Restriction endonuclease analysis of the genomes of virulent and avirulent Marek's disease viruses. Microbiol Immunol 25:671–681

Hirai K, Honma H, Ikuta K, Kato S (1984a) Genetic relatedness of virulent and avirulent strains of Marek's disease virus. Arch Virol 79:293–298

Hirai K, Ikuta K, Maotani K, Kato S (1984b) Evaluation of DNA homology of Marek's disease virus, herpesvirus of turkeys and Epstein-Barr virus under varied stringent hybridization conditions. J Biochem (Tokyo) 95:1215–1218

Hirai K, Nakajima K, Ikuta K, Kirisawa R, Kawakami Y, Mikami T, Kato S (1986) Similarities and dissimilarities in the structure and expression of viral genomes of various virus strains immunologically related to Marek's disease virus. Arch Virol 89:113–130

Hirai K, Yamada M, Arao Y, Kato S, Nii S (1990) Replicating Marek's disease virus (MDV) serotype 2 DNA with inserted MDV serotype 1 DNA sequences in a Marek's disease lymphoblastoid cell line MSB1-41C. Arch Virol 114:153–165

Hirai K, Sakaguchi M, Maeda H, Kino Y, Nakamura H, Zhu GS, Yamamoto M (1992) Construction of recombinant Marek's disease virus type 1 expressing the lacZ gene of *Eschericia coli*. 4th international Symposium on Marek's Disease, Proceedings of 19th World's Poultry Congress, Amsterdam, pp 150–155

Homma M, Ouchi M (1973) Trypsin action on the growth of Sendai virus in tissue culture cells. 3. Structural difference of Sendai viruses grown in eggs and tissue culture cells. J Virol 12:1457–1465

Igarashi T, Takahashi M, Donovan J, Jessip J, Smith M, Hirai K, Tanaka A, Nonoyama M (1987) Restriction enzyme map of herpesvirus of turkey DNA and its collinear relationship with Marek's disease virus DNA. Virology 157:351–358

Ikuta K, Nishi Y, Kato S, Hirai K (1981) Immunoprecipitation of Marek's disease virus-specific polypeptides with chicken antibodies purified by affinity chromatography. Virology 114:277–281

Ikuta K, Ueda S, Kato S, Hirai K (1983) Most virus-specific polypeptides in cells productively infected with Marek's disease virus or herpesvirus of turkeys possess cross-reactive determinants. J Gen Virol 64:961–965

Ikuta K, Ueda S, Kato S, Hirai K (1984) Processing of glycoprotein gB related to neutralization of Marek's disease virus and herpesvirus of turkeys. Microbiol Immunol 28:923–933

Iritani Y, Aoyama S, Takigami S, Hayashi Y, Ogawa R, Yanagida N, Saeki S, Kamogawa K (1991) Antibody response to Newcastle disease virus (NDV) of recombinant fowlpox virus (FPV) expressing a hemagglutinin-neuraminidase of NDV into chickens in the presence of antibody to NDV or FPV. Avian Dis 35:659–661

Isfort R, Jones D, Kost R, Witter R, Kung HJ (1992) Retrovirus insertion into herpesvirus in vitro and in vivo. Proc Natl Acad Sci USA 89:991–995

Isfort RJ, Witter R, Kung HJ (1994) Retrovirus insertion into herpesviruses. Trends Microbiol 2:174–177
Jackwood MW (1999) Current and future recombinant viral vaccines for poultry. Adv Vet Med 41: 517–522
Jakovleva LS, Mazurenko NP (1979) Increased susceptibility of leukemia-infected chickens to Marek's disease. Neoplasma 26:393–396
Jones D, Lee L, Liu JL, Kung HJ, Tillotson JK (1992) Marek disease virus encodes a basic-leucine zipper gene resembling the fos/jun oncogenes that is highly expressed in lymphoblastoid tumors [published erratum appears in Proc Natl Acad Sci USA 1993 Mar 15; 90(6):2556]. Proc Natl Acad Sci USA 89:4042–4046
Jones D, Isfort R, Witter R, Kost R, Kung HJ (1993) Retroviral insertions into a herpesvirus are clustered at the junctions of the short repeat and short unique sequences. Proc Natl Acad Sci USA 90:3855–3859
Jones D, Brunovskis P, Witter R, Kung HJ (1996) Retroviral insertional activation in a herpesvirus: transcriptional activation of US genes by an integrated long terminal repeat in a Marek's disease virus clone. J Virol 70:2460–2467
Kanamori A, Nakajima K, Ikuta K, Ueda S, Kato S, Hirai K (1986) Copy number of tandem direct repeats within the inverted repeats of Marek's disease virus DNA. Biken J 29:83–89
Kimman TG, de Wind N, Oei-Lie N, Pol JM, Berns AJ, Gielkens AL (1992) Contribution of single genes within the unique short region of Aujeszky's disease virus (suid herpesvirus type 1) to virulence, pathogenesis and immunogenicity. J Gen Virol 73:243–251
Kishi M, Bradley G, Jessip J, Tanaka A, Nonoyama M (1991) Inverted repeat regions of Marek's disease virus DNA possess a structure similar to that of the a sequence of herpes simplex virus DNA and contain host cell telomere sequences. J Virol 65:2791–2797
Kost R, Jones D, Isfort R, Witter R, Kung HJ (1993) Retrovirus insertion into herpesvirus: characterization of a Marek's disease virus harboring a solo LTR. Virology 192:161–169
Liu JL YY, Lee LF, Kung HJ (1998) Transforming potential of the herpesvirus oncoprotein MEQ: morphological transformation, serum-independent growth, and inhibition of apoptosis. J Virol 72:388–395
Longnecker R, Roizman B (1987) Clustering of genes dispensable for growth in culture in the S component of the HSV-1 genome. Science 236:573–576
Mackett M, Smith G, Moss B (1982) Vaccinia virus: a selectable eukaryotic cloning and expression vector. Proc Natl Acad Sci USA 79:7415–7419
Maotani K, Kanamori A, Ikuta K, Ueda S, Kato S, Hirai K (1986) Amplification of a tandem direct repeat within inverted repeats of Marek's disease virus DNA during serial in vitro passage. J Virol 58:657–660
Marshall DR, Reilly JD, Liu X, Silva RF (1993) Selection of Marek's disease virus recombinants expressing the *Escherichia coli* gpt gene. Virology 195:638–648
Mazariegos LA, Lukert PD, Brown J (1990) Pathogenicity and immunosuppressive properties of infectious bursal disease "intermediate" strains. Avian Dis 34:203–208
McMillen JK, Cochran MD, Junker DE, Reddy DN, Valencia DM (1994) The safe and effective use of fowlpox virus as a vector for poultry vaccines. Dev Biol Stand 82:137–145
Merz DC, Scheid A, Choppin PW (1981) Immunological studies of the functions of paramyxovirus glycoproteins. Virology 109:94–105
Meulemans G, Gonze M, Carlier MC, Petit P, Burny A, Long L (1986) Protective effects of HN and F glycoprotein-specific monoclonal antibodies on experimental Newcastle disease. Avian Pathology 15:761–768
Morgan RW, Gelb J Jr, Schreurs CS, Lutticken D, Rosenberger JK, Sondermeijer PJ (1992) Protection of chickens from Newcastle and Marek's diseases with a recombinant herpesvirus of turkeys vaccine expressing the Newcastle disease virus fusion protein. Avian Dis 36:858–870
Morgan RW, Gelb J Jr, Pope CR, Sondermeijer PJ (1993) Efficacy in chickens of a herpesvirus of turkeys recombinant vaccine containing the fusion gene of Newcastle disease virus: onset of protection and effect of maternal antibodies. Avian Dis 37:1032–1040
Morgan R, Anderson A, Kent J, Parcells M (1996) Characterization of Marek's disease virus RB1B-based mutants having disrupted glycoprotein C or glycoprotein D homolog genes. 5th International Symposium on Marek's disease, Michigan, pp 28
Nagai Y, Klenk HD, Rott R (1976) Proteolytic cleavage of the viral glycoproteins and its significance for the virulence of Newcastle disease virus. Virology 72:494–508
Nagy E, Krell PJ, Heckert RA, Derbyshire JB (1994) Vaccination of chickens with a recombinant fowlpox virus containing the hemagglutinin-neuraminidase gene of Newcastle disease virus under the

control of the fowlpox virus thymidine kinase promoter [published erratum appears in Can J Vet Res 1995 Jan; 59(1):25]. Can J Vet Res 58:306–308

Nakamura H, Sakaguchi M, Hirayama Y, Miki N, Yamamoto M, Hirai K (1992) Protection against Newcastle disease by recombinant Marek's disease virus serotype 1 expressing the fusion protein of Newcastle disease virus. 4th International Symposium on Marek's Disease, Proceedings of 19th World's Poultry Congress, Amsterdam, pp 332–335

Nazerian K, Lee LF, Yanagida N, Ogawa R (1992) Protection against Marek's disease by a fowlpox virus recombinant expressing the glycoprotein B of Marek's disease virus. J Virol 66:1409–1413

Nazerian K, Yanagida N (1995) A recombinant fowlpox virus expressing the envelope antigen of subgroup A avian leukosis/sarcoma virus. Avian Dis 39:514–520

Nazerian K, Witter RL, Lee LF, Yanagida N (1996) Protection and synergism by recombinant fowl pox vaccines expressing genes from Marek's disease virus. Avian Dis 40:368–376

Ogawa R, Yanagida N, Saeki S, Saito S, Ohkawa S, Gotoh H, Kodama K, Kamogawa K, Sawaguchi K, Iritani Y (1990) Recombinant fowlpox viruses inducing protective immunity against Newcastle disease and fowlpox viruses. Vaccine 8:486–490

Ono K, Takashima M, Ishikawa T, Hayashi M, Yoshida I, Konobe T, Ikuta K, Nakajima K, Ueda S, Kato S (1985) Partial protection against Marek's disease in chickens immunized with glycoproteins gB purified from turkey-herpesvirus-infected cells by affinity chromatography coupled with monoclonal antibodies. Avian Dis 29:533–539

Ono M, Katsuragi-Iwanaga R, Kitazawa T, Kamiya N, Horimoto T, Niikura M, Kai C, Hirai K, Mikami T (1992) The restriction endonuclease map of Marek's disease virus (MDV) serotype 2 and collinear relationship among three serotypes of MDV. Virology 191:459–1463

Panicali D, Paoletti E (1982) Construction of poxviruses as cloning vectors: insertion of the thymidine kinase gene from herpes simplex virus into the DNA of infectious vaccinia virus. Proc Natl Acad Sci USA 79:4927–4931

Parcells MS, Anderson AS, Cantello JL, Morgan RW (1994a) Characterization of Marek's disease virus insertion and deletion mutants that lack US1 (ICP22 homolog), US10, and/or US2 and neighboring short-component open reading frames. J Virol 68:8239–8253

Parcells MS, Anderson AS, Morgan RW (1994b) Characterization of a Marek's disease virus mutant containing a lacZ insertion in the US6 (gD) homologue gene. Virus Genes 9:5–13

Parcells MS, Anderson AS, Morgan TW (1995) Retention of oncogenicity by a Marek's disease virus mutant lacking six unique short region genes. J Virol 69:7888–7898

Parks RJ, Krell PJ, Derbyshire JB, Nagy E (1994) Studies of fowlpox virus recombination in the generation of recombinant vaccines. Virus Res 32:283–297

Peters WP, Kufe D, Schlom J, Frankel JW, Prickett CO, Groupe V, Spiegelman S (1973) Biological and biochemical evidence for an interaction between Marek's disease herpesvirus and avian leukosis virus in vivo. Proc Natl Acad Sci USA 70:3175–3178

Reddy SK, Sharma JM, Ahmad J, Reddy DN, McMillen JK, Cook SM, Wild MA, Schwartz RD (1996) Protective efficacy of a recombinant herpesvirus of turkeys as an in ovo vaccine against Newcastle and Marek's diseases in specific-pathogen-free chickens. Vaccine 14:469–477

Ricks CA, Avakian A, Bryan T, Gildersleeve R, Haddad E, Ilich R, King S, Murray L, Phelps P, Poston R, Whitfill C, Williams C (1999) In ovo vaccination technology. Adv Vet Med 41:495–515

Roizman B, Sears AE (1996) Herpes simplex viruses and their replication. In: Fields BN, Knipe DM, Howley PM (eds) Fields Virology. Lippincott-Raven Publishers, Philadelphia, pp 2231–2295

Ross LJ, Milne B, Biggs PM (1983) Restriction endonuclease analysis of Marek's disease virus DNA and homology between strains. J Gen Virol 64:2785–2790

Ross LJ, Binns MM (1991) Properties and evolutionary relationships of the Marek's disease virus homologues of protein kinase, glycoprotein D and glycoprotein I of herpes simplex virus. J Gen Virol 72:939–947

Ross LJ, Binns MM, Tyers P, Pastorek J, Zelnik V, Scott S (1993) Construction and properties of a turkey herpesvirus recombinant expressing the Marek's disease virus homologue of glycoprotein B of herpes simplex virus. J Gen Virol 74:371–377

Sakaguchi M, Urakawa T, Hirayama Y, Miki N, Yamamoto M, Hirai K (1992) Sequence determination and genetic content of an 8.9-kb restriction fragment in the short unique region and the internal inverted repeat of Marek's disease virus type 1 DNA [published erratum appears in Virus Genes 1993 Jun; 7(2):following 209]. Virus Genes 6:365–378

Sakaguchi M, Urakawa T, Hirayama Y, Miki N, Yamamoto M, Zhu GS, Hirai K (1993) Marek's disease virus protein kinase gene identified within the short unique region of the viral genome is not essential for viral replication in cell culture and vaccine-induced immunity in chickens. Virology 195:140–148

Sakaguchi M, Hirayama Y, Maeda H, Matsuo K, Yamamoto M, Hirai K (1994) Construction of recombinant Marek's disease virus type 1 (MDV1) expressing the *Escherichia coli* lacZ gene as a possible live vaccine vector: the US10 gene of MDV1 as a stable insertion site. Vaccine 12:953–957

Sakaguchi M, Nakamura H, Sonoda K, Matsuo K, Hirai K (1995) Marek's disease virus type 1 as a vaccine vector for poultry diseases. In: Chanock RM, Brown F, Ginsberg HS, Norrby E (eds) Vaccine 95. Cold Spring Harbor Laboratory Press, NewYork, pp 299–304

Sakaguchi M, Nakamura H, Sonoda K, Hamada F, Hirai K (1996) Protection of chickens from Newcastle disease by vaccination with a linear plasmid DNA expressing the F protein of Newcastle disease virus. Vaccine 14:747–752

Sakaguchi M, Sonoda K, Matsuo K, Zhu GS, Hirai K (1997) Insertion of tandem direct repeats consisting of avian leukosis virus LTR sequences into the inverted repeat region of Marek's disease virus type 1 DNA. Virus Genes 14:157–162

Sakaguchi M, Nakamura H, Sonoda K, Okamura H, Yokogawa K, Matsuo K, Hirai K (1998) Protection of chickens with or without maternal antibodies against both Marek's and Newcastle diseases by one-time vaccination with recombinant vaccine of Marek's disease virus type 1. Vaccine 16:472–479

Schat KA, Hooft van Iddekinge BJ, Boerrigter H, O'Connell PH, Koch G (1998) Open reading frame L1 of Marek's disease herpesvirus is not essential for in vitro and in vivo virus replication and establishment of latency. J Gen Virol 79:841–849

Sharma JM, Burmester BR (1982) Resistance to Marek's disease at hatching in chickens vaccinated as embryos with the turkey herpesvirus. Avian Dis 26:134–149

Sharma JM, Witter RL (1983) Embryo vaccination against Marek's disease with serotypes 1, 2 and 3 vaccines administered singly or in combination. Avian Dis 27:453–463

Sharma PN, Muneer MA, Cho Y (1989) Role of maternal antibodies in immunization of chicks against Newcastle disease virus. Vet Sci Zootec Int pp 51–55

Sheppard M, Werner W, Tsatas E, McCoy R, Prowse S, Johnson M (1998) Fowl adenovirus recombinant expressing VP2 of infectious bursal disease virus induces protective immunity against bursal disease. Arch Virol 143:915–930

Sheppard M (1999) Viral vectors for veterinary vaccines. Adv Vet Med 41:145–161

Silva RF, Witter RL (1985) Genomic expansion of Marek's disease virus DNA is associated with serial in vitro passage. J Virol 54:690–696

Sondermeijer PJ, Claessens JA, Jenniskens PE, Mockett AP, Thijssen RA, Willemse MJ, Morgan RW (1993) Avian herpesvirus as a live viral vector for the expression of heterologous antigens. Vaccine 11:349–358

Sonoda K, Sakaguchi M, Matsuo K, Zhu GS, Hirai K (1996) Asymmetric deletion of the junction between the short unique region and the inverted repeat does not affect viral growth in culture and vaccine-induced immunity against Marek's disease. Vaccine 14:277–284

Sonoda K, Sakaguchi M, Okamura H, Yokogawa K, Tokunaga E, Tokiyoshi S, Kawaguchi Y, Hirai K (2000) Development of an effective polyvalent vaccine against both Marek's and Newcastle diseases based on recombinant Marek's disease virus type 1 (MDV1) in commercial chickens with maternal antibodies. J Virol 74:3217–3226

Taylor J, Paoletti E (1988) Fowlpox virus as a vector in non-avian species. Vaccine 6:466–468

Taylor J, Edbauer C, Rey-Senelonge A, Bouquet JF, Norton E, Goebel S, Desmettre P, Paoletti E (1990) Newcastle disease virus fusion protein expressed in a fowlpox virus recombinant confers protection in chickens. J Virol 64:1441–1450

Taylor J, Christensen L, Gettig R, Goebel J, Bouquet JF, Mickle TR, Paoletti E (1996) Efficacy of a recombinant fowl pox-based Newcastle disease virus vaccine candidate against velogenic and respiratory challenge. Avian Dis 40:173–180

Tenser RB, Dunstan ME (1979) Herpes simplex virus thymidine kinase expression in infection of the trigeminal ganglion. Virology 99:417–422

Tripathy DN, Schnitzlein WM (1991) Expression of avian influenza virus hemagglutinin by recombinant fowlpox virus. Avian Dis 35:186–191

Tsukamoto K, Tanimura N, Kakita S, Ota K, Mase M, Imai K, Hihara H (1995) Efficacy of three live vaccines against highly virulent infectious bursal disease virus in chickens with or without maternal antibodies. Avian Dis 39:218–229

Tsukamoto K, Kojima C, Komori Y, Tanimura N, Mase M, Yamaguchi S (1999) Protection of chickens against very virulent infectious bursal disease virus (IBDV) and Marek's disease virus (MDV) with a recombinant MDV expressing IBDV VP2. Virology 257:352–362

Umino Y, Kohama T, Sato TA, Sugiura A (1990) Protective effect of monoclonal antibodies to Newcastle disease virus in passive immunization. J Gen Virol 71:1199–1203

Urakawa T, Sakaguchi M, Yamamoto M, Zhu GS, Hirai K (1994) Expression of a novel Marek's disease virus type 1 (MDV1)-specific protein p40 in insect cells by a baculovirus vector and in chick embryo fibroblasts infected with MDV1. Arch Virol 137:191–197

Van Zaane D, Brinkhof JM, Westenbrink F, Gielkens AL (1982) Molecular-biological characterization of Marek's disease virus. I. Identification of virus-specific polypeptides in infected cells. Virology 121:116–132

Weber PC, Levine M, Glorioso JC (1987) Rapid identification of nonessential genes of herpes simplex virus type 1 by Tn5 mutagenesis. Science 236:576–579

Webster RG, Kawaoka Y, Taylor J, Weinberg R, Paoletti E (1991) Efficacy of nucleoprotein and haemagglutinin antigens expressed in fowlpox virus as vaccine for influenza in chickens. Vaccine 9:303–308

Witter RL (1985) Principles of vaccination. In: Payne LN (ed) Marek's Diseases. Matinus Nijhoff Publishing, Boston, pp 203–250

Witter RL (1988) In: Kato S, Horiuchi T, Mikami T, Hirai K (eds) Marek's disease, prevention and control. 3rd International Symposium on Marek's disease, Advances in Marek's disease Research, Osaka, pp 389–397

Witter RL (1991) Attenuated revertant serotype 1 Marek's disease viruses: safety and protective efficacy. Avian Dis 35:877–891

Witter RL (1992) Recent developments in the prevention and control of Marek's disease. 4th international Symposium on Marek's Disease, Proceedings of the 19th World's Poultry Congress, Amsterdam, pp 298–304

Zelnik V, Darteil R, Audonnet JC, Smith GD, Riviere M, Pastorek J, Ross LJ (1993) The complete sequence and gene organization of the short unique region of herpesvirus of turkeys. J Gen Virol 74:2151–2162

Subject Index

Printing (Computer to Film): Saladruck, Berlin
Binding: H. Stürtz AG, Würzburg

Current Topics in Microbiology and Immunology

Volumes published since 1989 (and still available)

Vol. 213/II: **Günthert, Ursula; Birchmeier, Walter (Eds.):** Attempts to Understand Metastasis Formation II. 1996. 33 figs. XV, 288 pp. ISBN 3-540-60681-5

Vol. 213/III: **Günthert, Ursula; Schlag, Peter M.; Birchmeier, Walter (Eds.):** Attempts to Understand Metastasis Formation III. 1996. 14 figs. XV, 262 pp. ISBN 3-540-60682-3

Vol. 214: **Kräusslich, Hans-Georg (Ed.):** Morphogenesis and Maturation of Retroviruses. 1996. 34 figs. XI, 344 pp. ISBN 3-540-60928-8

Vol. 215: **Shinnick, Thomas M. (Ed.):** Tuberculosis. 1996. 46 figs. XI, 307 pp. ISBN 3-540-60985-7

Vol. 216: **Rietschel, Ernst Th.; Wagner, Hermann (Eds.):** Pathology of Septic Shock. 1996. 34 figs. X, 321 pp. ISBN 3-540-61026-X

Vol. 217: **Jessberger, Rolf; Lieber, Michael R. (Eds.):** Molecular Analysis of DNA Rearrangements in the Immune System. 1996. 43 figs. IX, 224 pp. ISBN 3-540-61037-5

Vol. 218: **Berns, Kenneth I.; Giraud, Catherine (Eds.):** Adeno-Associated Virus (AAV) Vectors in Gene Therapy. 1996. 38 figs. IX,173 pp. ISBN 3-540-61076-6

Vol. 219: **Gross, Uwe (Ed.):** Toxoplasma gondii. 1996. 31 figs. XI, 274 pp. ISBN 3-540-61300-5

Vol. 220: **Rauscher, Frank J. III; Vogt, Peter K. (Eds.):** Chromosomal Translocations and Oncogenic Transcription Factors. 1997. 28 figs. XI, 166 pp. ISBN 3-540-61402-8

Vol. 221: **Kastan, Michael B. (Ed.):** Genetic Instability and Tumorigenesis. 1997. 12 figs.VII, 180 pp. ISBN 3-540-61518-0

Vol. 222: **Olding, Lars B. (Ed.):** Reproductive Immunology. 1997. 17 figs. XII, 219 pp. ISBN 3-540-61888-0

Vol. 223: **Tracy, S.; Chapman, N. M.; Mahy, B. W. J. (Eds.):** The Coxsackie B Viruses. 1997. 37 figs. VIII, 336 pp. ISBN 3-540-62390-6

Vol. 224: **Potter, Michael; Melchers, Fritz (Eds.):** C-Myc in B-Cell Neoplasia. 1997. 94 figs. XII, 291 pp. ISBN 3-540-62892-4

Vol. 225: **Vogt, Peter K.; Mahan, Michael J. (Eds.):** Bacterial Infection: Close Encounters at the Host Pathogen Interface. 1998. 15 figs. IX, 169 pp. ISBN 3-540-63260-3

Vol. 226: **Koprowski, Hilary; Weiner, David B. (Eds.):** DNA Vaccination/Genetic Vaccination. 1998. 31 figs. XVIII, 198 pp. ISBN 3-540-63392-8

Vol. 227: **Vogt, Peter K.; Reed, Steven I. (Eds.):** Cyclin Dependent Kinase (CDK) Inhibitors. 1998. 15 figs. XII, 169 pp. ISBN 3-540-63429-0

Vol. 228: **Pawson, Anthony I. (Ed.):** Protein Modules in Signal Transduction. 1998. 42 figs. IX, 368 pp. ISBN 3-540-63396-0

Vol. 229: **Kelsoe, Garnett; Flajnik, Martin (Eds.):** Somatic Diversification of Immune Responses. 1998. 38 figs. IX, 221 pp. ISBN 3-540-63608-0

Vol. 230: **Kärre, Klas; Colonna, Marco (Eds.):** Specificity, Function, and Development of NK Cells. 1998. 22 figs. IX, 248 pp. ISBN 3-540-63941-1

Vol. 231: **Holzmann, Bernhard; Wagner, Hermann (Eds.):** Leukocyte Integrins in the Immune System and Malignant Disease. 1998. 40 figs. XIII, 189 pp. ISBN 3-540-63609-9

Vol. 232: **Whitton, J. Lindsay (Ed.):** Antigen Presentation. 1998. 11 figs. IX, 244 pp. ISBN 3-540-63813-X

Vol. 233/I: **Tyler, Kenneth L.; Oldstone, Michael B. A. (Eds.):** Reoviruses I. 1998. 29 figs. XVIII, 223 pp. ISBN 3-540-63946-2

Vol. 233/II: **Tyler, Kenneth L.; Oldstone, Michael B. A. (Eds.):** Reoviruses II. 1998. 45 figs. XVI, 187 pp. ISBN 3-540-63947-0

Vol. 234: **Frankel, Arthur E. (Ed.):** Clinical Applications of Immunotoxins. 1999. 16 figs. IX, 122 pp. ISBN 3-540-64097-5

Vol. 235: **Klenk, Hans-Dieter (Ed.):** Marburg and Ebola Viruses. 1999. 34 figs. XI, 225 pp. ISBN 3-540-64729-5

Vol. 236: **Kraehenbuhl, Jean-Pierre; Neutra, Marian R. (Eds.):** Defense of Mucosal Surfaces: Pathogenesis, Immunity and Vaccines. 1999. 30 figs. IX, 296 pp. ISBN 3-540-64730-9

Vol. 237: **Claesson-Welsh, Lena (Ed.):** Vascular Growth Factors and Angiogenesis. 1999. 36 figs. X, 189 pp. ISBN 3-540-64731-7

Vol. 238: **Coffman, Robert L.; Romagnani, Sergio (Eds.):** Redirection of Th1 and Th2 Responses. 1999. 6 figs. IX, 148 pp. ISBN 3-540-65048-2

Vol. 239: **Vogt, Peter K.; Jackson, Andrew O. (Eds.):** Satellites and Defective Viral RNAs. 1999. 39 figs. XVI, 179 pp. ISBN 3-540-65049-0

Vol. 240: **Hammond, John; McGarvey, Peter; Yusibov, Vidadi (Eds.):** Plant Biotechnology. 1999. 12 figs. XII, 196 pp. ISBN 3-540-65104-7

Vol. 241: **Westblom, Tore U.; Czinn, Steven J.; Nedrud, John G. (Eds.):** Gastroduodenal Disease and Helicobacter pylori. 1999. 35 figs. XI, 313 pp. ISBN 3-540-65084-9

Vol. 242: **Hagedorn, Curt H.; Rice, Charles M. (Eds.):** The Hepatitis C Viruses. 2000. 47 figs. IX, 379 pp. ISBN 3-540-65358-9

Vol. 243: **Famulok, Michael; Winnacker, Ernst-L.; Wong, Chi-Huey (Eds.):** Combinatorial Chemistry in Biology. 1999. 48 figs. IX, 189 pp. ISBN 3-540-65704-5

Vol. 244: **Daëron, Marc; Vivier, Eric (Eds.):** Immunoreceptor Tyrosine-Based Inhibition Motifs. 1999. 20 figs. VIII, 179 pp. ISBN 3-540-65789-4

Vol. 245/I: **Justement, Louis B.; Siminovitch, Katherine A. (Eds.):** Signal Transduction and the Coordination of B Lymphocyte Development and Function I. 2000. 22 figs. XVI, 274 pp. ISBN 3-540-66002-X

Vol. 245/II: **Justement, Louis B.; Siminovitch, Katherine A. (Eds.):** Signal Transduction on the Coordination of B Lymphocyte Development and Function II. 2000. 13 figs. XV, 172 pp. ISBN 3-540-66003-8

Vol. 246: **Melchers, Fritz; Potter, Michael (Eds.):** Mechanisms of B Cell Neoplasia 1998. 1999. 111 figs. XXIX, 415 pp. ISBN 3-540-65759-2

Vol. 247: **Wagner, Hermann (Ed.):** Immunobiology of Bacterial CpG-DNA. 2000. 34 figs. IX, 246 pp. ISBN 3-540-66400-9

Vol. 248: **du Pasquier, Louis; Litman, Gary W. (Eds.):** Origin and Evolution of the Vertebrate Immune System. 2000. 81 figs. IX, 324 pp. ISBN 3-540-66414-9

Vol. 249: **Jones, Peter A.; Vogt, Peter K. (Eds.):** DNA Methylation and Cancer. 2000. 16 figs. IX, 169 pp. ISBN 3-540-66608-7

Vol. 250: **Aktories, Klaus; Wilkins, Tracy, D. (Eds.):** Clostridium difficile. 2000. 20 figs. IX, 143 pp. ISBN 3-540-67291-5

Vol. 251: **Melchers, Fritz (Ed.):** Lymphoid Organogenesis. 2000. 62 figs. XII, 215 pp. ISBN 3-540-67569-8

Vol. 252: **Potter, Michael; Melchers, Fritz (Eds.):** B1 Lymphocytes in B Cell Neoplasia. 2000. XIII, 326 pp. ISBN 3-540-67567-1

Vol. 253: **Gosztonyi, Georg (Ed.):** The Mechanisms of Neuronal Damage in Virus Infections of the Nervous System. 2001. approx. XVI, 270 pp. ISBN 3-540-67617-1

Vol. 254: **Privalsky, Martin L. (Ed.):** Transcriptional Corepressors. 2001. XIV, 190 pp. ISBN 3-540-67611-2

CPSIA information can be obtained
at www.ICGtesting.com
Printed in the USA
LVHW081140121122
732998LV00005B/84

9 783540 677987